The Peregrine Profession

Studies in Global Social History

VOLUME 36

Studies in Global Migration History

Series Editor

Dirk Hoerder (*University of Arizona, Phoenix, AZ, USA*)

VOLUME 12

The titles published in this series are listed at *brill.com/sgmh*

The Peregrine Profession

Transnational Mobility of Nordic Engineers and Architects, 1880–1930

By

Per-Olof Grönberg

BRILL

LEIDEN | BOSTON

Cover illustration: The Swedish engineer Wilhelm Unge (right) together with Alfred Nobel (left) in front of the 'Villa Nobel' in San Remo, 1896.
SOURCE: The Nobel Foundation.

The Library of Congress Cataloging-in-Publication Data is available online at http://catalog.loc.gov
LC record available at http://lccn.loc.gov/2018964047

Typeface for the Latin, Greek, and Cyrillic scripts: "Brill". See and download: brill.com/brill-typeface.

ISSN 1874-6705
ISBN 978-90-04-36647-3 (hardback)
ISBN 978-90-04-38520-7 (e-book)

This book is printed on acid-free paper and produced in a sustainable manner.

Printed by Printforce, the Netherlands

Contents

Acknowledgements VII
List of Illustrations IX

1 **Introduction** 1
1 Background, Research Areas, and the Organisation of This Study 4
2 Major Concepts and Definitions 5
3 Theoretical Framework 12
4 Previous Research 26
5 Methodology 58
6 Sources, Selections, and Cohorts 74

2 **A Peregrine Profession** 82
1 Aspects of Nordic Technicians' Transnational Mobility 84
2 Summary 100

3 **The Choice of Destinations** 103
1 Domination of the German-Speaking Countries and North America 103
2 Characteristics of Destination Choices 107
3 Summary 121

4 **To Study and to Practise in German-Speaking Europe** 127
1 Different Types of Learning Mobility 130
2 For Nordic Companies in Austria and Germany 150
3 Summary 152

5 **Journeymen and Traditional Emigrants to North America** 156
1 Aspects of Nordic Technical Migration to North America 161
2 Nordic Technicians as Traditional Transatlantic Emigrants 178
3 Summary 183

6 **A Worldwide Labour Market** 187
1 Intra-Nordic Studies and 'Expertise' Migration 187
2 Study Travelling Dominated the Mobility to Continental Europe 195
3 The Old Industrial Empire Still Attracted 210
4 Among Nobel Employees and Finnish Technicians in Russia 213
5 Danish Domination in Asia and Oceania 217

6 Infrastructure, Mining, and Sugar Plantations in Latin America and the Caribbean 225
7 Summary 231

7 **Burning No Bridges Behind Them** 237
1 What Technicians Returned? 239
2 More Than Every Third Nordic Technician Had Foreign Experience 249
3 Geographical Dispersion of Returnee Technicians 265
4 Summary 293

8 **Summary and Concluding Discussion** 300

Appendices 325
Appendix 1: International destinations of Nordic technical school graduates, 1880–1930 (graduation 1880–1919) 325
Appendix 2: Highest share per country of mobile engineers and architects from Sweden, Denmark, Norway and Finland, 1880–1930 329
Appendix 3: Destinations of Nordic mechanical and electrical engineers and naval architects, 1880–1930 329
Appendix 4: Destinations of Nordic civil and construction engineers, 1880–1930 330
Appendix 5: Destinations of Nordic chemical engineers, 1880–1930 330
Appendix 6: Destinations of Nordic mining engineers and metallurgists, 1880–1930 331
Appendix 7: Destinations of Nordic architects (educated at technical schools), 1880–1930 331
Sources and References 332
Index 359

Acknowledgements

I am finally at the end of the 'long and winding road' leading to this book. Writing these words is a challenge as some water has flowed under the bridge. I just hope that no one feels forgotten when reading these acknowledgements.

This road started in Trondheim: at the Norwegian University of Science and Technology's (NTNU) Department of History and Classical Studies. My first thanks go to Håkon With Andersen. This book would never have been written without his professional and friendly support. Håkon backed my application and was my advisor during my stay at NTNU, sharing his profound knowledge of the history of technology. Håkon and many of his colleagues contributed to my impression of NTNU and the department as welcoming environments. Pål Thonstad Sandvik read several versions of the manuscript and gave valuable suggestions, based in his deep insights in Norwegian, Nordic, and international economic history. Pål also included me in several conference sessions he arranged, giving me the opportunity to present this research to a wider audience. The walking encyclopaedia, Svein Henrik Pedersen, who took me to almost every restaurant in the city, taught me a lot about Norway and Norwegian history. Pål and Svein Henrik, together with their respective wives/partners Margrethe and Hege, have become very close friends. Margrethe's father, the late Gudmund Stang, whom I never had the chance to meet, pioneered historical studies of transnational technical mobility at NTNU. Gudmund Stang called nineteenth and early twentieth century Scandinavian engineers a *peregrine profession*. The title of this book is a tribute to him. Hans Otto Frøland, Jan Thomas Kobberød, and Staffan Wahlgren also showed interest in my research project and assisted in different ways. As head of department, Staffan re-invited me to Trondheim to give seminars and work with the project.

I was also warmly welcomed to Denmark by Henrik Harnow and Lars Heide, and to Finland by Susanna Fellman, Marjatta Hietala, Sampsa Kaataja and Panu Nykänen. Seminars at the Copenhagen School of Business, the universities of Helsinki and Tampere, Chalmers in Gothenburg, Halmstad University College, Umeå University, University of Stavanger, as well as the Norwegian School of Management in Oslo gave me valuable suggestions and deeper insights into the history of each of the Nordic countries. There are also many archivists and librarians around the Nordic countries who deserves my deepest gratefulness. This is not least true for Tove Dahl Johansen at the National Library in Oslo, who went to work a few hours earlier than required to make sure that I was able to see and photograph everything before the library closed early for Easter. Tove Dahl Johansen's help was essential to the inclusion of the pre-1900 graduates from the technical school in Kristiania in the dataset.

Recently, I cooperated closely with Fay Lundh Nilsson from the Department of Economic History at Lund University in research projects on twentieth century labour immigration to Sweden as well as nineteenth century Swedish technical education. Fay has been a wonderful colleague and our cooperation has developed into deep friendship. This cooperation also inspired some of the concepts used in this book. Another valued colleague and friend, Paul Puschmann at Radboud Universiteit Nijmegen and Katholieke Universiteit Leuven, has given me particularly valuable suggestions and readings on migration theory. Many of the colleagues I have met in places like Umeå, Leuven and Stavanger, both before and after I began to write this book, have also developed into close friends, like Soili-Maria Olli, who always has been very supportive and encouraging.

The same is valid for my family, not least my sister Anna-Karin Sandström and her husband and daughters. My sister, as an English teacher, has also provided some linguistic suggestions. While visiting seminars and doing archival research in Finland, my aunt Maire Martikka let me stay in her apartment, and my cousin Anu Martikka helped me with several practical matters. I was also reintroduced to my family in Helsinki and in Karelia, wherefrom my mother Anja came to Sweden as a war child. Together with my late father, Bert, my mother is essential to this book. Although going to university studies was not self-evident for me, my parents always believed in me taking the step into academia. This step, of course, is crucial to this book.

The 'long and winding road' ends here and now, in the interesting, friendly and welcoming research environment provided by the Division of Social Sciences at the Luleå University of Technology. You could not ask for better colleagues and friends than the present and former members of our little history unit: Carina, Curt, Dolly, Josefin, Kristina, Lars, May-Britt, Roine, and Tore!

My contacts at Brill have also been essential to finishing the book. I am deeply grateful to Wendel Scholma, Evelien van der Veer, Malathy Chandrasekaran, and not least Gerda Danielsson Coe, who assisted me tremendously in these final stages. As for making my English readable, I like to thank Anders Michael Nielsen, who has checked the entire manuscript, and my room neighbour, Gregory Poelzer, who helped me with minor language checks in the very final stages. Both Anders Michael Nielsen and Gregory agreed on very short notice.

Julia has shown all the patience an author and researcher in a relationship could ever ask for! You often believed more than I did that this book would see the day. Thank you for being you! Soon, there will be both time and place for us!
Luleå in June 2018.

Per-Olof Grönberg

Illustrations

Figures

1 Birthplaces of Nordic technical school graduates, 1880–1919 62
2 Distribution of specialisation among Nordic technical school graduates, 1880–1919 66
3 Distribution per decade of graduation for Nordic technicians, 1880–1919 68
4 Transnational mobility before 1930 among Nordic engineers and architects leaving school 1880–1919 83
5 Distribution between 'migration' and 'study travel' before 1930 among transnationally mobile Nordic technical school students graduating 1880–1919 85
6 Transnational mobility before 1930 per graduation age among Nordic engineers and architects leaving school 1880–1919 94
7 Transnational mobility before 1930 per social status among Nordic engineers and architects leaving school 1880–1919 94
8 Transnational mobility before 1930 per birthplace among Nordic engineers and architects leaving school 1880–1919 95
9 Transnational mobility before 1930 per specialisation among Nordic engineers and architects leaving school 1880–1919 98
10 Transnational mobility before 1930 per graduation year among Nordic engineers and architects leaving school 1880–1919 99
11 Migration and study travelling before 1930 per graduation year among Norwegian and Finnish engineers and architects leaving school 1880–1919 100
12 Distribution on destination (in per cent) before 1930 per social status among transnationally mobile Nordic technical school graduates 1880–1919 110
13 Distribution on destination in per cent per decade of graduation among transnationally mobile Nordic technical school graduates 1880–1919 going abroad 1880–1930 118
14 Distribution of destination in per cent per age at graduation among transnationally mobile Nordic technical school graduates 1880–1919 and going abroad 1880–1930 119
15 Distribution on destination in per cent per time interval between graduation and migration among Nordic technical school graduates 1880–1919 migrating abroad 120
16 Distribution between 'migration' and 'study travel' before 1930 per destination among transnationally mobile Nordic technical school graduates graduating 1880–1919 121

17 Pre-1930 return migration and permanent migration in numbers and percentages of Nordic engineers and architects educated 1880–1919 (migrants only) 238
18 Duration of stay abroad (first intermission) among engineers and architects educated in the Nordic countries 1880–1919, who returned before 1930 240
19 Pre-1930 return migration per social class in percentage of transnational migration Nordic engineers and architects educated 1880–1919 241
20 Pre-1930 return migration per TIME BETWEEN GRADUATION AND MIGRATION in percentage of transnational migration of Nordic engineers and architects educated 1880–1919 244
21 Pre-1930 return migration per SPECIALISATION in percentage of transnational migration of Nordic engineers and architects educated 1880–1919 245
22 Pre-1930 return migration per DECADE OF DEPARTURE in percentage of transnational migration of Nordic engineers and architects educated 1880–1919. SOURCES: see Figure 1. 244 246
23 Pre-1930 return migration per FIRST DESTINATION in percentage of transnational migration of Nordic engineers and architects educated 1880–1919 246
24 Percentages of engineers and architects with foreign experience in the Nordic countries 1900, 1910, and 1920. Total experience (Tot) and experience from migration (Mig) 250
25 Percentages of foreign experiences per 'area of experience' among engineers and architects in the Nordic countries 1900, 1910, and 1920 254
26 Percentages of foreign experiences per area of experience among engineers and architects in Sweden 1900, 1910, and 1920 255
27 Percentages of foreign experiences per 'area of experience' among engineers in Denmark 1900, 1910, and 1920 255
28 Percentages of foreign experiences per area of experience among engineers and architects in Norway 1900, 1910, and 1920 256
29 Percentages of foreign experiences per 'area of experience' among engineers and architects in Finland 1900, 1910, and 1920 256

Tables

1 Registered positions of Nordic technicians and upward mobility 1900, 1910, and 1920 70
2 Engineers and architects with at least three years of education at technical schools in the Nordic countries 1880–1919 in per cent per educational institute 76

3 Distribution on destination (numbers-per cent) before 1930 among transnationally mobile Nordic technical school graduates 1880–1919 104
4 Number and percentage of Nordic technical school graduates, 1880–1919, migrating or study travelling to Germany, Switzerland, and Austria before 1930 129
5 Number and percentage of Nordic technical school graduates, 1880–1919, migrating or study travelling to the United States and/or Canada before 1930 161
6 Median and average duration of stay (first intermission) abroad among engineers and architects educated in the Nordic countries 1880–1919, who returned before 1930 239
7 Pre-1930 return migration per age of graduation in percentage of transnational migration among Nordic engineers and architects educated 1880–1919 243
8 Pre-1930 return migration per country of birth in percentage of transnational migration among Nordic engineers and architects educated 1880–1919 244
9 Percentages of migrants, study travellers, and graduates with no foreign experience who had experienced upward occupational mobility among Nordic engineers and architects in 1900, 1910, and 1920 248
10 Percentages of technical school graduates with foreign experience (total) in the Nordic countries, 1900–1910–1920. Total experience above, placements abroad below 252
11 Experience from migration in German-speaking Europe in numbers and percentages among engineers and architects in the Nordic countries in 1900, 1910, and 1920 259
12 Foreign experience per specialisation among graduates from technical schools in the Nordic countries working in Sweden, Denmark, Norway, and Finland in the years 1900, 1910, and 1920 261
13 Foreign experience per region among graduates from technical schools in Sweden 1880–1919, working in Sweden in the years 1900, 1910, and 1920 266
14 Foreign experience per region among graduates from technical schools in Denmark 1880–1919, working in Denmark in the years 1900, 1910, and 1920 275
15 Foreign experience per region among graduates from technical schools in Norway 1880–1919, working in Norway in the years 1900, 1910, and 1920 282
16 Foreign experience per region among graduates from technical schools in Finland 1880–1919 working in Finland in the years 1900, 1910, and 1920 288

CHAPTER 1

Introduction

Civil engineer Herluf Forchhammer graduated in Copenhagen in 1898 and was employed by two railroad companies in New York in 1903. Upon his return to Denmark, he became an engineer with building contractors Christiani & Nielsen, headed their offices in Aarhus and Copenhagen, served periods in London and Hamburg and headed the company's department in Turkey.[1] His colleague from Finland, Fjalar Witting, went abroad immediately upon graduation. Specialising in dye-chemistry, he worked close to Frankfurt-am-Main and made study trips to Berlin in 1905. After an intermission in Saint Petersburg, Witting was briefly employed in Tampere and Pori before he became head of a large company's chemical department. Witting also had his own business in Helsinki before he was employed as a dye-chemist in Tampere in 1921.[2] Forchhammer was in the United States and Witting in Germany, whereas their Norwegian colleague Haakon Hauan went to both the leading industrial countries after his 1889 graduation as a chemical engineer in Trondheim. He studied at the renowned technical university in Charlottenburg near Berlin and received a grant to study the European grease industry. After a visit to the The World's Columbian Exhibition of 1893 in Chicago, Hauan worked one year in New Jersey before returning. In 1905, he became director of an oil-refining company and was later appointed minister of industrial supply.[3] Electrical engineer Julius Körner also visited the United States and Germany. Upon his 1902 graduation from the Royal Institute of Technology in Stockholm, Körner was employed with the electrotechnical company ASEA. He stayed one year at General Electric in 1906—on a 'mission' from ASEA—and reported about spearhead technology. He visited Switzerland, France, and Germany on his way back and became responsible for railway electrification

1 Aage Hannover (ed.), *Dansk Civilingeniørstat 1942, biografiske Oplysninger om Polytekniske Kandidater 1829–1941* (København 1942) 90.

2 Sulo Heiniö (ed.), *Matrikel öfver Polytekniska institutets i Finland lärare och elever 1898–1908* (Hämeenlinna 1918) 189; 'Fjalar Witting', *Tammerfors Aftonblad*, 30 April 1935.

3 Otto Delphin Amundsen, *Den kongelige norske Sankt Olavs orden* (Oslo 1947) 83; Bjarne Bassøe (ed.), *Ingeniørmatrikkelen: norske sivilingeniører 1901–55 med tillegg* (Oslo 1961) 168–169; O. Alstad, (ed.), *Trondhjemsteknikernes matrikel: biografiske meddelelser om samtlige faste og hospiterendeelever av Trondhjems tekniske læreanstalt 1870–1915: med ca.1300 ungdomsportrætter* (Trondhjem 1916) 74.

 | DOI:10.1163/9789004385207_002

in northernmost Sweden. In 1917, he became the editor of Sweden's leading technical journal.[4]

The four graduates went abroad during a period of international upheaval, also reaching our planet's northernmost located group of nations. Norway achieved full sovereignty through the dissolution of the union with Sweden in 1905, and Finland became fully independent from Russia in 1917. Democratic reforms such as universal suffrage for men and women were introduced. The period from 1880 to 1930 also encompassed large-scale industrialisation, and the four countries experienced one of the world's fastest economic growths: The Nordic gross domestic product increased by 285 per cent; Sweden and Finland were close, while the Danish growth was somewhat stronger and the Norwegian a little weaker.[5] Denmark's industrial breakthrough can be placed in the years 1880–1900, Sweden's 1890–1910, Norway's in the first two decades of the twentieth century, and Finland's from 1920 to 1938.[6] Economic growth was accompanied by population growth. The Nordic population increased by 52 per cent, from between ten and eleven million in 1880 to somewhat over sixteen million in 1930. Denmark and Finland had stronger population growth; Norway lay somewhat below the Nordic average and Sweden a little lower. Swedish and Norwegian population growth was in part hampered by a significantly strong transatlantic emigration. Per capita, only Ireland exceeded the two Scandinavian neighbours in the mass movement to America. Danish and Finnish overseas migration was considerably more moderate.[7] In Finland, Russia in general, and especially Saint Petersburg played partly the role that America had elsewhere in Europe. As in most parts of Europe, however, the Nordic countries witnessed an even stronger movement of people from rural to urban areas.

---

4 Govert Indebetou and Erik Hylander (eds.), *Svenska teknologföreningen 1861–1936: biografier* (Stockholm 1937) 626; Per-Olof Grönberg, *Learning and Returning: return migration of Swedish engineers from the United States, 1880–1940* (Umeå 2003) 136–138; Roine Viklund, *Den första statsbaneelektrifieringen. Ett pionärprojekt i subarktisk miljö* (Gävle 2015).

5 Angus Maddison, *The world economy: historical statistics* (Paris 2003).

6 Olle Krantz, 'Industrialisation in Three Nordic Countries: A Long-Term Quantitative View', in: Hans Kryger Larsen (ed.), *Convergence?: aspects on the industrialisation of Denmark, Finland and Sweden 1870–1940* (Helsinki 2001) 52–55; Christian Venneslan, 'Electrification and industrialisation: An assessment of the industrial breakthrough in Norway', *Scandinavian Economic History Review* 57:2 (2009) 124–155.

7 Sten Carlsson, 'Chronlogy and Composition of Swedish Emigration to America', in: Harald Runblom and Hans Norman (eds.), *From Sweden to America: a history of the migration* (Uppsala 1976) 129.

Denmark was the most urbanised Nordic country, but the process was substantial in all countries.[8]

Urbanisation led to programmes designed to improve water and sewerage systems as well as electrification and the erection of houses, roads, bridges, and harbours. The Nordic region, thus, saw a development that created a stronger need for theoretically and practically trained technicians. The beginnings of higher technical education in the Nordic countries have been placed in 1829 in Denmark, 1848 in Sweden, 1879 in Finland, and 1910 in Norway. Despite being considered an institute of higher education, the Finnish Polytechnic Institute had a more limited curriculum and a lower level of teaching compared to technical universities in countries such as Germany and Sweden.[9] This implied different needs to go abroad to obtain or complete technical education in the four countries.

Recent decades have witnessed a discussion about the danger of brain drain in Western as well as in developing countries. Some have argued that the economic incentives to 'stay at home' must be strengthened, while others have claimed that the worries are exaggerated, and the out-migration rates are moderate. However, the movement abroad of educated people is not exclusive to our time. Middle Age scholars from peripheral Europe explored 'enlightened' countries such as Britain, France, Germany, Switzerland, and Italy. Students from several European countries were enrolled at universities in, for example, Bologna, Siena, Prague, and Paris as early as the fourteenth century. Journeymen became important from the sixteenth century: Artisans generally could not become masters without spending time abroad, and this tradition remained almost until 1900. Returning journeymen and immigrants brought new technology and other skills. The dependency on artisan migration has been discussed, for instance in studies of Denmark-Norway and pre-industrial Germany. Artisan technology was continuously updated. In the nineteenth century, technical school graduates took over this role to a large extent.[10]

---

8 Jonas Ljungberg and Lennart Schön, 'Domestic markets and international integration: paths to industrialisation in the Nordic countries', *Scandinavian Economic History Review* 61:2 (2013) 108.

9 Timo Myllyntaus, 'Foreign Models and National Styles in Teaching Technology in the Nordic Countries', in: Irina Gouzévitch, André Grelon and Anousheh Karvar (eds.), *La Formation des ingénieurs en perspective. Modèles de référence et réseaux de médiation—XVIIIe-XXe siècles* (Rennes 2004) 149–151.

10 Pär Eliasson, 'Svenska studenter i Tyskland 1372–1800', in: Kurt Genrup (ed.)*,'Förtyskningen' av Sverige* (Umeå 1994) 43–65; Eric Engström, *Bokbindargesällen Karl Stellan Söderströms gesällvandring 1843–1858: lärande i skråväsendet speglat i personliga dokument* (Stockholm 1995) 93; Tom Ericsson, *Mellan kapital och arbete: småborgerligheten i Sverige 1850–1914* (Umeå 1988) 108–133; Vello Helk, *Dansk-norske studierejser fra reformationen*

Boel Berner has emphasised that extensive geographic mobility distinguished nineteenth- and early twentieth-century engineers from other academics.[11] This book is about Julius, Haakon, Fjalar, Herluf, and more than 6,500 of their colleagues who were educated in the Nordic countries between 1880 and 1919, and who went abroad before 1930. The overall purpose is to shed light on technicians who Norwegian historian Gudmund Stang calls a 'peregrine profession'.[12]

1 Background, Research Areas, and the Organisation of This Study

This study is located near the point of intersection between the history of migration and the history of science and technology. The study connects to medium-range theories and deals with one of several important processes when the Nordic countries were transformed into modern industrialised welfare societies. It is one of many contributions to the study of this transformation; together with studies in, for example, economic, social, and political history, the history of ideas as well as studies in other subjects.

The author is Swedish, has some Finnish family background, and has conducted a lot of the study in a Norwegian research environment. This background can be regarded as suitable for a comparative Nordic study, but the author's knowledge of Sweden is still better than his knowledge of the rest of

til enevælden 1536–1660: med en matrikel over studerende i udlandet (Odense 1987); Vello Helk, *Dansk-norske studierejser: 1661–1813* (Odense 1991); Steve Hochstadt, 'Migration in Preindustrial Germany', *Central European History* 16:3 (1983) 195–224; Even Lange, *Norske ingeniører i Amerika 1900–1950: en moderne svennevandring* (Bekkestua 1988) 1–2; Timo Myllyntaus, '"The Best Way to Pick Up a Trade", Journeys Abroad by Finnish Technical Students 1860–1940', *Icon (London)* 2 (1996) 141–142; Jussi Nuorteva, *Suomalaisten ulkomainen opinkäynti ennen Turun akatemian perustamista 1640* (Helsinki 1997) 56; Panu *Käyttännön ja teorian välissä: teknillisen opetuksen alku Suomessa* (Espoo 1998) 80–82; Reinhold Reith, 'Einleitung: Elitenwanderung und Wissentransfer', in: Dittmar Dahlmann and Reinhold Reith (eds.), *Elitenwanderung und Wissenstransfer im 19. und 20. Jahrhundert* (Essen 2008) 10–14; Ana Simões, Ana Carneiro and Maria Paula Diogo, 'Travels of Learning. Introductory Remarks', in: Ana Simões, Ana Carneiro and Maria Paula Diogo (eds.), *Travels of learning: a geography of science in Europe* (Dordrecht 2003) 1–18; Sverker Sörlin, *De lärdas republik: om vetenskapens internationella tendenser* (Malmö 1994) 121–132.

11 Boel Berner, *Sakernas tillstånd: kön, klass, teknisk expertis* (Stockholm 1996) 43.

12 Gudmund Stang, 'A measure of relative development?: a note on the education and dispersion of Scandinavian engineers 1870–1930', in: Bjørn L. Basberg, Helge W. Nordvik and Gudmund Stang (eds.), *I det lange løp. Essays i økonomisk historie tilegnet Fritz Hodne* (Bergen-Sandviken 1997) 93.

the Nordic area. The author is also mainly an empirically rooted migration historian, and this may be transparent. Readers may object to the presence of many details and names in this book. However, these details and names may form a base for future researchers, who would like to go deeper into certain individuals. The educational background of the technicians mentioned in the text will be given in footnotes.

Comparative Nordic perspectives will run as a main thread through the study, but comparisons of technicians based on their social, geographical, and educational background will also be performed. As this study deals with four countries, it has not been possible to conduct in-depth studies. Examples are based mostly on secondary material. The database consisting of 12,376 engineers and architects educated in the Nordic countries from 1880 to 1919 makes possible a high degree of generalisation.

This chapter will deal with definitions and limitations, historical backgrounds, theoretical aspects, previous research, sources, and methods. Chapter two deals with the mobility out of the Nordic countries: its magnitude, and what characterised the graduates who departed. Chapter three discusses the choice of destination; where the graduates went, why, and what governed their choices. Chapter four takes a more in-depth look at the activities of the Nordic engineers and architects in their main destination area, the German-speaking countries of Europe, that is, Germany, Switzerland, and Austria. Chapter five does the same for the second largest destination area, North America, the United States and Canada. Chapter six follows the Nordic technical school graduates to other destinations around the world: from those who crossed a border to another Nordic country to the ones in Argentina and the Pacific. Chapter seven joins the graduates on their journeys back home, discussing the magnitude of return migration, who returned, the presence in the Nordic countries of technicians with foreign experience, and their careers. Chapter eight, finally, is a concluding discussion.

2 Major Concepts and Definitions

Below, we will describe some major concepts and definitions. They are related to technicians and technology, mobility and migration as well as the Nordic countries.

2.1 *Concepts and Definitions Related to Technicians and Technology*

The modern engineer can be described as a person with deep technological-scientific knowledge who has undergone an education at a technical university or lower level technical school.[13] Partly for reasons related to the available source material, this definition will also be used in this study. In addition, from a narrow social science terminology, a profession can be viewed as an occupation whose authority and status are grounded in a formally high education. Nevertheless, we should underline that people performed 'engineering' tasks long before the establishment of educational institutions.[14] Also during our studied period, other 'engineers' co-existed with the formally educated ones, for example, bricklaying masters, entrepreneurs, and autodidactic master mechanics.[15] Some were skilled and not necessarily less important than educated engineers, but it is not within the scope of this study to include them all.

An architect can be described as a person who is educated to design buildings and housing areas. He or she should perform these tasks from esthetical and functional points of departure.[16] Education to become an architect was provided by technical schools and art schools, but there were also several autodidactic 'architects'. This study includes educated architects, but only graduates from technical schools. This selection implies a relatively coherent study, with a 'technological' emphasis and, when discussing the entire Nordic group of engineers and architects or 'national' groups, we will sometimes use *technicians* as a term. This is relevant as they all were educated at technical schools. However, to vary the language, we will sometimes also use *graduates*. Thus, these words will basically be used as synonyms.

Apart from the technical school connection, the reasons to include architects relate to the fact that they have much in common and often performed similar tasks as civil and construction engineers, especially in Norway. Also, excluding architecture graduates would have made the very low number of female graduates in this study even lower and the difference between the already significantly smallest Finnish cohort and the other countries even larger. An inclusion of art school architects had, of course, been interesting from many points of departure; we would have been able to include Danish architects (educated at the Copenhagen Academy of Arts), and we would almost

13 Henrik Harnow, *Viden om—den danske tekniske rådgivnings historie 1850 til i dag* (København 2004) 14; Karl-Erik Michelsen, *Viides sääty: insinöörit suomalaisessa yhteiskunnassa* (Helsinki 1999) 8.

14 Henrik Harnow, *Den danske ingeniørs historie 1850–1920: danske ingeniørers uddannelse, professionalisering og betydning for den danske moderniseringsproces* (Århus 1998) 13–14; Bosse Sundin, *Den kupade handen: historien om männsiskan och* tekninken (Stockholm 2006) 243.

15 Harnow, *Den danske ingeniørs historie*, 13; Berner, *Sakernas tillstånd*, 121.

16 Björn Linn, 'arkitekt', *Nationalencyklopedin*. n.d. http://www.ne.se/uppslagsverk/encyklopedi/lång/arkitekt (29 June 2017).

certainly have reached a higher female share. However, as the source material is less comprehensive, we have excluded them.

Technology has often been defined as 'activities aimed at satisfying human wishes using physical artefacts'. This definition is relatively workable and covers activities such as scientific management as well as innovation and development of artefacts. The construction or improvement of a machine aiming to manufacture something new or better is based on human wishes to start or improve such production. Scientific management can be defined as technology since it is a way to satisfy wishes of higher productivity and includes artefacts.[17] In everyday speech, technology and technique are often used synonymously. Nordic sources and secondary literature often use words that linguistically are better translated into 'technique' than into 'technology': *teknik* (Danish and Swedish), *teknikk* (Norwegian), and *tekniikka* (Finnish). We are using these words synonymously for 'technology'.

Technology transfer is not necessarily the movement of technology over national borders, but this is what we deal with in this study. Cross-national technology transfer embraces the movement of established or new technologies from industrialised to developing countries as well as between industrialised nations. This study deals primarily with transfer between established and leading industrial countries to a region of nations that underwent their large-scale industrialisation process during the studied period.

A general definition of technology transfer may also include transfer from one industry or activity to another, but this phenomenon is more often called 'diffusion of innovation' and is not the prime topic of this study.[18] This study does not claim that a graduate who returned and used practical knowledge acquired abroad necessarily was the first to use it in the country. Questions revolve more around how practises were implemented in several places and over time.

2.2 *Concepts and Definitions Related to Migration and Mobility*

Transnational mobility has been a widely used concept, not least in the European Union, since the Treaty of Rome in 1957. Søren Kristensen has, however, pointed to problems with the concept; it has been widely used to cover many different activities, and the precise meaning has, therefore, become unclear. Kristensen has developed a taxonomy that resembles the concepts in this

17 Svante Lindqvist, 'Vad är teknik?', in: Bosse Sundin and Boel Berner (eds.), *I teknikens backspegel: antologi i teknikhistoria* (Stockholm 1987) 32–33.

18 Svante Lindqvist, 'Social and cultural factors in technology transfer', in: Kristine Bruland (ed.), *Technology transfer and Scandinavian industrialisation* (New York NY 1991) 15.

study. The lines are drawn between mobility for recreation, labour-market purposes, and as a learning process.[19] The technicians in our study followed two main lines; they performed transnational mobility related to labour-market issues and—to an even larger extent—they crossed international borders to learn.

Kristensen also makes use of the concept 'placements abroad', which shares many features with the patterns of many technicians in this study. This is not the same as ordinary employment in other countries since the placements need to be consciously organised for learning purposes. Kristensen defines the concept as follows: 'A shorter or longer period spent abroad in a public or private enterprise, which has been consciously organised for learning purposes, which implies active involvement in concrete work processes, and which can be paid or unpaid'.[20] Today, it is more common that placements abroad are organised by companies, schools, and authorities, but Kristensen underlines that they also can be self-organised. This was the case for a majority, albeit not all of them, of our late nineteenth- and early twentieth-century technicians. The fact that placements abroad should involve concrete work processes excludes several other activities; one is study trips, where the participants observe the work process; they normally are not involved in it.

This study uses the concepts outlined in Krsitensen's study. We will deal with transnational mobility for labour-market purposes, that is, Nordic technicians who supposedly went abroad because of difficulties in finding employment in the native country and the like. This mobility was not primarily for learning, but this is not to say that learning did not happen. On the contrary, Kristensen underlines that most learning is unconscious, and it may very well have been the case that labour-market mobile graduates returned, made use of their experiences, and became important in the Nordic countries.

Transnational mobility for learning purposes will be discussed in terms of two sub-categories: placements abroad and study trips/study travel. The reason for separating them thus relates to the assumption that the latter did not involve in real work processes. Journeys that, for example, aimed at visiting mechanical workshops, ironworks, and the like, often only for a few days, cannot be regarded as placements abroad, nor can visits to fairs like the ones in

19 Søren Kristensen, 'Support for transnational mobility for young people', in: Alexandros Tassinopoulos, Heinz Werner and Søren Kristensen, *Mobility and migration of labour in the European Union and their specific implications for young people* (Thessaloniki 1998) 99, 102.

20 Søren Kristensen, *Learning by leaving: placements abroad as a didactic tool in the context of vocational education and training in Europe* (Luxembourg 2004) 16.

Chicago in 1893, Paris in 1900, or St. Louis in 1904. This type of mobility implies stays at hotels, with friends, colleagues, or the like. Nevertheless, this division is not a completely lucky one. Some technicians, especially in Finland, may have stated study trip, although they really were on what we define as 'placements abroad'. However, there is a need to distinguish between technicians who provided information on foreign workplaces as well as time periods and those who only stated that they had travelled in other countries.

In this study, we do include enrolment at foreign educational institutes in the *placement abroad* concept; as opposed to Kristensen. One reason is that these technicians mostly spent at least one semester, often longer, at these institutes; they had to reside abroad during such periods; they hardly stayed in hotels, and so forth. Therefore, we will regard university activities like lectures, reading in the library, and exams as the university's *concrete work processes.* Hence, we will regard studies as a kind of work.

'Migration', 'migrants', and so forth will be used as generic terms for *labour-market mobility* and *placements abroad* together. Migration embraces permanent and temporary residential changes as well as moves with open time perspectives, but hence not tourism and other types of temporary travelling that do not imply residential change.[21] In chapter 5, dealing with the Nordic engineers' and architects' mobility to North America, we defined the transatlantic mass migration from Europe as *traditional emigration.* This term mainly implies permanent settlement in Nordic/Scandinavian-American areas, i.e. to some extent based on the popular view of the transatlantic emigrant conveyed in literature like Norwegian-American author Ole Edvart Rølvaag's novel *Giants in the Earth* (1927) and Swedish author Wilhelm Moberg's novel series *The Emigrants* (1949–1959). The term is not unproblematic, however, as the transatlantic mass migration was extraordinary manifold, even when engineers and architects are excluded. Having stated that, the term is practical for comparing the technicians' North American migration patterns with the general ones. It also serves a purpose for a discussion on whether some of the technicians were emigrants in the traditional sense or going temporarily to America to acquire useful experiences for a career back home.[22]

21 Dirk Hoerder, Jan Lucassen, and Leo Lucassen, 'Terminologien und Konzepte in der Migrationsforschung', in Klaus J. Bade, et. al (eds.), *Enzyklopädie Migration in Europa: vom 17. Jahrhundert bis zur Gegenwart* (Paderborn 2008) 36.

22 Per-Olof Grönberg, 'Journeymen or traditional emigrants? Norwegian and Swedish engineers and architects in North America, 1880–1930', in: Philip J. Anderson and Dag Blanck (eds.), *Norwegians and Swedes in the United States. Friends and neighbors* (St. Paul, MN 2011) 197–218.

As for *mobility*, however, migration and travelling as well as commuting and other types of daily movement can be embraced by the term. This study, however, uses *total transnational mobility* as a generic term for all the registered cross-border moves of these technicians, be it migration or study trips. We are studying a *transnational mobility* that is connected to these individuals' professional lives, but we cannot state for sure that the change of workplace always was the main reason for migration; it could have been the longing to live in Paris, love, or something else. We will never know for sure the true purposes of all transnational mobility.

As for study travellers, it is inbuilt that they are returning. Once they settle permanently abroad, they become migrants. *Return migration* is, therefore, limited to migrants. As a term, it can be defined as people's return to their country, place of origin, or an earlier place of residence after having lived elsewhere for some time.[23] In this context, we are speaking of return migration back to the *country of education*, not necessarily the country of birth. This migration was not necessarily directed to a previous place or region of residence in that country, that is, a technician who went abroad immediately after graduation in Trondheim is regarded a returned migrant to Norway even if the person was employed in Bergen after the intermission abroad. When we are discussing the experience of returned migrants and study travellers together, we will use the term *foreign-experienced*.

2.3 *Concepts and Definitions Related to the Nordic Area and Individual Countries*

The Nordic area, that is, Denmark with the Faroe Islands and Greenland, Finland with the Åland Islands, Iceland, Norway, and Sweden, has historically been a relatively homogenous geographic entity. The countries have had strong political ties. The extraordinary mutually intelligible Scandinavian languages make the area very suitable for comparative studies. Finnish is completely different, but Swedish is also an official language in Finland, and native Finnish speakers generally possess knowledge of Swedish. Language has, nevertheless, been used as a reason to neglect Finland in comparative Nordic studies. The inclusion here is partly possible since the author has some knowledge of Finnish. Source material, written documents, and several historical reference works are, however, often available in Swedish, not least in studies of higher professional groups.

---

23 Russell King, 'Generalizations from the History of Return Migration', in: Bimal Ghosh (ed.), *Return migration: journey of hope or despair?* (Geneva 2000) 8.

We will use the term Finnish to describe technicians educated in Finland. This is based on citizenship and birth country rather than native tongue. Most technicians were Swedish speaking, that is, Finland-Swedes.[24] Finnish as a generic term is often used also by scholars from Finland. Many place names in Finland are in both languages. This study follows the recommendations to use the place's majority language when writing in English.[25]

Scandinavia is an often-used generic term for all the countries mentioned above and dependencies, especially internationally. In this study, however, we will use the Nordic area to describe this entity, whereas Scandinavia sometimes will be used as a term embracing Denmark, Norway and Sweden. The reason for this selection is partly practical. We will discover some typically 'Finnish' traits in this study, and it is useful to discuss them in terms of Finnish-Scandinavian differences.

During the period under study, the borders of two of the Nordic countries differed from today. Parts of what was then Finland's Viipuri County, located on the Karelian Isthmus, belong today to Russia. Petsamo, east of today's Norwegian-Russian border belonged to Finland from 1920 to 1947. Denmark settled its present borders in 1920 when northern Schleswig (southernmost Jutland) was reunited with the country. For forty years of our time period, today's southernmost Jutland was, thus, outside our definition of the Nordic area. This study follows the national borders of the studied period.

The political status of the countries diverged. Sweden and Norway were in a union between 1814 and 1905. The Swedish king was head of state in both

---

24 Susanna Fellman, *Uppkomsten av en direktörsprofession: industriledarnas utbildning och karriär i Finland 1900–1975* (Helsingfors 2000) 60; Pasi Tulkki, *Valtion virka vai teollinen työ?: insinöörikoulutus sosiaalisena ilmiönä 1802–1939* (Turku 1996) 173. The 'Swedish' administration remained intact, during the Grand Duchy period, higher education was for long offered only in Swedish and the language dominated private industry. Native Finnish speaking engineers often worked in administration. Swedish speakers constituted a significantly higher share of the technical school graduates compared to the population at large. It is difficult to reach an exact measurement. Many Finnish speakers had on the one hand Swedish family names, while some Swedish speakers translated their names to Finnish in the light of the growing national consciousness. Names are only a rough indication, but it is reasonable to assume that Swedish speakers constituted about every second graduate, albeit the Finnish share increased over time. An inclusion of the technical school in Tampere, whose first graduates left in 1916, would probably to a small extent have levelled out these differences.

25 For instance, as Finnish is the majority language in Finland's former capital, Turku will be used and not the Swedish form *Åbo*. As Swedish is the spoken language in the Finnish supremacy island group between Turku and Stockholm, the Åland Islands will be used instead of the Finnish name *Ahvenanmaa*.

countries, and the ministry for foreign affairs in Stockholm ran a mutual foreign policy. Norway, however, had its own laws, parliament, and institutions. Consequently, we regard Sweden as a fully independent country and Norway as a semi-independent one. Mobility over the intra-union border is regarded as transnational. The same goes for pre-1917 mobility over Finland's eastern border. Finland was a Grand Duchy under Russia until it reached full independence in 1917 but had its own institutions, senate, national bank, currency, and the Finnish people could, to a high degree, govern themselves.[26] Max Engman criticises the term 'Russian parenthesis' and emphasises the period as important for the constitution of Finland as a nation. The relative autonomy made it, for example, possible to introduce universal suffrage for men and women in 1906.[27] Hence, Finland is also regarded as a semi-independent country.

Denmark was the only Nordic country possessing dependencies. Furthest away lay the Caribbean islands St. Croix, St. John, and St. Thomas, constituting the Danish West Indies. Greenland and Iceland were also under Danish supremacy, albeit Iceland from 1874 stepwise became more autonomous until the island nation reached full independence in 1944. Iceland's industrialisation can be dated to around 1910 and was dominated by the steam trawler industry.[28] The Faroe Islands lay closer but were nevertheless clearly separated from Denmark proper. The Faroe legislative position of the time was, however, on a par with the mainland regions; the island cluster constituted an *amt,* a Danish county. We will not consider mobility to the Faroe Islands as transnational, but to the Danish West Indies, Greenland, and Iceland.

3 Theoretical Framework

Migration/mobility and technology transfer are research agendas in which a myriad of theoretical approaches have been evident for a long time. It would take more than one book to survey all of them. This survey focuses on some approaches that are relevant for this study: theories of migration and mobility as well as return migration and technology transfer.

---

26 Riitta Hjerppe and Jukka Jalava, 'Economic Growth and Structural Change—A Century and a Half of Catching-up', in: Jari Ojala, Jari Eloranta and Jukka Jalava (eds.), *The road to prosperity: an economic history of Finland* (Helsinki 2006) 34.

27 Max Engman, *Lejonet och dubbelörnen: Finlands imperiella decennier 1830–1890* (Stockholm 2000) 9–37.

28 Magnús S. Magnússon, *Iceland in transition: labour and socio-economic change before 1940* (Lund 1985) 19–20.

3.1 Migration and Mobility

Mobile technicians constitute a microscopic share of human activity with a very long history. The great movement of tribes over Europe before the Early Middle Ages is usually labelled the *migration period*, and people in the pre-industrial society changed residence only two times less than we do today.[29] We have touched upon early students, journeymen, mass movement over the Atlantic as well as the even more significant relocation of people from rural to urban areas. Post-war European nations became countries of labour and refugee immigration, and movements within and over borders continue to be one of the most momentous human activities. Migration's diversity and complexity have made some scholars predict the eternal absence of a general migration theory.[30] Some concepts have, nevertheless, remained widely used over time, like Ravenstein's *laws of migration* and the *push and pull* factors.

Migration models deriving from neo-classic economics tended to focus on economic causes and viewed migrants as rational individuals whose choices were grounded in cost-benefit calculations and based on the access to full information and free choices. Theories from the *new economics of migration* criticised this perspective and wanted to focus on the importance of families and households as the context in which individual migration decisions were taken. This is an overly family-oriented approach for a study of professional mobility. *Historical-structural theories* also emerged as a reaction against the neoclassical ones and interpreted migration as a manifestation and reinforcement of unequally distributed economic and political power. This happened in peasant societies, and Immanuel Wallerstein's *world-system theory* applied them on a global scale. Peripheral areas were undermined and suffered from brain-drain through the gradual incorporation in the global capitalist system. These models were criticised for rigidity and for viewing migrants as helpless 'victims' of structures. Different countries have experienced diverging economic development after their incorporations in more 'globalised' economic systems, and remittances and return migration have often had positive long-term effects.

One idea in this study is that the return of Nordic technicians was positive. An application of general historical-structural models is therefore difficult, but models from neoclassical economics are also problematic. The widely used push and pull concept resembles, mostly on an individual level, neoclassical theories. The model is based on the coincidence between repulsive factors in

29 Martin Dribe and Maria Stanfors, *Demografins grunder* (Lund 2015).

30 Hein De Haas, *Migration and Development in Southern Morocco. The Disparate Socio-Economic Impacts of Out-Migration on the Todgha Oasis Valley* (Nijmegen 2003) 17.

the area of origin and attractive ones at potential destinations. Everett S. Lee emphasised these factors alongside intervening obstacles and personal factors. Migration is a response to better opportunities elsewhere and the flow of information about those opportunities. Migration is selective; individuals with dissimilar characteristics respond differently to positive and negative factors at the places of origin and destination and diverge in their abilities to cope with intervening obstacles. Migrants are, therefore, rarely representative of their area of origin.

One problem with the *push and pull* concept is its insufficiency to explain why people return to the areas from where they were earlier 'pushed' away. Almost every place and area experiences out- and in-migration at the same time.[31] Timo Myllyntaus uses the model in an article on Nordic engineering students in Zurich. He claims that technicians or technical students may leave their home countries because of political circumstances, shortages in the domestic educational systems, such as poor teaching quality, lack of laboratories and research facilities, and limited employment possibilities. These factors cannot, however, explain the choice of destinations. Students may leave despite the availability of an appropriate domestic education if a potential host country offers an attractive and interesting educational system with possibilities in various disciplines, high-quality teaching, access to up-to-date technology, good opportunities for financial support, a pleasant atmosphere, and career opportunities. Myllyntaus argues that it 'could be claimed that the Swiss Federal Institute of Technology educated a technical elite for the Nordic countries and that its graduates acted as important agents of technology transfer and industrial development in their home countries'.[32] This is true, but it also points to the difficulties of using *push and pull* factors in this context. One might argue that the newly acquired degrees changed the situation, but this would have required a 'reversed' application of the model and a discussion of why this new prestige suddenly made the native countries attractive again. When analysing migration decisions in which returns to the areas of departure are embedded, it is difficult to use a model where deficiencies in exactly these areas constitute a focal point.

In this study, we need concepts that can cope with return migration. We have already touched upon the framework borrowed from Kristensen's division

31 De Haas, *Migration and Development*, 24.

32 Timo Myllyntaus, 'Discovering Switzerland. Internationalisation among Nordic Students of Technology prior to World War II' in: Ana Simões, Ana Carneiro, and Maria Paula Diogo (eds.), *Travels of learning: a geography of science in Europe* (Dordrecht 2003) 299–328, quote from 325.

of transnational mobility into a *learning process* and a *labour-market process.* The reasons for leaving the country of education are, thus, connected to these traits; individuals go abroad because they want to learn new things, implying that a return is predetermined. In the next step, 'learning mobility' is divided into *placements abroad* and *study travelling,* the former implying involvement in concrete work-processes, whereas the latter indicates that the participants 'only' were observers. However, our technicians also leave because domestic labour markets do not satisfy their ambitions. Sometimes, it may be difficult even to find employment. At other times, accessible positions might not be satisfactory in terms of assignments, wages, and career possibilities. These are, of course, in a way, *push* factors, but the concept still does not explain their choice of destination.

The *placements abroad* concept has a lot in common with *target migration:* a concept based on, for example, studies conducted by Frank Bovenkerk.[33] Bovenkerk divided migrants into four categories based on intentions and outcome: (1) migrants with return intentions, who did in fact return; (2) migrants with return intensions, who changed their minds and settled permanently; (3) migrants with permanent settlement intentions, who changed their minds and returned; (4) migrants with permanent settlement intentions, who did, in fact, settle permanently. The first category embraces *target migrants* who move with specific aims in mind: to accumulate money, knowledge, or skills through education and practise and to utilise real and/or human capital upon a return. Technicians leaving for self-organised *placements abroad* belong to this category of migrants, but *labour-market mobility* may also be a part of it. A technician may work abroad to avoid unemployment, low payment, or uninteresting assignments, but intend to return home when times are better. Such a migration is not necessarily primarily for learning purposes but may imply 'unconscious' learning. Similarly, the intended permanent migrant may start to consider return as an alternative, since he or she 'unconsciously' has learned something useful. Of course, an intended temporary migration, a self-organised *placement abroad,* may also result in permanent out-migration. A projected returnee technician may very well find the working conditions and other aspects of life in the host country satisfactory enough to change his or her mind and settle for good.

As for today's *placements abroad,* it is more common that they are organised by schools, companies, authorities, and the like. This implies that

33 Frank Bovenkerk, *The sociology of return migration: a bibliographic essay* (The Hague 1974) 9–19; King, 'Generalizations from the History of Return Migration', 11–12.

destinations often are governed by external forces. The technicians who went abroad in the decades around 1900 usually organised their foreign intermissions themselves; thereby, they also decided where to go. What governed their destinations? Generally, a choice of destination cannot be explained solely by economic forces. Today's migrants from Morocco, for example, often prefer France and French Canada even if wages generally are lower compared to, for example, Germany. Colonial ties and language similarities have a profound influence on the shaping of migration patterns. Direct labour recruitment is sometimes important, but more important are the movements that originate in settlement by an initial group of migrants. Distance is another factor; it is no coincidence that contemporary Moroccans go to Spain or that most of the international migrants in the late nineteenth- and early twentieth-century Stockholm came from Finland and, in addition, from the nearby Swedish-speaking regions. A group establishing a migrant community at a destination increases the likelihood of subsequent migration between the same places; this has been called *chain migration* or, in more recent terminology, *network migration*. These networks tie people together; they may be kin, friends, or share a community of origin or the same profession. If migrant networks grow to a critical size, they become self-perpetuating for migration. These networks form a social capital that migrants can use, together with real capital and human capital, to find employment. Social capital thereby enables migration as do the other capital forms. Networks are most important in transnational migration since transnational migration generally involves higher costs and risk. These networks and social capital offset other migrant obstacles.

Migration systems can be viewed as a set of geographical places linked by flows and return-flows of people, in addition to goods, service, and information. These processes tend to facilitate further migration between the places. If information about favourable conditions is received at the place of origin, new and sometimes organised migration occurs. Ravenstein's law that a migration stream from one place to another always generates a counter-stream is not contradictory; the back-migration becomes, instead, an integrated part of the system. Internationally, migration systems may consist of countries—or places within the countries—that are characterised by similar feedback mechanisms. Migration systems link people together in transnational communities resulting in relatively clear migration structures; they are encouraging migration along certain pathways and discouraging it along others. We can find clusters of people with the same origin in certain areas, cities, and even city quarters. Macro-level migration systems tend to strengthen ties between countries and regions and include, for example, today's migrant workers from the eastern and southern Mediterranean in the European Union and Central Americans in the United States.

From the middle ages to around 1800, several regional migration systems had emerged in Europe. One was the so-called 'Baltic system' that existed between the 1650s and 1720s, a period when Sweden's power had reached its zenith. An extensive administrative apparatus, exploitation of natural resources like iron and copper, and construction of castles and residences resulted in a demand of skills and in-migration. German, Dutch, Walloon, French, Jewish, and Scottish specialists migrated and often hired masters and journeymen from their respective places of origin.[34] Around 1800, there were seven migration system in Europe, of which the largest was the North Sea system, primarily providing the Netherlands with seasonal labour. Other systems centred on southeast England, Paris and its surroundings, Madrid, Castile, the coastal regions between Catalonia and Provence, and the Po Valley in Italy. The pull of cities was very important. Two parallel migration systems emerged when intercontinental migration resumed after the 1815 Congres of Vienna. The eastward Russo-Siberian system brought migrants to industrialising cities such as Saint Petersburg and Moscow as well as to the Donets basin, fertile regions along the Volga, and to the building of the Trans-Caspian and Trans-Siberian railways respectively. Dirk Hoerder identifies a border line between the systems stretching from Lake Peipus—between today's Estonia and Russia—through Smolensk to Odessa,[35] but it seems clear that Finland for a long time also was included in the eastern system.

A western, Atlantic migration system also emerged in the nineteenth century, based on the demand for labour, but also in the search for cultivable lands. The local and regional migration patterns became increasingly more international and not least intercontinental: about 50 million people left Europe from 1800 until World War I, and a clear majority went to North America. The United States and Canada became part of the labour importing Atlantic core, along with European countries like Britain, France, Germany, and Switzerland. At the same time, Britain and Germany exported labour and settlers to North America, whereas the French and Swiss transatlantic migration was considerably lower. The labour-exporting European peripheries included nations like Ireland, Italy, eastern Austria-Hungary, Poland, and Scandinavia. The Scandinavian countries exported vast numbers of workers and settlers to

---

34 Dirk Hoerder, 'Migration in the Atlantic economies: Regional European origins and worldwide expansion', in: Dirk Hoerder and Leslie Page Moch, *European migrants: Global and local perspectives* (Boston, MA 1996) 27.

35 Hoerder, 'Migration in the Atlantic economies', 34–36.

North America, and a more limited number went to Germany. The core attracted labour from the periphery. Around 1900, the Atlantic migration system included the importing Atlantic core in North America and western Europe, and an eastern, southern and northern labour-exporting European periphery. The western parts of North America constituted an importing periphery. Hoerder calls the rest of the world—the largely politically independent Latin American countries and the still colonised areas in Africa, Asia and Oceania—an 'appendix' to the core nations. They attracted settlers, labour, and often investments.[36]

The mobility of the Nordic technicians will be discussed with these migration systems as background, but we will investigate the technicians' own migration systems, or rather *mobility systems* on a micro-level.

3.2 *Impact of Return Migration and Transfer of Technology and Ideas*

There are many concepts in technology transfer. Early influential scholars include Everett M. Rogers, whose book *Diffusion of Innovasions* and vocabulary—innovators, early adopters, laggards, opinion leaders and change agents—influenced almost all researchers in the 1960s and 1970s.[37] Historians of technology and economic historians are, of course, only two groups among a multitude of scholars who paid attention to technology transfer.[38] Economic historian Nathan Rosenberg's conclusions that the machine tool industry was pivotal to technical diffusion in nineteenth century United States constitute some of the most influential historical studies.[39] Scholars like Vernon W. Ruttan and Yujiro Hayami, as well as John M. Staudenmaier, focused the importance of support networks in technology adaptation.[40] In his influential *Networks of Power,* Thomas P. Hughes argues that late nineteenth and early twentieth century electrical power systems evolved through five phases, where technology transfer was one. Hughes does not emphasise mobility, but one of the components in the transfer of Edison's system was that people met at exhibitions like the one in Paris in 1881.[41] People meeting each other is important for technology transfer. Bruce Edsall Seely refers to several studies on the

---

36 Hoerder, 'Migration in the Atlantic economies', 36–39.

37 Everett M. Rogers, *Diffusion of Innovations* (New York, NY 1962).

38 Bruce Edsall Seely, 'Historical Patterns in the Scholarship of Technology Transfer', *Comparative Technology Transfer and Society*, 1:1 (2003) 7–48.

39 Nathan Rosenberg, *Perspectives on technology* (New York, NY 1976); Nathan Rosenberg, *Inside the black box: technology and economics* (Seoul 1982).

40 Cf. Seely, 'Historical Patterns', 22.

41 Thomas P. Hughes, *Networks of Power. Electrification in Western society, 1880–1930* (Baltimore, MD 1983) 47–78.

diffusion of steam engines, iron, and textile manufacturing to nineteenth century United States and concludes: 'These studies emphasized that successful transfers required moving people, not just hardware'.[42]

However, mobile technicians constitute of course not the only channel for technology transfer. Six channels for nineteenth-century technology transfer to peripheries have been identified: (1) import of foreign machinery and equipment; (2) foreign direct investments; (3) foreign licenses and patents; (4) recruiting skilled workers, engineers, artisans, teachers, and consultants from abroad or permitting mass immigration; (5) journeys abroad by the country's citizens to study at educational institutions, train in factories, offices, workshops, and so forth, or visits to fairs and other places for exchange of technological knowledge; (6) 'natural and low-cost diffusion', that is, dispersion of know-how through scientific publications, trade, analysis of foreign products, the establishment of research institutes, and so on.

Previous research has argued that technological know-how transferred through the first channels has had a quicker impact on the recipient countries' economies. It has been stated that all channels were utilised across the world, but the importance of them has differed between different countries.

The United States relied heavily on mass immigration, while machinery import, natural diffusion, and licenses and patents were less important or became important in a later stage. Russia acquired technology through foreign direct investments and skilled immigration. Japan attracted minor foreign investments and relied first on machinery import and natural diffusion. In a later stage, recruitment of foreign specialists and Japanese technicians returning from abroad became important. Foreign licenses and patents were most important in Germany. In Sweden, all the channels were important except foreign direct investments, but natural diffusion was most important. Finland's most significant channel was citizens coming back from studies abroad, but machinery import, immigration, and natural diffusion also contributed.[43]

This study focuses on the mobility of persons, that is, it does not set out to discuss all six channels and compare their importance. Such a study would have required a different source material and different research questions, and this is also in part already discussed by earlier researchers. Hence, we are dealing primarily with channel five and, to a smaller extent, channel four. Mobility leading to direct personal contact has been viewed as important for historical

42 Seely, 'Historical Patterns', 21.

43 Timo Myllyntaus, *The gatecrashing apprentice: industrialising Finland as an adopter of new technology* (Helsinki 1990) 100–105.

technology transfer, for example, by Seely and in a study of Scandinavian industrialisation by Kristine Bruland.[44] Seely writes:

> Most importantly, historical analysis emphasizes that successful transfers rest on the exchange of people, not just machines, drawings, blueprints, patents or other technical literature.[45]

Hence, mobility has been focused as a major source of technology transfer. Technical societies were important as they spurred mobility by sponsored visits to foreign countries and dissemination of information on foreign technical development, for instance, through journals. However, it was more difficult for technical journals, drawings, and the like to play independent roles.[46] The importance of personal contact for technology transfer may have been more obvious during the pre-industrial and early industrial eras when formalised knowledge made up only a small part of all knowledge transferred. Formalised knowledge became more extensive during the industrialised era, but we will claim that humans continued to be essential in the transfer process.

The relationship between migration and development has been an ongoing scientific debate. Migration as positive for both the sending and receiving areas was the dominant perspective from the 1950s to the early 1970s: development was strongly linked to return migration. The early 1970s to the early 1990s was dominated by the 'development of underdevelopment' perspective, while the issue was nearly completely neglected in the 1990s. The first decade of the third millennium has turned the tide towards optimism again; return migration and remittances are viewed as potential development forces and

44 Kristine Bruland, 'Skills, learning and the international diffusion of technology: a perspective on Scandinavian industrialization', in: Maxine Berg and Kristine Bruland (eds.), *Technological revolutions in Europe* (Cheltenham 1998) 176–183; Lange, *Norske ingeniører*, 1–2; Ole Hyldtoft, 'Foreign Technology and the Danish Brick and Tile Industry, 1830–1870', in: Kristine Bruland (ed.), *Technology transfer and Scandinavian industrialisation* (New York, NY 1991) 226; John M. Staudenmaier, *Technology's storytellers: reweaving the human fabric* (Cambridge, MA 1985) 124, 218; Göran Ahlström, 'Världsutställningar och teknikspridning 1850–1914', in: Bengt Berglund, Per-Olof Grönberg and Tomas Nilson (eds.), *Historiska perspektiv på tekniköverföring 1800–2000* (Göteborg 2006) 75–76.

45 Seely, 'Historical Patterns', 22.

46 Henrik Björck, 'Bilder av maskiner och ingenjörskårens bildande: tekniska tidskrifter och introduktion av ny teknik i Sverige, 1800–1870', *Polhem* 5 (1987) 267–310, 294–295; Bruland, 'Skills, learning', 176–183; Lange, *Norske ingeniører*, 1–2.

the prospects to turn 'brain drain' into 'brain gain' are once again seen as fully feasible.[47]

There are numerous studies and debates on the effects of 'brain drain' and whether it can be turned into a 'brain gain'; in developing as well as developed countries. Bimal Ghosh concludes that the countries of origin can expect development from return migration if three criteria are fulfilled. Migrants need to have acquired more advanced knowledge and skills compared to what is already possible in the native country, this knowledge and these skills must be relevant for the native country, and the returnees must be willing and must be given the opportunity to use it. Return migration's impact has diverged from one country to another.[48] India, and especially her software industry, has been described as one 'gainer'. Uwe Hunger has noted that a very large proportion of the key positions around the turn of the millennium were filled by software engineers who had been in the United States, read Silicon Valley. This migration, Hunger claims, increased the amount of human capital. Upon return, this capital was invested in India's development and the initial 'brain drain' was thus turned into a 'brain gain'. There are two things to point out: (1) The success of India's software industry was also dependent on other factors, and (2) the impact was in a limited sector of the economy.[49]

One thing worth pointing out is that individual returnees' economic prosperity does not necessarily imply aggregate contribution of return migration on the local, regional or national level.[50] There is a need to discuss different impacts of return migration to acknowledge the contextual influence on the possibility to make an impact or not. Bovenkerk has outlined a framework, emphasising a few influential factors, which is useful for this study. The first emphasised factor relates to the *quantity of returnees,* where a larger number may provide a critical mass for change, whereas a smaller number may have little or no impact. Substantial real numbers and percentages are both important, even if there might be a risk of overstraining at the other end. Some studies have concluded that return migration hardly had more than a local influence. However, if returning migrants act similarly in many different places

---

47 Hein De Haas, 'Migration and Development: A Theoretical Perspective', *International Migration Review* 44:1 (2010) 230.

48 Bimal Ghosh, 'Return Migration. Reshaping Policy Approaches', in: Bimal Ghosh (ed.), *Return migration: journey of hope or despair?* (Geneva 2000) 187–189.

49 Uwe Hunger, The 'Brain Gain' Hypothesis: Third World Elites in Industrialized Countries and Socioeconomic Development in their Home Country (San Diego, CA 2002). Working Paper 47. The Center for Comparative Immigration Studies, University of California, San Diego.

50 King, 'Generalizations from the History of Return Migration', 23–24.

in a country or area as well as over time, the influence may be substantial also on a wider geographical scale. This was assumed in a Norwegian project on returnees from America; by applying new solutions, the returnees became *living examples of innovation.* They treated innovative methods and ideas as a part of everyday life and 'forced their neighbours to evaluate the old and traditional against the new and to make decisions about what to accept and what to reject'.[51] Knut Djupedal continues: 'If we treat the returnees as parts of a great flow of information stretching over decades and now, centuries, perhaps they did have a heavy influence on their surroundings'.[52] This is an interesting perspective. The share of returnees from America among the Norwegian population in general was about one percent in 1900. If such a small segment could exert considerable influence on a whole society, what influence could not considerably higher shares of foreign-experienced technicians exert on industrial and technological development?

Duration of absence is also important and 'somewhere in between' is believed to have the strongest impact. A too short sojourn may not have enabled a returnee with the appropriate knowledge and skills to act in the innovative way revealed above. This is perhaps more relevant for 'ordinary' returnees compared to foreign-experienced technicians; the ones studying at renowned technical universities abroad were generally not away for very long, but they still returned with valuable formal and informal assets. Nevertheless, as for *working practise,* the likelihood of acquiring more knowledge and skills increased the longer the time abroad. Presumably, this was an especially relevant difference between *placements abroad* and *study trips:* the former implying involvement in real work processes, the latter implying only observation. Anders Lundgren has studied a Swedish chemical engineer, Cronquist's, travels in a few European countries in the early 1870s. He was important, but this importance was *not* primarily based on his experience as a traveller. Cronquist stayed only for shorter periods at the work places he visited. Lundgren asserts that he thereby lacked the continuous work experience necessary to acquire the knowledge and skills to make an impact.[53] However, a too long intermission may also have less impact. A return at a higher age may imply fewer ambitions to act innovatively, and a long

51 Knut Djupedal, 'Some thoughts on the influence of returned migrants in Norway', *Göteborgs-emigranten* 6 (1997) 207.

52 Djupedal, 'Some thoughts', 208.

53 Anders Lundgren, 'Kunskap och resande. Några aspekter på den kemiska industrins tidiga utveckling i Sverige', in: Bengt Berglund, Per-Olof Grönberg and Tomas Nilson (eds.), *Historiska perspektiv på tekniköverföring 1800–2000* (Göteborg 2006) 90–92.

absence may also indicate that the returnee has become alienated from the home society.

Bovenkerk identifies two categories that can be viewed as very similar to each other: the *difference between the home environment and the adopted environment* and *quality of the training and skills* acquired abroad. Therefore, we merge them into this study. Ingrid Semmingsen pointed to this problem when comparing Norwegian return from North America with Italian return. Rural Norwegians crossed the Atlantic to work as farm labourers, whereas rural Italians worked in urban industries. Both groups returned to their native areas, but the Italians failed to copy city life and were often looked upon as 'strange' The Norwegians, however, could act innovatively.[54] The training, as Ghosh also pointed out, must be appropriate for the home country. This implies, on the one hand, that the training cannot be too advanced; this means that the technician returns to a country where he, or in a few cases she, cannot apply the acquired knowledge because the native country is not ready. On the other hand, the quality cannot be too low. This implies that the foreign intermission, at least from a technical point of departure, is 'useless'. The technician could, basically, have received the same training at home. In our context, the differences between the Nordic countries and the technicians' 'adopted' countries are essential. Receptiveness has been emphasised as important in several technology transfer studies: the ability to integrate the transferred technology into the technical infrastructure is, to a large extent, decisive regarding whether it will be persistent.[55]

Göran Ahlström uses the often award-winning Swedish products at international fairs in the latter part of the nineteenth century to conclude that the country was at the international forefront in areas such as mining, machinery, and industrial chemistry. Ahlström emphasises the early professionalisation of Swedish engineering, almost equalling the German pattern, as important for the country's relative successfulness.[56] Henry Nielsen and Michael F. Wagner argue that Denmark's unusualness about technology was the filtering and implementation of some technologies to make them appropriate in Denmark at different times.[57] Thus, Denmark's main contribution lies in the adaptation of

54 Ingrid Semmingsen, *Veien mot vest. 2, Utvandringen fra Norge 1865–1915* (Oslo 1950).

55 Yujiro Hayami and Vernon W. Ruttan, *Agricultural development: an international perspective* (Baltimore, MD 1971); Nathan Rosenberg, 'Economic Development and the Transfer of Technology: Some Historical Perspectives', *Technology and Culture* 11:4 (1970) 550–575.

56 Göran Ahlström, *Technological development and industrial exhibitions 1850–1914: Sweden in an international perspective* (Lund 1995) 124, 210.

57 Henry Nielsen and Michael F Wagner, 'Technology in Denmark', in: Jan Hult and Bengt Nyström (eds.), *Technology & industry. A Nordic Heritage* (Canton, MA 1992) 1.

technology, in the adding of a special design and the unique style given to various products, processes, and organisations. This is valid for all Nordic countries. Riitta Hjerppe and Jukka Jalava describe Finland:

> Finland is not known for major or pioneering inventions, but clearly Finnish entrepreneurs have been effective borrowers of technology, thus actually saving themselves the development costs of the pioneers.[58]

David Kirby also observes that the Grand Duchy was receptive to new ideas and characterised by an ability to apply them quickly within important areas.[59] According to Gunnar Nerheim, Norwegians have, over the course of history, successfully adapted foreign technologies to national and regional contexts.[60]

Social class also matters. Returning professionals—such as engineers and architects—are more likely to make an impact than manual labourers. The willingness to return is generally higher among migrants originating from the elites of their countries, as the chances of obtaining higher positions are greater. This is clearly connected to the fifth category, the (5) motives and how migration and return are organised. A back-and-forth migration made to acquire skills and knowledge—a more or less self-organised *placement abroad*—to transform into prestige and a good career upon return is, of course, more likely to make an influence than a return migration performed because of disappointment, homesickness, or the like. Furthermore, if the migration is a part of a programme including grants for studying abroad and so forth, the implication is a strategy related to economic and technological development and the development potential of the returnee is stronger.

The *view of the adopted country in the home environment* is also important. If the adopted country is viewed as a good example and a forerunner in the home environment—by authorities, politicians, industrialists, but also in the public conscience—returnees have more development potential than returnees from countries viewed as hostile or insignificant. There are studies underlining the importance of, primarily, American[61] and

58 Hjerppe and Jalava, 'Economic Growth', 55.

59 David Kirby, *Östersjöländernas historia. 1772–1993* (Stockholm 1996) 310.

60 Gunnar Nerheim, 'Patterns of Technological Development in Norway', in: Jan Hult and Bengt Nyström (eds.), *Technology & industry. A Nordic Heritage* (Canton, MA 1992) 53; Pål Thonstad Sandvik, *Mekanisk industri i en europeisk periferi: Fabrikken ved Nidelven 1843–76* (Oslo 1994).

61 Henrik Björck, *Staten, Chalmers och vetenskapen: forskningspolitisk formering och sociala ingenjörer under Sveriges politiska industrialisering 1890–1945* (Nora 2004) 69; Nils Runeby, 'Americanism, taylorism and social integration: action programmes for Swedish

German[62] influences on the Nordic countries. Tomas Nilson, for example, states that engineers and technicians recommended Swedish industry to follow German examples to make 'everything' better. Sweden could take several 'industrial' steps forward by imitating German organisation and technology.[63] Switzerland was another model: characterised by beauty, freedom of press and associations as well as neutrality. Her independence was an example for Finland and Norway. Swiss economic and technical development, not least her Polytechnic school in Zurich, were admired.[64] The United States was also admired. Norway's commissar at the 1893 Chicago fair described the country as the most advanced country in manufacturing, and nine years later, America was described as superior to Europe in a Norwegian lecture. Stories on the marvels of electricity and modern tools arrived in Europe, including its northernmost part.[65]

Different groups had, of course, different sources of inspiration; architects diverged following Lisa Brunnström as they were more inspired by Europe and Germany than by the United States, although a few Swedish architects were influenced by architects like Richardson and Sullivan.[66] The Mediterranean was another source of inspiration. Italy was the classical destination of architects.[67] Around the turn of the nineteenth century, many northern European architects 'believed that direct contact with the classical remains was indispensable, so that the study tour of Italy assumed great importance as a formative

---

industry at the beginning of the twentieth century', *Scandinavian journal of history* 3:1–4 (1978) 21–46.

62 Bernd Henningsen and Birgitta Högvall (eds.), *Skandinavien och Tyskland: 1800–1914: möten och vänskapsband* (Stockholm 1997); Jarle Simensen and Ole Kristian Grimnes (eds.), *Tyskland—Norge: den lange historien* (Oslo 1999); Edgar Hösch (ed.), *Deutschland und Finnland im 20. Jahrhundert* (Wiesbaden 1999).

63 Tomas Nilson, "Vacker, föredömlig, rationell'—bilder av tysk teknik i Teknisk Tidskrift 1890–1914', in: Bengt Berglund, Per-Olof Grönberg and Tomas Nilson (eds.), *Historiska perspektiv på tekniköverföring 1800–2000* (Göteborg 2006) 66–67.

64 Myllyntaus, 'Discovering Switzerland', 301–303.

65 Sigmund Skard, *USA i norsk historie: 1000–1776–1976* (Oslo 1976) 180; Ole Hyldtoft, 'Perioden 1896–1930', in: Ole Hyldtoft and Hans Chr. Johansen, *Teknologiske forandringer i dansk industri 1896–1972* (Odense 2005) 15–241.

66 Lisa Brunnström, *Den rationella fabriken: om funktionalismens rötter* (Umeå 1990) 202; Göran T. Rygert, *Svenska arkitekter i USA 1846–1930: forskningsprojekt 1993–1996* (Stockholm 1996) 26–28.

67 Hilding Ekelund, Eva Ekelund and Kim Björklund, *Italia la bella: arkitekterna Hilding och Eva Ekelunds resedagbok 1921–1922* (Helsingfors 2004); Nicola Flora, Paolo Giardiello and Gennaro Postiglione, 'Journey to Italy', in: Nicola Flora, Paolo Giardiello, and Gennaro Postiglione (eds.), *Sigurd Lewerentz: 1885–1975* (Milano 2002).

experience'.[68] The main road usually went through Paris to Italy. Sometimes, the architects continued to Greece and/or Spain. Travel grants from the Swedish Academy of Arts usually required that the holders of the scholarships spend a significant share of the three years abroad in Italy.[69] France was also an inspirational source; garden cities in Britain were another. Nordic neighbouring countries also were attractive. Danish architecture was recognised for its use of bricks.[70] Sweden also attracted visitors, especially among Swedish speaking architects in Finland. Swedish landscape architecture and buildings like the Stockholm City Hall and public library exerted influence on other Nordic architects.[71]

4 Previous Research

This overview will focus on what Ana Simões, Ana Carneiro, and Maria Paula Diogo among others, call *travels of learning*. We will begin with academic travel from the Middle Ages to industrialisation, continue with the crossings of European borders among specialists and journeymen as well as transatlantic connections and return migration before surveying transnational mobility of technicians on to around the 1930s.

4.1 *Academic Travel from the Middle Ages to Industrialisation*

Terri Kim claims that Confucian universities in China received many students from nearby countries and that ancient Islamic and medieval European universities were more 'transnational' than many universities today. Students from peripheral Europe travelled to do all or parts of their studies abroad. Going to places like Bologna was often to qualify for professions in law, medicine, and teaching or to become a church servant. Transnational student travel was often a gateway to a lucrative career in a time when borders and immigration regulations were non-existent.[72] Jussi Nuorteva discovered, for example, documents of Finnish students at the University of Paris in 1313. The French

68 Cf. Flora, Giardiello and Postiglione, 'Journey to Italy', 35.

69 Anders Bergström, *Arkitekten Ivar Tengbom: byggnadskonst på klassisk grund* (Stockholm 2001) 35, 43.

70 Bergström, *Arkitekten Ivar Tengbom*, 35–38.

71 Joakim Hansson, 'The architectural-historical journeys of Hilding Ekelund and Eva Kuhlefelt and their early links with Sweden', in: Timo Tuomi, Kristina Paatero, and Eija Rauske (eds.), *Hilding Ekelund: (1893–1984) arkkitehti = arkitekt = architect* (Helsinki 1997) 61–63.

72 Terri Kim, 'Shifting patterns of transnational academic mobility: a comparative and historical approach', *Comparative Education* 45:3 (2009) 387–403.

capital continued to be a main destination for Swedish and Finnish students during the Middle Ages, but students from Sweden also found their way to the universities in Italy. The first German university, in Prague, became important in the fourteenth century. From the late Middle Ages until the nineteenth century, German language universities were the most important ones for this mobility.[73]

Transnational student mobility was, however, hampered after the Renaissance. This was connected to the formation of European nation-states, and universities thereby became parts of a nation state's institutional framework.[74] Travel during the Enlightenment established to some extent the boundaries between 'enlightened spaces' of Europe and peripheries that lagged behind in terms of science and technology. The so-called Grand Tour included France, the Netherlands, Switzerland, and parts of today's Germany and Italy. This tour was—quoting Simões, Carneiro, and Diogo—done in 'search of enlightenment and culture both in its academic form and in terms of social rules, manners and taste'. The aim was to reinforce the traveller's status in the Republic of Letters, and the tours combined pleasure with learning. Visited places were chosen by artists, men of letters, and men of science. Narratives became popular; they fulfilled people's curiosity and were a means to obtain information and knowledge on the road towards enlightenment. Both travellers and non-travellers read the narratives and engaged in discussions about what they had learned from them.[75]

Simões, Carneiro, and Diogo describe a new kind of travel emerging from the Grand Tour. It had similar educational purposes, but showed more specificity when it came to purposes and destinations. Intellectuals and scientists from the European periphery went to the continent's major centres in different fields. The Portuguese historians ascribe these journeys immense importance for circulation, diffusion, and appropriation of knowledge in Europe. These tours ranged from the search for education or training to missions that can be characterised as espionage. This, more professionalised travel pattern was consolidated in the nineteenth century. Travel was now rarely the former combination of pleasure and learning, but divided into tourism and journeys for professional purposes. Professional travel became increasingly more specialised; the choice of destinations was governed by specific aspects in scientific and technological fields that had developed in certain areas or places.[76]

---

73 Nuorteva, *Suomalaisten ulkomainen*, 523–524; Eliasson, 'Svenska studenter', 43.

74 Kim, 'Shifting patterns', 388; Kristensen, 'Support for transnational mobility', 111.

75 Simões, Carneiro, and Diogo, 'Travels of Learning. Introductory Remarks', 1–2, quote from 2.

76 Simões, Carneiro, and Diogo, 'Travels of Learning. Introductory Remarks', 4.

In architect education, study tours were emphasised at an early stage. French *Académie Royale d' Architecture* awarded, for example, their best students grants to spend several years in Rome from the early eighteenth century and onwards. The *Royal Academy* in Sweden began to give grants for architecture studies in France and Italy in 1788. Going abroad was also essential for Finnish architects. Before architecture education was established at the technical junior secondary school in Helsinki in 1863, some studied abroad, often in Stockholm, but also in, for example, Paris and Munich. Later, these trips to the European continent continued, sometimes for formal studies but also freely to discover famous and beautiful buildings. Professors and other teachers emphasised the necessity to study classical architecture 'on the spot'. Every famous architect had undertaken a longer journey in other countries, that is, travelled around in Europe.[77]

Towards the end of the nineteenth century, German research centres had won worldwide recognition and scholars from all over Europe as well as the United States and Japan travelled there.[78] In the earliest years of the twentieth century, students from the Russian Empire (excluding Finland and Poland) constituted the largest group and accounted for 41 per cent of the foreign students in Germany. Students from the Austro-Hungarian double monarchy, including Romania and Serbia, accounted for 22 per cent. Switzerland and Bulgaria accounted for about six percent each, the Americas for five percent, Asia for three percent, and Britain and Greece for about two percent each. Norway was the strongest represented Nordic country.[79]

4.2 *Specialists and Journeymen Crossing European Borders*

Pre-industrial Europe also saw other types of travel. Migration of specialists was essential, so was journeyman or artisan migration.[80] Religious schisms often precipitated skilled refugee migration. The textile workers fleeing from the Catholic Spanish Netherlands (today's Belgian regions of Flanders and Brabant), who brought modern products such as new draperies and luxury goods as well as up-to-date means of production to sixteenth and seventeenth

77 Kim Björklund, 'En sentida Grand Tour. Resan 1921–1922 och bildningsresornas tradition', in: Hilding Ekelund, Eva Ekelund, and Kim Björklund, *Italia la bella: arkitekterna Hilding och Eva Ekelunds resedagbok 1921–1922* (Helsingfors 2004) 15.

78 Kim, 'Shifting patterns', 390.

79 Wolfgang König, 'Technical education and industrial performance in Germany: a triumph of heterogeneity', in: Robert Fox and Anna Guagnini (eds.), *Education, technology and industrial performance in Europe, 1850–1939* (Cambridge 1993) 65–87; Myllyntaus, 'The Best Way', 145 –146.

80 Hoerder, 'Migration in the Atlantic economies', 24.

century England, constitute one example.[81] The Huguenot refugees from France, contributing to a long-term productivity increase in the Prussian textile industry after their settlement in 1685, constitute another example.[82]

In the Nordic area, it was impossible for an artisan to become a master without spending time on the European continent. The 'journeyman was subsequently expected to seek to improve his education by moving from place to place, working for longer or shorter periods in workshops abroad',[83] writes Poul Strømstad. Returning journeymen were active in Danish piano manufacturing, bronze casting, printing, and brewing. The founder of Copenhagen's Carlsberg brewery was a returned journeyman from Munich.[84] Germany's influence on the food and stimulant industry has also been revealed for Sweden, where many brewing masters and machines were German.[85] We can note two vehicles of transfer of brewing technology; through returning natives and direct immigration. Stephan R. Epstein writes that the main vehicles of technology transfer were 'permanent emigration of master artisans and the temporary migration of journeymen'.[86] Reinhold Reith argues that 'brain-drain', 'brain-gain' and 'brain-circulation' can be viewed as modern conceptualisations of traditions with deep historical roots. Travel by textile experts and miners, watchmakers, glass-blowers, and several other groups of skilled workers were customary in the Middle Ages and the early modern period; employers and authorities often recruited them.[87] It was common that an 'initiator' returned to start activities back home, and the recruitment of skilled foreign workers followed this.

81 Raingard Esser, 'Südniederländische Textilarbeiter im England des 16. und 17. Jahrhunderts', in: Klaus J. Bade, et. al. (eds), *Enzyklopädie. Migration in Europa. Vom 17. Jahrhundert bis zur Gegenwert* (Paderborn 2007) 1029–1031.

82 Erik Hornung, 'Immigration and the Diffusion of Technology: The Huguenot Diaspora in Prussia', *American Economic Review,* 104:1 (2014) 84–122.

83 Poul Strømstad, 'Artisan Travel and Technology Transfer to Denmark, 1750–1900', in: Kristine Bruland (ed.), *Technology transfer and Scandinavian industrialisation* (New York, NY 1991) 136.

84 Strømstad, 'Artisan Travel', 135–136; For an example of brewing technology transfer from Germany to Scotland, see: Stefan Manz, 'Technologietransfer und spezialistenwanderung: eine Augsburger lagerbrauerei in Glasgow, 1889–1959', *Zeitschrift für Unternehmensgeschichte* 45:2 (2000) 225–247.

85 Torsten Gårdlund, *Industrialismens samhälle* (Stockholm 1942) 252–255.

86 Stephan R. Epstein, 'Craft guilds, apprenticeship, and technological change in preindustrial Europe', *Journal of Economic History* 58:3(1998) 684–713, 702.

87 Reinhold Reith, 'Circulation of skilled labour in late Medieval and early Modern Europe', in: Stephen R. Epstein and Maarten Prak (eds), *Guilds, innovation and the European economy, 1400–1800* (Cambridge 2009) 114–142; Reith, 'Einleitung', 10–13.

There are myriad international studies of these phenomena. Here, we will mention some dealing with the Nordic countries. Göran Rydèn and Chris Evans have written on forging experts from Liege, Germany, and the Spanish Netherlands in sixteenth- and seventeenth-century Swedish iron trade.[88] Bjørn Ivar Berg and Gunnar Nerheim have revealed how skilled workers from Saxony arrived in Norwegian mining districts in the same centuries.[89] Rolv Petter Amdam has written on the arrival of glass-blowers from Germany, France, and Britain in Norway some century later.[90] Torbjörn S. Fogelberg's and Friedrich D. Holl's study of the Swedish glass-industry also shows an ongoing dependency on skilled foreign labourers from the sixteenth to the early twentieth century. Whereas glass-blowers from Italy, France, and Britain had some marginal importance, the Swedish dependency on German specialists was immense.[91] Nielsen and Wagner have identified Dutch, British, and Swiss experts as essential for eighteenth-century Danish dairy development.[92]

Ole Hyldtoft has revealed how British engineers supervised the erection of mid-nineteenth-century Danish brickworks and supplied technology. The Netherlands, France, and Germany were other technical sources for Denmark's large brick and tile industry.[93] Reverberatory and cupola furnaces were, as Martin Fritz has shown, transferred to the Swedish foundry industry around 1800 through British foundry workers.[94] Rydèn and Claus Wohlert have emphasised Swedish travelling interchange with Britain as important for the dispersion of, for example, charcoal refining, Lancashire forging and ingot steel methods.[95]

---

88 Chris Evans and Göran Rydén, 'Kinship and the transmission of skills: bar iron production in Britain and Sweden, 1500–1860', in: Maxine Berg and Kristine Bruland (eds.), *Technological revolutions in Europe* (Cheltenham 1998) 190–193.

89 Bjørn Ivar Berg, 'Bergseminaret på Kongsberg (1757–1814) og annen tidlig kunnskapsformidling i bergfag', in: Anne Kristine Børresen and Astrid Wale (eds.), *Vitenskap og teknologi for samfunnet?: bergfagene som kunnskapsfelt* (Trondheim 2005) 101–111; Nerheim, 'Patterns of Technological Development'.

90 Rolv Petter Amdam, 'Industrial Espionage and the Transfer of Technology to the Early Norwegian Glass Industry', in: Kristine Bruland (ed.), *Technology transfer and Scandinavian industrialisation* (New York, NY 1991) 73–93.

91 Torbjörn S. Fogelberg and Friedrich D. Holl, *Wanderungen Deutscher Glashüttenleute und Schwedens Glasindustrie In den letzten fünf Jahrhunderten* (Växjö 1988) 9.

92 Nielsen and Wagner, 'Technology in Denmark', 2–3, 8–9.

93 Hyldtoft, 'Foreign Technology'.

94 Martin Fritz, 'British Influence on Developments in the Swedish Foundry Industry around the turn of the Eighteenth Century', in: Kristine Bruland (ed.), *Technology transfer and Scandinavian industrialisation* (New York, NY 1991) 59–72.

95 Göran Rydén, 'Bergsbruket, moderniteten och den första svenska industrispionen', in: Bengt Berglund, Per-Olof Grönberg, and Tomas Nilson (eds.), *Historiska perspektiv på tekniköverföring 1800–2000* (Göteborg 2006); Claus Wohlert, 'The Introduction of the

Torsten Gårdlund describes how British workers introduced new methods in nineteenth-century Swedish porcelain making and were important for the cement industry.[96] British immigrants, technology, and machines also had a strong impact on the Nordic textile industry; this has, for example, been shown by Myllyntaus for Finland,[97] Strømstad for Denmark,[98] Kristine Bruland and Trine Parmer for Norway,[99] and Bengt Berglund for Sweden.[100] Similar patterns can be observed at nineteenth-century engineering workshops, as revealed, for example, in Norway by Bruland and Pål Thonstad Sandvik.[101] Shipbuilding was one area in Norway where British influences remained important for a long period, as Håkon With Andersen has shown.[102] Denmark, too, imported shipbuilding technology from Britain in the mid-nineteenth century; the technical advisor for the Copenhagen dry dock and many skilled workers came from Southampton.[103] The Fiskars workshop in Finland also had managers from Britain, but some also from Sweden.[104] Swedish workshops also often had British managers and experts.[105]

All travelling interchange was not with Britain; some was intra-Nordic. Norwegian immigrant entrepreneurs and workers helped modernise the Finnish sawmill industry and set up machines at Swedish counterparts. Danish skills were important for Sweden's cement industry. A more pronounced pattern is, however, the increasing interest for Germany. We have already

Bessermer Process in Sweden', in: Kristine Bruland (ed.), *Technology transfer and Scandinavian industrialisation* (New York, NY 1991) 295–306.

96 Gårdlund, *Industrialismens samhälle*, 256–257.

97 Timo Myllyntaus, 'Technological Change in Finland', in: Jan Hult and Bengt Nyström (eds.), *Technology & industry. A Nordic Heritage* (Canton, MA 1992) 37–39.

98 Strømstad, 'Artisan Travel'.

99 Kristine Bruland, *British technology and European industrialization: the Norwegian textile industry in the mid-nineteenth century* (Cambridge 1989); Trine Parmer, 'How Industrial Technology First Came to Norway', in: Kristine Bruland (ed.), *Technology transfer and Scandinavian industrialisation* (New York, NY 1991) 37–57.

100 Bengt Berglund, 'Från förlag till automat. Textilindustrins utveckling och organisation i Sjuhäradsbygden 1750–2000', in: Bengt Berglund, Per-Olof Grönberg, and Tomas Nilson (eds.), *Historiska perspektiv på tekniköverföring 1800–2000* (Göteborg 2006) 189–195.

101 Kristine Bruland, 'The Norwegian Mechanical Engineering Industry and the Transfer of Technology, 1800–1900', in: Kristine Bruland (ed.), *Technology transfer and Scandinavian industrialisation* (New York, NY 1991) 252–262; Sandvik, *Mekanisk industri*, 107–108.

102 Håkon With Andersen, *Fra det britiske til det amerikanske produksjonsideal: forandringer i teknologi og arbeid ved Aker mek. verksted og i norsk skipsbyggingsindustri 1935–1970* (Trondheim 1986) 422–424.

103 Bruland, 'Skills, learning', 172–173.

104 Myllyntaus, 'Technological Change in Finland', 38.

105 Gårdlund, *Industrialismens samhälle*, 242.

revealed how German education and universities developed to world-class centres at the end of the nineteenth century and earlier influences on brewing and the like. There was interesting technological development going on in Germany; the inventor of the sulphite technology, the Swedish engineer C. D. Ekman, had, for example, worked with German technicians. Germans also erected several wood grinderies in Sweden before 1870.[106] Some wood-grinding machines used in Finnish paper mills in the 1860s and 1870s were designed by German immigrant engineers, and technologies were, following Eli Moen, also introduced at paper factories in Kristiania and nearby Drammen.[107]

4.3 *Transatlantic Connections and Return Migration*

This was also a time when different groups seriously started to 'discover' development in North America. In the 1860s, milling machines and drills were, for example, brought back by Swedish engineers with special missions whereupon 'the American system' was introduced in some mechanical workshops, as Mats Fagerberg has identified.[108] Gårdlund claims that visits at world exhibitions were important for the reorientation away from Britain and towards Germany and America.[109] The Philadelphia fair in 1876 provided an opportunity for European nations to acquaint themselves with American technology. In Sweden, this resulted in the admiration of American efficiency and a fear of the country's competitive potential.[110] In the early 1890s, the United States was established as the main model in Swedish technical journals. It was a progressive country, a picture that was echoed in Denmark, Norway, and many other countries.[111]

This new admiration for America cannot be separated from the great exodus of Europeans over the Atlantic, which was one of the most significant

106 Gårdlund, *Industrialismens samhälle*, 247–250.

107 Eli Moen, 'Norway's Entry into the Age of Papers: The Development of the Pulp and Paper Industry in the Drammen District', in: Kristine Bruland (ed.), *Technology transfer and Scandinavian industrialisation* (New York, NY 1991) 377–386.

108 Mats Fagerberg, 'Det amerikanska systemets införande i den svenska järnmanufaturindustrin. Diskussionen i svenska tidskrifter under 1880-talet', in: Bengt Berglund, Per-Olof Grönberg, and Tomas Nilson (eds.), *Historiska perspektiv på tekniköverföring 1800–2000* (Göteborg 2006) 161–188.

109 Gårdlund, *Industrialismens samhälle*, 241–243.

110 Ingrid Jansson, *Svensk rapportering av amerikansk teknologi på världsutställningen i Philadelphia 1876* (Stockholm 1980) 16–17.

111 Marie-Louise Bowallius, *Den förändrade synen på amerikansk teknologi: rapportering och värdering av amerikansk teknologi i Teknisk Tidskrift 1870–1893* (Stockholm 1980).

transplantations of people in history. For a long time, it was less emphasised that many emigrants returned to Europe.

However, return migration can hardly be described as a neglected field any longer. This part will focus on technical artefacts and so forth, but mention some other aspects as well. Return migration was often a movement of unmarried and unskilled male workers up to the age of 45 settling in their native districts. Upward social mobility during the time abroad was uncommon, partly because the foreign intermissions rarely exceeded five years.[112] This is, however, not to say that social mobility never occurred after return.[113] Studies of the impact on European societies have reached different conclusions. Some scholars argue that the impact of return from North America was generally minor and only local; similar conclusions have been drawn on post-war return to southern Europe and northern Africa.[114]

Mark Wyman describes the emigration era as a time of progress for the people and nations of Europe. Taking his point of departure in several European studies, Wyman concluded that returnees brought ideas in a variety of fields. Many reacted against the conditions in the home societies: illiteracy, the lack of rights for women, and the lack of political rights in general were leading concerns. Proponents of female and political rights in Norway as well as prime ministers in Norway, Finland, and Latvia were returnees from the United States, so were many 'challengers' of monopolised religion. Dressing, dining, and languages are other fields.[115] Long before Wyman, however, there were positive pictures of American influences in Europe, for example in a study by Halvdan Koht.[116] Theodore Saloutus concluded in 1956 that returnees brought back important ideas for the Greek economy and often went into business; a

---

112 Leslie Page Moch, *Moving Europeans: migration in Western Europe since 1650* (Bloomington, IN 2003).

113 Magnus Persson, *Coming full circle? Return migration and the dynamics of social mobility on the Bjäre peninsula 1860–1930* (Lund 2007).

114 For example: Dino Cinel, *The national integration of Italian return migration, 1870–1929* (Cambridge 1991); Arnold Schrier, *Ireland and the American emigration 1850–1900* (Minneapolis, MN 1958); Semmingsen, *Veien mot vest. 2*; Adam Walaszek, 'Perserving or Transforming Role? Migrants and Polish Territories in the Era of Mass Migrations', in: Dirk Hoerder and Jörg Nagler (eds.), *People in transit: German migrations in comparative perspective, 1820–1930* (Cambridge 1995) 101–124; Keijo Virtanen, *Settlement or return: Finnish emigrants (1860–1930) in the international overseas return migration movement* (Turku 1979).

115 Mark Wyman, *Round-trip to America: the immigrants return to Europe, 1880–1930* (Ithaca, NY 1993).

116 Halvdan Koht, *Amerika i Europa: impulser från väster i teknik, politik, kultur* (Stockholm 1950).

warehouse, a dairy, and the first modern hotel in the country are examples of returnee businesses.[117]

The Norwegian returned emigrant project also gave a positive picture.[118] In Sweden, the first academic studies were more or less demographic and written in the emigration project at Uppsala University in the 1960s and 1970s.[119] Return migration's impact was, however, considered in 1938 by 'popular educator' E. H. Thörnberg, who claimed that returnees mediated American impulses, especially in religion.[120] Hans Lindblad, a former member of Parliament, brings forward thoughts that returnees were crucial when Sweden was transformed into a democracy and a modern welfare state as well as for the country's economic development. Popular movements like the free churches and the temperance movement took several influences from the returnees. Lindblad also argues that the impact of returnees has been underestimated in his own Liberal Party as well as in the Social Democratic movement. Two groups were especially important in the flow of ideas between America and Sweden: pastors and engineers.[121]

We will return to this. Below, we will give some 'technology-related' examples from Europe, albeit not necessarily mediated through educated engineers. Wyman writes on the interchanges of various technical artefacts and knowledge.

> These ranged from bringing back American phonographs, Singer sewing machines, bicycles, hatchets and doubled-bitted axes, among smaller items, to developing iron rolling mills. New logging procedures were introduced in Finland and new fishing methods in Yugoslavian coastal waters, and in the Apennines of northern Italy the first regular rural bus service in Lucca was launched by a remigrant. Not far away in Fornaci di Barga, aging brick and cement kilns were transformed by a group of returnees into an industrial establishment that subsequently attracted large paper and textile mills and a munitions factory.[122]

117 Theodore Saloutos, *They Remember America; the story of the repatriated Greek-Americans* (Berkeley, CA 1956).

118 Skard, *USA i norsk historie*, also gives a positive of return migration's impact in Norway.

119 Lars-Göran Tedebrand, *Västernorrland och Nordamerika 1875–1913: utvandring och återinvandring* (Stockholm 1972); Bo Kronborg and Thomas Nilsson, *Stadsflyttare: industrialisering, migration och social mobilitet med utgångspunkt från Halmstad 1870–1910* (Uppsala 1975).

120 E. H. Thörnberg, *Sverige i Amerika, Amerika i Sverige* (Stockholm 1938).

121 Hans Lindblad, 'Impulser som förändrade Sverige', in: Ingvar Henricsson and Hans Lindblad, *Tur och retur Amerika: utvandrare som förändrade Sverige* (Stockholm 1995) 101–272.

122 Wyman, *Round-trip to America*, 147.

Hence, farming and fishing were influenced; returnees introduced tomatoes in Finland, lettuce in Ireland, tobacco as a crop in several countries, new types of nets for herring fishing, and one returnee cultivated swampy lands in northern Norway, earlier considered impossible for farming. Returnees were also behind road building into peripheral European mountain regions. Other businesses included 'Yankee' pubs in Ireland, barber shops and watchmaker shops in Italy, several banks in Romania, American-type stores with fixed prices and no bargaining possibilities in Hungary, and film companies, shoe-making business, mechanical workshops, and hauliers in Poland. In the Nordic countries, we can point to potteries, furniture factories, blacksmith's workshops, car repair shops, manufacture of electrical motors, and village shopkeepers in Sweden; electrical workshops and canneries in Denmark, clipper ship projects, painting and construction businesses as well as furnishing and camera shops in Norway. In Finland, we can point to a machine shop manufacturing the first American-style harrows, a plant making petrol-fuelled motors, a brickworks, an asphalting firm, and the Finnish-American Mining Company.[123]

4.4 *Transnational Mobility of Technicians to the 1930s*

Engineers and technicians became one of the major travelling groups in this context; technicians overtook to a certain extent the journeyman role as technology carrier. The mobile engineers or technicians have been the subjects of studies in several countries, not least studies that deal with today's mobility. This survey, however, is limited to studies focusing on this phenomenon until around the 1930s.

Discussing the construction of a professional identity among eighteenth century Russian engineers, Dmitri Gouzévitch focuses on the physical, transnational mobility as an important part.[124] This 'bird-of-passage identity' followed technicians to the twentieth century. The Swedish-American engineer Lawrence E. Widmark wrote the following in 1938:

> Is it the lure of gold or other riches that causes a man to leave his native country and spend the rest of his life in some remote parts of the globe,

---

123 Ulf Beijbom, *Amerikaminnen: berättelser i utvandrarbygd* (Stockholm 1996) 279–280; Dorothy Burton-Skårdal, 'Rapport om prosjektet: Hjemvendte emigranter i Agder', in: Hans Try and Berit Eide Johnsen (eds.), *Arbeidsvandringar på Agder* (Kristiansand 1990) 113; Hyldtoft, 'Perioden 1896–1930', 45–46, 98–99; Reino Kero, *Suureen länteen: siirtolaisuus Suomesta Yhdysvaltoihin ja Kanadaan* (Turku 1996) 261–266; Lindblad, 'Impulser', 216–221; Wyman, *Round-trip to America*, 136–148.

124 Dmitri Gouzévitch, 'The rise of state engineering administrations in eighteenth century Russia', *Engineering Studies* 3:3 (2011) 195–213.

> enjoying and gloating over an ever-increasing hoard of accumulated wealth? No! It is hardly that. A man born with the inclination and capacity for solving technical problems, would never be truly satisfied until he reached the proper field of activity for his particular capacity. If this is not available in his own country, he had to seek it outside of its borders'.[125]

At first glance, Widmark's statement appears relevant: no other professional group in the nineteenth and early twentieth century crossed national borders more often than engineers. However, that does not present the whole picture of the 'transnational technician'. Håkon With Andersen argues for a dualism among engineers: they base their legitimacy and credibility on international experience but are also among the professional groups most likely to participate in the race between nations on technical and economic development and prosperity.[126] Henrik Björck emphasised that an engineer's interest to introduce, and work with, new technology can be connected to an interest in the technology as such, a will to raise the status of the engineering profession as well as the individual position.[127] A fourth criterion can be added: a will to contribute to the economic, industrial and technical development of one's native country.[128] This will is reflected in the high number of technicians returning home after acquiring knowledge and experience abroad.

Engineer travel and exchange have been featured in many countries; they have been a geographically very mobile profession, which is—of course—not to say that members of other professions did not move and/or travel. Not all engineer travels were, of course, *placements abroad*, some were basically *labour-market mobility*. Engineers and technicians moved because of at least the perception of better options elsewhere. Often, they acted as agents of transfer of technology and ideas in the 'other' direction. This will partly be focused here, but the prime objective is the influence on the countries of origin. The order of presenting studies of countries and earlier researchers is done only to make better changeovers between different countries and topics; it does not mirror opinions on importance.

---

125 Lawrence E. Widmark, 'Engineers', in: Adolph B. Benson and Naboth Hedin (eds.), *Swedes in America 1638–1938* (New Haven, CT 1938) 409–410.

126 Håkon With Andersen, 'Den nye vitenskapsprofesjonen: teknologene', in: Anton Fredrik Andersen and Guri Hjeltnes (eds.), *Universitet, samfunn og politikk: 18 innlegg om universitets- og vitenskapshistorie* (Oslo 1997) 172–173.

127 Björck, 'Bilder av maskiner', 298.

128 Per-Olof Grönberg, 'The engineer paradox: international migration as a patrotic act', in: K. G. Hammarlund (ed.), *Borders as experience* (Halmstad 2009) 101.

There are some previous studies of transnational mobility of technicians. American engineers and their role in transfer of technology and other ideas to South Africa have been studied by John Higginson and Jessica Teisch.[129] Engineers and other technicians sometimes also travelled to colonies. According to C. R. Day, nineteenth-century French engineers travelled relatively little, especially considering the colonies and France's heavily foreign investments. About every sixth engineer went abroad; the colonies are included in this share. Many of them were on missions from French companies, but Day underlines how industrialists complained about the difficulty of finding qualified persons for duty outside the country.[130] Daniel M. Ringrose reveals, however, a French pattern of 'sending' technicians to the colonies after 1900. However, these persons did not necessarily possess engineering degrees. Ringrose's study of early twentieth-century Indochina suggests a transnationally mobile group of French technical workers, often with experience from other parts of the empire. They brought considerable experience and skills, but few of them made Indochina a permanent residence. The diffusion of technology has been considered essential for empire-building, but many projects have been studied in isolation and in-migration of technical experts from the motherland has been underestimated.[131]

There are a few studies on how technicians from different countries went abroad to study and to return with valuable experience. Czechoslovak engineering studies abroad after the independence in 1918 and how the experiences were utilised in all industrial branches upon return has been a focus of Marcela Efmertowa.[132] Greek studies of this topic have been conducted by Yiannis Antoniou, Fotini Assimacopoulou, Konstantinos Chatzis, and Anna Mahera. Nineteenth- and early twentieth-century teaching staff at Athens Polytechnic were often trained in French engineering schools and

---

129 John Higginson, 'Privileging the Machines: American Engineers, Indentured Chinese and White Workers in South Africa's Deep-Level Gold Mines, 1902–1907', *International Review of Social History* 52:1 (2007) 1–34; Jessica B. Teisch, ' "Home is not so far away": Californian Engineers in South Africa, 1868–1915', *Australian economic history review* 45:2 (2005) 139–160.

130 C. R. Day, 'The Making of Mechanical Engineers in France: The Ecoles d'Arts et Métiers, 1803–1914', *French Historical Studies* 10:3 (1978) 458.

131 Daniel M. Ringrose, 'Nomadic Technicians and Migration in the Francophone World', *Proceedings of the Western Society for French History* 32 (2004) 308–327.

132 Marcela Èfmertovà, 'Les professeurs èlectrotechniques tchèques dans le monde: formation et impact des travaux scientifiques dans les annèes 1918–1938', in: Ana Cardoso de Matos et al.(eds.), *The Quest for a Professional Identity: Engineers between Training and Action* (Lisboa 2009) 513–523.

constituted driving forces in establishing modern Greek engineering education and were in the forefront of modernisation in general.[133] Especially from the 1880s onwards, many Greeks went abroad to study engineering. Many of them still had state grants, but an increasingly higher share travelled on their own initiative; 'undoubtedly a break with the past'.[134] Most went to France, but Switzerland and later also Germany and Belgium became major destinations. A French-trained engineer introduced reinforced concrete in the country. Early twentieth-century Greek industry generally employed few qualified engineers, but a small group of returnees from German technical schools and, mainly, the *Eidgenössische Technische Hochschule* (ETH) in Zurich were important for industrialisation as they were involved in industrial establishments in various fields: chemistry, electrotechnology, cement industries, and the construction business. Many of them were also inspired by technocratic ideals: rationality in the spirit of Taylor and Ford.[135]

Portugal is another country where French schools were important. Ana Cardoso de Matos points to the insufficiency in numbers as well as skills when it came to engineers who could take part in mid-nineteenth-century railway and road construction and hydraulic engineering. Parisian *École des Ponts et Chausses* became essential; engineers educated at this school did not constitute a large group, but designed many of the important public works in the latter part of the nineteenth century. In addition, many non-migrant engineers learned from them, some as employees at the public works they directed and others as students at Portuguese technical schools that had hired them as teachers.[136] In another article, Cardoso de Matos and Diogo conclude that going abroad to improve one's education or to visit and 'study' became increasingly more common among engineers from the mid-nineteenth century onwards. This was important for Portugal's appropriation of foreign technical

133 Fotini Assimacopoulou et al., 'Elève en France, enseignant en Grèce. Les ensignants de l'Ecole polytechnique d'Athènes (1837–1912) formès dans des ècoles d'ingènieurs en France', in: Ana Cardoso de Matos, et. al. (eds.), *The Quest for a Professional Identity: Engineers between Training and Action* (Lisboa 2009) 25–41.

134 Yannis Antoniou et al., 'Greek Engineers: two Centuries of History (19th-20th Centuries)', in: Ana Cardoso de Matos, et. al. (eds.), *The Quest for a Professional Identity: Engineers between Training and Action* (Lisboa 2009) 387.

135 Antoniou et al., 'Greek Engineers', 386–395.

136 Ana Cardoso de Matos, 'Asserting the Portuguese Civil Engineering Identity: the Role Played by the École des Ponts et Chaussées, 1825–1866', in: Ana Cardoso de Matos et. al.(eds.), *The Quest for a Professional Identity: Engineers between Training and Action* (Lisboa 2009) 177–208.

knowledge, but also for the creation of a professional identity. France continued to play a major role and received about 70 per cent of the Portuguese engineering students abroad, whereas about 15 per cent went to Belgium and 15 per cent to Germany. Another pattern was travels to study railways, harbours, and factories. These journeys were mostly destined to France, Belgium, the Netherlands, and Spain and often were financed by the government. The acquired technical knowledge was primarily important for the development of public works and urban infrastructure, but there are a few 'industrial' examples like the renewal of the machinery at a textile factory and the introduction of modern machines and methods in cement making. Going to world fairs and other international meetings constituted a third mobility pattern, especially the five fairs in Paris from 1855 to 1900 were frequently visited by Portuguese engineers. Fairs and other meetings were arenas of exchange, and Portuguese engineers reported what they had seen and heard as well as compared foreign technological development to the less lively industrial environment back home. When engineers gathered at meetings and fairs around 1900, training of engineers as well as agricultural, electrotechnical and chemical development were in focus. Other points of interest were for example steam machines and mechanical tools. The geographic pattern of Portuguese engineer travels clearly ascribed France the strongest role.[137]

Basque engineers have been the subject of Aitor Anduaga. He focuses on engineering studies in Belgium between 1850 and 1914. The number of Basque graduates was remarkably high. Wealthy Basque families sent their sons to what were considered the world's most important schools. Social, geographical, and economic factors lay behind the choice. The University of Liege was viewed as a 'better' place to study for future mining engineers; Spanish technical education had deficiencies. Liege had less stringent requirements than other foreign schools to be accepted as a student. The profitable and productive mining environment around Liege provided valuable work experiences. A Basque student either returned immediately upon graduation or remained a couple of years abroad working in the Belgian steel and iron industry or in nearby establishments in the neighbouring countries. A typical returnee joined the family business and was likely to engage in technology import. He had a degree and prestige enough to be influential in

137 Ana Cardoso de Matos and Maria Paula Diogo, 'Bringing it all back home: Portuguese engineers and their travels of learning', *Journal of History of Science and Technology* 1 (2007) 155–181.

such matters.[138] 'The engineers educated abroad had two main advantages. One was the social credibility gained from the cosmopolitan experience and higher education in elite engineering centres. The other was their privileged status in communication, due to their position midway between supplier agents and host companies' boards of managers'.[139] The engineer was a 'linking agent'. Briefly, the engineer established a link between the Basque company he was affiliated with and a Belgian technical spearhead company.[140]

Russia used engineer travel as an important part of tsarist polices in the early nineteenth century. Irina Gouzévitch and Dmitri Gouzévitch state that French technical education constituted an important model. Under Alexander I, students were sent abroad to study engineering, especially in France, but also in Germany. The tsar invited French engineers to set up higher technical schools. Nicholas I reduced the number of foreign experts, whereas the number of engineering students abroad increased. A network of engineers serving Russian interests in Western Europe was established: they had diplomatic status or represented ministries. Sometimes, they performed a kind of industrial espionage. Many travelling engineers wrote books, which served as guides for potential travellers.[141]

Engineer travel to, and back and forth to, the country was also important in Japan. Erich Pauer reveals that about 3,000 engineers, technicians, and scientists were employed in the country between 1868 and 1912; a majority arrived before the late 1880s and came from Britain, the United States, France, Germany, and the Netherlands. Few were very highly qualified; many were 'generalists', but it was 'enough' to acquaint the Japanese with some Western technologies. French technicians played a role in shipbuilding. Public resources were long used to support Japanese students abroad, that is, to Europe. Most returnee students were less successful, but the ones who combined studies with practise often showed better results. Pauer estimates the number of Japanese engineering students abroad between 1860 and 1900 at somewhere between 600 and 1,500; most of them came from the upper societal layers and travelled to Germany, at least after 1880. Pauer argues that almost all technology transfer

---

138 Aitor Anduaga, 'The engineer as a 'linking agent' in international technology transfer: the case of Basque engineers trained in Liège', *Engineering Studies* 3:1 (2011) 45–70, 46–53.

139 Anduaga, 'The engineer as a "linking agent"', 66.

140 Anduaga, 'The engineer as a "linking agent"'.

141 Irina Gouzévitch and Dmitri Gouzévitch, 'Travelling Interchanges between the Russian Empire and Western Europe: The Travels of Engineers during the first half of the nineteenth century', in: Ana Simões, Ana Carneiro, and Maria Paola Diogo (eds.), *Travels of learning: a geography of science in Europe* (Dordrecht 2003) 213–231.

prior to 1850 came to Japan through different kinds of written documents, and that this channel also dominated from 1850 to 1870. The next fifteen years saw a combination of immigrant engineers, imported machinery, and written documents, while the period to 1900 was dominated by the import of machinery. The importance of foreign engineers decreased.[142]

The travel to and life in North America by immigrant engineers and technicians have also been the subject of studies outside the Nordic countries. We can point to biographies of successful and famous foreign-born technicians in North America, such as alternating-current pioneers Nikola Tesla from Serbia[143] and Charles P. Steinmetz from Germany.[144] There has also been a focus on technical exchange between Europe and America. Hans-Joachim Braun has written about technology transfer from North America to Germany. Engineering visits to exhibitions and contacts with American engineers were important for the introduction of American measuring instruments, water turbines and steam pumps into Germany.[145] Mary Nolan writes on interwar Germany:

> Germany's acute economic problems provided the impetus for industrialists and engineers, professors and newspaper editors, industrial psychologists and social workers, socialists and intellectuals, trade union functionaries and factory workers to make the pilgrimage to America in the mid-1920s. Certainly their German experiences, social position, political proclivities, and individual aspirations shaped their perceptions of America; but America also changed their understanding of Germany—what it was and what it could become. [...] America did provide a working vision of modernity from which Germans could pick and choose different elements as they strove to imagine not an ideal future, but at least an updated and improved one. America, Americanism, and Fordism provided not only a model to emulate or modify, but a vivid, colorful, and controversial language in which to debate modernity.[146]

142 Erich Pauer, 'Technologietransfer und industrielle Revolution in Japan, 1850–1920', *Technikgeschichte* 51:1 (1984) 34–54.

143 Robert Lomas, *The man who invented the twentieth century: Nikola Tesla, forgotten genius of electricity* (London 2000).

144 Ronald R. Kline, *Steinmetz: engineer and socialist* (Baltimore, MD 1992).

145 Hans-Joachim Braun, 'Franz Reuleaux und der Technologietransfer zwischen Deutschland und Nordamerika am Ausgang des 19. Jahrhunderts', *Technikgeschichte* 48:2 (1981) 112–130.

146 Mary Nolan, *Visions of modernity: American business and the modernization of Germany* (New York, NY 1994) 8–9.

Thus, the transatlantic crossings were made not only by Nordic technicians, nor were technicians the sole group in this interchange. The Weimar travellers were not the first from Germany, but they were more numerous and had different objectives. They arrived to learn from America to a greater extent than earlier sojourners. Engineers constituted a favoured group in this context; they could offer detailed assessments of technologies in, for example, coal mines and railroads as well as steel and iron, but also on Fordism, Taylorism, and labour-management relations. The motives were usually connected to the study of these phenomena. The route of the typical German traveller began in New York, wherefrom he[147] proceeded, via a stop at Niagara Falls, to the Ford works in Detroit. After visiting this, to use Nolan's words, 'Mecca', he usually continued to Chicago or Pittsburgh before he returned east to begin the journey back to Germany. Very few travellers went outside this circle. It was a journey through the heartland of America's second industrial revolution, to vast and astoundingly productive steel and ironworks, to mechanical industries with far-reaching and meticulously subdivided labour processes, and to industries with developed labour-management relations. This focus on factories and technology was a new trait in the 1920s.

The time spent in the United States was now also shorter; the trip usually lasted from a few weeks to some months. This was a change from the pre-war pattern with longer intermissions that included employment and higher degrees of acquaintance with American life in general.

Nolan notes that there were visitors from elsewhere in Europe. She claims, however, that debates on Taylor and Ford stood at the very core of German society, whereas they were more marginal in countries like France and Italy. Travel reports, often exaggeratingly positive, formed a base for conceptions and debates of reforms.[148]

Of course, European-born technicians such as Steinmetz and Tesla also made valuable contributions to the American economy and technology. Braun has emphasised transfer of mechanical technology from Germany to the United States between 1870 and 1939. Germany and the United States were basically on the same technical level in mechanical engineering, and the main incentive for the transfer of German technology to the United States was a significantly larger American market. Dealing with diesel engines, steam pumps, steam engines and tubes, Braun underlines that personal transfer by travels of German

147 Nolan states that there were very few women among the travellers in general; among engineers there were hardly any female travellers at all.

148 Nolan, *Visions of modernity*, 17–30.

engineers was decisive, but engineering journals and other media had a function in the initial stages.[149]

Grönberg has written a few comparative articles on transnational mobility in the period 1880–1930 including the four countries and concludes that German-speaking Europe and North America dominated the destination pattern, but that Nordic technicians were dispersed to around 100 countries all over the world. More than every second Nordic technician went abroad; more often from Finland and Norway than from Denmark and Sweden. Swedish and Danish technicians chose North America as their 'first' destination, whereas Norwegian and Finnish more often went to German-speaking Europe. In Finland, of course, Russia also played a major role. The Danes tended, however, to move more outside of these two 'main' areas. Most technicians were 'target migrants', meaning that they returned home. Upon return, they were important for technical development. Return was much stronger from German-speaking countries than from North America. German returnees made up considerable clusters of knowledge and must have had an impact, for example as managers of major companies. Some also returned to teach at the technical universities, especially in Finland and Norway.[150] Arguing that engineers often were international and patriotic at the same time, Grönberg identifies their will to acquire valuable experience and knowledge to be able to contribute to technical development in their native countries as one reason for transnational mobility. The engineer often identified a technological gap in the native country, whereupon he went to study this at one of the places where the development had reached furthest. 'While their actions and proposals were made to promote

149 Hans-Joachim Braun, 'Technologietransfer im Maschinenbau von Deutschland in die USA 1870 bis 1939', *Technikgeschichte* 50:3 (1983) 238–252; Hans-Joachim Braun, 'A Technological Community in the United States: The National Association of German-American Technologists, 1884–1941', *Amerikastudien: American Studies* (1985) 447–463.

150 Per-Olof Grönberg, 'Internationale Migration und Rückwanderung von nordischen Ingenieuren, 1880–1930', *Technikgeschichte* 73: 3/4 (2006) 169–206; Per-Olof Grönberg, 'Den vandrande professionen. Nordiska ingenjörers internationella mobilitet runt förra sekelskiftet', in: Jan Frode Hatlen and Pål Thonstad Sandvik (eds.), *En sann historiker. Festskrift til Svein Henrik Pedersen* (Trondheim 2007) 89–96; Per-Olof Grönberg, 'To study or to work? A Comparative Perspective on Nordic Engineer Migration to German-speaking Europe, 1880–1930', in: Karl Gunnar Hammarlund and Tomas Nilson (eds.), *Technology in time, space, and mind: aspects on technology transfer and diffusion* (Halmstad 2008) 71–96; Per-Olof Grönberg, 'Die internationale Migration und die Rückwanderung skandinavischer Ingenieure zwischen 1880 und 1930', in: Dittmar Dahlmann and Reinhold Reith (eds.), *Elitenwanderung und Wissenstransfer im 19. und 20. Jahrhundert* (Essen 2008) 121–157.

their own careers, the belief that their countries needed new technology and ideas was also important'.[151]

Grönberg has focused solely on migration to North America among Norwegian and Swedish engineers and architects. Asking the question whether these technicians were 'journeymen' or 'traditional emigrants', Grönberg concludes that both nationalities primarily belonged to the former category. They differed from 'ordinary' emigrants when it came to the choice of destinations and extraordinary high return rates. The Norwegians were somewhat more 'traditional' than their Swedish colleagues; this may be explained by Norway's slower industrialisation. Norwegian technicians settled permanently to a greater extent, and the ones who came back had also often been away for somewhat longer and were a little older, albeit a lower average graduation age in Norway. Norwegian technicians also more often went to 'Scandinavian' areas.[152]

Important studies on transatlantic mobility of Scandinavian engineers have been conducted by the earlier-mentioned Stang. He studied mobility to North and South America from 1870 to 1930 concluding that engineers constituted a *peregrine profession.* Stang explains the worldwide dispersion as a combination between individual choices and non-favourable domestic labour-market conditions. On the one hand, the *fresh engineer's wish to gain some extra qualifying experience, or the pursuing of a promising future in a transnational company.* On the other hand, *periods in all three countries* with *tendencies to an oversupply of qualified engineers.*[153] Technicians crossing the Atlantic gained experience, but it had different impacts. Employments in a drawing office at a larger American company or in the municipal bodies of a remote South American town were often well-paid, but not necessarily good career strategies for potential returnees. Some gained experience as representatives of domestic companies, a significant pattern among Danish engineers and a dominant trait for their Swedish colleagues by the time of World War I. This was weaker for Norway, but the country's engineers still had the strongest transnational migration propensity, a difference that seems more dramatic considering the Danish West Indies. Transatlantic engineer migration had connections to general emigration. Denmark's comparably moderate patterns were due to relatively good economic development. Sweden and Norway were among the major emigration countries in Europe, but economic and industrial development also weakened Swedish emigration incentives after the turn of the century; this was reflected in engineer migration. The Norwegian industrial economy developed slower

151 Grönberg, 'The Engineer Paradox', 98–105, quote from 105.
152 Grönberg, 'Journeymen or Traditional Emigrants?', 197–218.
153 Stang, 'A measure of relative development?', 96.

and, in this light, Norwegian youngsters showed an *unreasonably high propensity* to get a technical education. The dispersion was cyclical and followed general emigration patterns in all three countries, but Norway's relatively high number of graduates *opened for the possibility of a large emigration.*[154] In the 1920s, Norwegian transatlantic engineer migration *approached epidemic proportions.* At some points, it exceeded the yearly graduation. Stang underlines that his studies only cover the Americas, but that Australia and South Africa became increasingly important towards 1930.[155]

Myllyntaus has dealt with engineering students from the four countries at Zurich's ETH before World War II. He observes a minor peak in the early 1860s and peaks in the early 1870s, around 1905, around 1920, and after 1935. The highest number of Nordic students arrived in 1904, the lowest in 1885. Other 'low' years were 1864, the early 1910s, and the mid-1920s. The numbers often decreased with economic depressions and increased with booms, wars, and other military offensives. Norwegians constituted nearly half the Nordic students, Swedes about one-fourth, Danes a rough sixth, and Finns about one-eighth. This was partly due to slower development of higher technical education in Norway, and the pattern was also echoed at German technical universities. ETH's high-quality teaching and good reputation attracted students from several countries, but similarities in terrain made Swiss solutions in bridge, road, and railway building, building of hydroelectric power plants and development of the chemical industry easily applicable in Norway. Independent, self-reliant, and industrialising Switzerland was a model; experience from this country was prestigious. This was relevant in countries striving towards and achieving independence, like Norway and Finland, but also elsewhere. Personal ties were also important; many former students recommended ETH to friends and relatives, which launched a chain reaction. This was more common among Norwegians than among other Nordic technicians.

ETH was an attractive institution for the study of technology. Nordic students studied on average two years in the 1860s, two and a half years in the 1870s and 1880s, and one and a half years in 1890s. After that, it gradually increased to almost four years in the 1930s. Danes and Norwegians stayed, on average, longer than Finns, whereas Swedes had the shortest duration of stay. Somewhat more than half the Nordic students studied mechanical and electrical engineering, while civil engineering also was more popular among the Nordic students than among other ETH students in general. Chemistry was studied by about

154 Stang, 'A measure of relative development?', 93–96. Quote from 96.

155 Stang, 'A measure of relative development?', 97.

every seventh ETH student, Nordic or not. Architecture and forestry were less popular subjects among Nordic students than among ETH's students in general. Civil engineering was the most popular 'Nordic' subject until about 1885 when mechanical and electrical engineering overtook this position. Chemistry rose in popularity from around 1890. Norwegian students focused on civil engineering, while most of their Swedish counterparts studied mechanical and electrical engineering. The Finns were more evenly divided between these two disciplines and were more dispersed than the other Nordic nationalities in general. Chemistry was studied more by Norwegians and Swedes than Finns and Danes, but Myllyntaus gives no further information on Danish specialities.

The main reasons to go to Zurich included taking courses from famous professors and becoming familiar with the Swiss technical educational system, rather than obtaining a graduate engineer degree. These modest objectives were partly determined by too short scholarships. Students with other native tongues also often lacked German language proficiency. Many Nordic students did not possess appropriate knowledge of mathematics and some other important subjects. After the turn of the century, however, and especially in the interwar years, obtaining the degree of a graduate engineer became more common. Norwegian students most frequently obtained these degrees (35 per cent). It was also more common among Danes (26 per cent) than among Swedes (13 per cent) and Finns (10 per cent). Myllyntaus refers to the Norwegian and Finnish preference to send students abroad instead of establishing their own technical universities, but Finnish scholarships were generally too short. Myllyntaus does not explain the strong Danish tendency and the difference compared to Sweden. Doctoral degrees were obtained by only three Finnish and three Norwegian students; this embraced less than one percent of all Nordic and Norwegian students, but nearly 4 per cent of the Finnish ones.[156]

ETH was not only a renowned educational institution but also a famous research centre and a very good place to establish contacts in different countries. After employment in Switzerland, Germany, the United States, France, and other countries many ETH graduates made successful careers. Myllyntaus argues that ETH educated technical elites for the Nordic countries and that the school's returning graduates acted as agents for industrial development and technology transfer. Some served at railway buildings and the state railways, some became managers in the private sector, a few became entrepreneurs, whereas others were given teaching positions at technical universities, and some became professors.

---

156 Myllyntaus, 'Discovering Switzerland', 320–323.

Transnational mobility of technicians has been focused in Finland. Max Engman, Susanna Fellman, Marjatta Hietala, Karl-Erik Michelsen, Myllyntaus, and Pasi Tulkki have dealt with the theme. One reason for this interest may be that study tours of nations have been identified as the most important channel of technology transfer.[157] In another article, Myllyntaus concludes:

> At the turn of the [nineteenth] century, almost two generations of Finnish engineers had been educated in a cosmopolitan spirit. For them, studying abroad and going on 'educational business trips' was not just a fashion of the time. For some devoted professionals, studying at renowned foreign universities of technology and later regularly travelling abroad turned into an obsession when they claimed that without this, an engineer's competence was questionable. In any case, these two generations had managed to create a popular image that engineers were knowledgeable persons versed in languages and internationally oriented. The image of the engineer began to change after World War II when the profession gradually came to consist of the first generation of engineers to have been educated almost entirely within the Finnish educational system.[158]

Journeys to more advanced Western nations became customary and essential for technology transfer in nineteenth-century Finland, 'the modern form of a medieval tradition: for craftsmen to have training abroad under foreign masters'.[159] Authorities awarded grants and required travel reports and the like to use for their own technical development. Journeys were also financed by private firms, sometimes connected to the purchase of machinery. These trips were nevertheless unable to 'provide up-to-date expertise in advanced technology or to educate Finns to become highly qualified specialists'.[160] Many considered investments in domestic higher technical education as too costly. For a long time, certain technological fields could be studied only to a limited extent; it was, for example, impossible to become a certified electrical engineer in Finland before 1911. The solution was to award grants to study at renowned foreign technical universities. Teachers at Finnish technical schools also often inspired students to specialise abroad. This was the most common pattern, although some youngsters chose to do all their post-high-school studies abroad.

157 Myllyntaus, *The gatecrashing apprentice*, 128.
158 Myllyntaus. 'The Best Way', 160.
159 Myllyntaus, 'The Best Way', 140.
160 Myllyntaus, 'The Best Way', 142.

This strategy was satisfactory; the first book of reinforced concrete in Finnish was, for example, based on notes taken at lectures at *Technische Hochschule Berlin-Charlottenburg* (THBC), and there are several other examples.

Despite the Grand Duchy status, Russia proper received few Finnish students as their 'technical education had several disadvantages, some related to politics, other to the low prestige of the schools, their admission requirements, and to the content and level of education'.[161] Sweden gradually rose in popularity over the course of the nineteenth century, but Switzerland was thus also 'discovered'. Germany overtook as the most popular destination towards the turn of the century, in the light of the rapid development of *Technische Hochschulen.* Finnish students nevertheless constituted a rather small share of all the foreign students in Germany, and Norwegians were the most represented Nordic nationality. They rarely moved abroad because of political or ethnic discrimination, but rather to pursue educational goals and professional careers. Students from Finland returned more frequently than colleagues from Eastern Europe and Norway. Sometimes, their grants obliged them to come back and take up, for example, teaching appointments. Somewhat above a third of the engineers in Finland in the mid-1910s had studied abroad. They 'rose to key posts in the Finnish economy and joined the leaders of the new establishment when Finland became independent.'[162]

Foreign intermissions of Finnish technical students decreased in the interwar years, more dramatically than among Norwegians. Myllyntaus relates to better opportunities to study technology at home, riskier to go abroad in the light of the hyperinflation in Germany and a new policy emphasising the national education system. Study tours were no longer viewed as equally important for technology transfer, foreign experience was not as rewarding in a job application process, and degrees from foreign universities were less prestigious. Nevertheless, foreign studies remained important in non-technical fields and medical sciences overtook technology as the main discipline. German-speaking Europe had lost some prestige, but was still the most popular destination for all foreign as well as Finnish, students in the mid-1920s. Among interwar Finnish technical students, Sweden held its position well; the United States rose as a 'new' destination, whereas the international tendency to study in France and Britain did not really catch on.[163]

Fellman has also emphasised foreign education and experience in her thesis on the education and careers of industrial managers in Finland, 1900–1975.

161 Myllyntaus, 'The Best Way', 143.
162 Myllyntaus, 'The Best Way', 151.
163 Myllyntaus, 'The Best Way'.

Managers born from 1875 to 1904 typically specialised in certain fields at foreign universities, but also practised and undertook study trips, and this was a break with earlier patterns. It was more common that engineers graduated from foreign universities—mostly German, but also Swedish, Swiss, Austrian, and British—compared to managers with other types of education. Future managers also often completed domestic education as well as practising abroad, regardless of technical or mercantile background. Heirs, that is, sons of managers, had better possibilities to go on longer foreign intermissions. For the less fortunate, employment abroad was essential, but grant possibilities improved over time. Engineer-managers were primarily in Germany, Switzerland, France, Britain, and Scandinavia before World War I. Sweden 'kept' this position in the interwar years, while North America became a more popular destination. Many had also worked and made study trips in Russia, but few had enrolled there. It is difficult to ascertain the importance of foreign experience on managerial careers; many would probably have had a good career regardless of any foreign intermission.[164]

A few other Finnish scholars have also studied mobile technicians; study trips to Europe by city planners, architects, and engineers have been discussed by, for example Hietala.[165] The Helsinki City Council was keen on acquiring foreign know-how in areas like hygiene and healthcare, city planning, education, libraries, social policy, labour protection, and energy supply. This can be viewed in the light of Russification, and the need to employ Finnish experts with local knowledge. Inspiration came from other Nordic capitals, but also from cities such as Berlin, Paris, Vienna, Budapest, London, Hamburg, Dresden, and Zurich. Each sector had their particular 'model cities'. Several experts who had studied abroad—the engineers had studied electro-technology, engineering, and chemistry at German universities—were employed. As city officials, they went abroad again; almost 400 study tours were made between the mid-1870s and 1917. Some travellers took the initiative themselves; many officials felt the pressure to keep up with the latest developments; others were prompted and

---

164 Fellman, *Uppkomsten*, 207–220, 226.

165 Kirsi Ahonen, Marjaana Niemi, and Jaakko Pöyhönen, *Tietoa, taitoa, asiantuntemusta: Helsinki eurooppalaisessa kehityksessä 1875–1917. 3, Henkistä kasvua, teknistä taitoa* (Helsinki 1992); Marjatta Hietala, 'The diffusion of infrastructural innovations in Finland', in: Pekka Ylä-Anttila and Synnöve Vuori (eds.), *Mastering technology diffusion: the Finnish experience* (Helsinki 1992); Marjatta Hietala, 'Finnische Wissenschaftler in Deutschland 1860–1950. Allgemeine bemerkungen mit besonderer Berücksichtigung medizinischer Kontakte', in: Edgar Hösch (ed.), *Deutschland und Finnland im 20. Jahrhundert* (Wiesbaden 1999) 373–394; Jussi Kuusanmäki, *Tietoa, taitoa, asiantuntemusta: Helsinki eurooppalaisessa kehityksessä 1875–1917. 2, Sosiaalipolitiikkaa ja kaupunkisuunnittelua* (Helsinki 1992).

financed (almost half of the tours) by the city, while a fourth group received invitations to congresses and fairs. Finnish study trips increased in the 1880s and 1890s, but even more during the economic growth period in the early years of the twentieth century when more travelling grants could be awarded. These trips went to various European cities. Helsinki's waterworks engineer, for example, travelled for three months in 1889 through Stockholm and Copenhagen via Hamburg, Bremen, and the Netherlands to Sheffield and Liverpool. On his way back from London, he passed Antwerp, Frankfurt-am-Main, Leipzig, and Berlin before he embarked on a boat back to Finland in Stettin. All in all, this travelling created an internationally experienced body of public officials in Helsinki; they also possessed an extensive contact network abroad.[166]

However, all transnational mobility of technicians from Finland was not necessarily for learning purposes. Michelsen has pointed to limited volumes in late nineteenth century industry, and this did not create a need for theoretically trained technicians. Many had to look for employment abroad.[167] Tulkki has also noted this pattern and estimates that about two-thirds of those who were educated as mechanical engineers from the Polytechnic Institute in Helsinki before 1900 had their first employment abroad.[168] Engman has shown that Russia proper often became a solution, especially the nearby imperial capital of Saint Petersburg. Some worked for Finnish companies, others for mechanical workshops like the Swedish Nobel brothers' establishments in Saint Petersburg. Railway and canal building were other fields, so was Nobel's oil business around Baku. Many of these engineers returned to Finland, particularly around the revolution, and made remarkable careers and valuable contributions. The lighthouse service was, for example, developed by a few engineers who had been in Russia; the directors general of the state railways and the customs were also engineers with experience from Russia and so were many leading industrialists.[169]

Like in Finland, technicians from Norway were also frequently going abroad, and this has been focused on in books and articles. Bruland has already been mentioned. Håkon With Andersen, Kenneth O. Bjork, Even Lange, Nerheim, Skard, and Stang as well as Sandvik have written on the topic as a part of other studies or exclusively. Andersen, Nerheim, and Stang have written about journeys to German-speaking countries to study technology. Andersen writes: 'During the second half of the nineteenth century the German institutes

166 Hietala, 'The diffusion of infrastructural innovations in Finland', 269–278.
167 Michelsen, *Viides sääty*, 166.
168 Tulkki, *Valtion virka*, 149, 175.
169 Engman, *Lejonet och dubbelörnen*, 145–164.

(Hochschulen) were the ultimate centres for higher technological education in Norway'. Andersen underlines that Germany educated many Norwegian engineers; she was the model for education, provided textbooks and later professors.[170] Nerheim identifies five forms of technology transfer between Germany and Norway; import of skilled German specialists, import of machinery and other technical equipment, purchase of rights to patents and the like, establishment of subsidiaries of German companies, and the travel of Norwegians to Germany to learn about a new technology or to acquire competence in a certain technical field.[171] Stang states that several engineers found their way to Europe rather than cross the Atlantic. These young men wanted primarily to complete their engineering education. Besides Germany, they went to schools in places such as Vienna, Prague, Paris, Zurich, Edinburgh, and Gothenburg. One reason was their will to get a more thorough education than in Norway, but even more important was the fact that the 'formal' competence from the domestic schools was not appropriate enough to get employment in Norwegian public engineering. Private employers also raised doubts. The stream of fresh engineers travelling to German *Hochschulen* continued year after year, but Stang raises questions about whether these journeys really brought that much competence. Most engineers just followed a limited number of courses; few graduated.[172]

One alternative to Germany was the United States; others were, for example, Argentina or South Africa. Some years in America were, following Stang, a good investment in a purely 'economic' perspective because of relatively good wages, but also to acquire competence in the world's most technology-oriented society. This competence could be used upon return; it equalled some years in Germany. Many had contacts in America and arrived at companies where many Norwegian technicians already were employed, like the Northern Pacific Railway Company. Many engineers awaited possibilities in Norway, returned and evaluated the situation, and went back to America if the possibilities were not good enough. One engineer called the Norwegian migration *a subsidy of American industry*. Nevertheless, some engineers wanted to return,

170 Håkon With Andersen, 'Germany and the education of Norwegian engineers. With some reflections on the role of engineers as a social group', in: *Bürgertum und Bürokratie im 19. Jahrhundert: Technologie, Innovation, Technologietransfer: Bericht über das 2. deutsch-norwegische Historikertreffen in Bonn, Mai1987* (Oslo 1987) 100–114.

171 Gunnar Nerheim, 'Tysklands rolle i norsk industrialisering frem til andre verdenskrig' in: Jarle Simensen (ed.), *Tyskland—Norge: den lange historien* (Oslo 1999) 130–131.

172 Gudmund Stang, 'Ble det for mange ingeniører?', in: *Trondheim ingeniørhøgskole 1912–1987: festskrift til jubileumsfeiringen 31. oktober 1987* (Trondheim 1987) 35–36.

and to succeed in Norway. The country got a reservoir of high-quality technical expertise through this engineer mobility. Once Norway could utilise these experiences through return migration, it was one the most important reasons explaining the country's comparably rapid modernisation.[173]

There have also been a few studies focusing migration of Norwegian engineers to and back from the United States. Skard includes a part about engineers and industrial influences in Norway in his book on Norwegian-American relations. Norwegians wrote travel reports from the fairs in America in the decades around 1900; the United States was described as technologically outstanding. The transatlantic migration of engineers was a very distinctive characteristic. No other educated group migrated to such a high extent and several Norwegian-born engineers became important in America. Nevertheless, it was not what we would call a 'brain drain'. Some engineers returned; others 'commuted' over the Atlantic, and many wrote in Norwegian technical journals. The engineers thereby contributed to technological development at home: pioneers in radio, shipbuilding, oil, tin-can and metal industry as well as mining had all been in North America. Harbour, railway, and road building were other fields where returned engineers were influential, so were military technology, water power, and the electrotechnical field at large. Two hydro-electric pioneers, Jens Bache-Wiig and Ole S. Bragstad, had worked in the large American electrotechnical industries.[174]

Return was also one trait mentioned in the classical *Saga in Steel and Concrete* from 1947 by the Norwegian-American scholar Bjork. The tendency to regard America as a temporary place of residence gave the life of Norwegian engineers in the United States an 'orientation that was different from that of the main body of Norwegian Americans'.[175] One trait was their pattern of destinations: larger cities and industrial centres rather than the traditional areas of settlement for Norwegian immigrants. Hence many went on what we might characterise as self-organised *placements abroad,* but economic motives and an 'urge to see a little of the world and to try out our strength in a bigger field'[176] also contributed. Norwegian engineers had, according to Bjork, a good reputation in the United States. However, he also describes returns that were grounded in disillusion and reveals that the 'Norwegian-Swedish' state of Minnesota received relatively many of these technicians: a pattern that, following Sten

173 Stang, 'Ble det for mange ingeniører?', 37–40.

174 Skard, *USA i norsk historie*, 180–184.

175 Kenneth O. Bjork, *Saga in steel and concrete: Norwegian engineers in America* (Northfield, MN 1947) 36.

176 Bjork, *Saga in steel and concrete,* 37.

Carlsson, was less relevant for Swedish colleagues. The bulk of Bjork's study is development in the United States. Almost all aspects of American technology received a Norwegian contribution: the spanning of rivers, tunnel, road and skyscraper construction, metallurgic development and even 'Taylorist-Fordist-type' organisation and management. Nevertheless, a few returnees are mentioned by Bjork. Nicolay Fougner, who produced the first seagoing concrete-built ship in Norway in 1917, is one. Anton Martin Grønningsæter, who became a pioneer in the Norwegian nickel industry as manager of the plant in Kristiansand, is another. Grønningsæter's achievements have also been described by Sandvik.[177]

Lange has also discussed engineers in the United States and their impact upon return, especially in the first half of the twentieth century. He connects to artisan travel and a traditionally cosmopolitan orientation among Norwegian engineers, because they 'needed' to study abroad because of the prolonged lack of higher technical education and an international outlook of nineteenth- and twentieth-century Norway in general. Transatlantic migration and connections to the United States became important, also for technology. American industries and research environments started to 'challenge' their German counterparts as models for Norwegian technicians in the early years of the twentieth century. Some years, the migration of engineers was so strong that certain specialisations at the Norwegian Institute of Technology (NTH) were given the nickname *The America Line.* After World War I, the United States started to overtake Germany as source of technical inspiration, and American universities such as the Massachusetts Institute of Technology (MIT) and Carnegie-Mellon in Pittsburgh attracted an increasingly higher number of Norwegian students. In 1919, every fifth foreign-born MIT-student was Norwegian, and this was a higher share than the one for nearby Canada. Nevertheless, Lange assumes that returning engineers educated in Norway or elsewhere, with working experience from American industries, were more important than students. Even if some settled in America, more than every second engineer returned. Lange estimates that roughly 13 per cent of the engineers working in Norway from around the turn of the century to the mid-1950s had experience from North America, about every third engineer with foreign experience. They were equally dispersed over the different specialisations. Lange proceeds to look at the period from 1925 to 1935, a period characterised by unemployment,

177 Bjork, *Saga in steel and concrete*; Pål Thonstad Sandvik, *Falconbridge nikkelverk 1910–1929–2004: et internasjonalt selskap i Norge* (Kristiansand 2004); Sten Carlsson, 'Swedish engineers in Chicago', in: Philip J. Anderson and Dag Blanck (eds.), *Swedish-American life in Chicago: cultural and urban aspects of an immigrant people, 1850–1930* (Uppsala 1991) 182.

economic crisis and transformation. He argues that the transfer of technology through these returnees was important for Norway's economic and industrial development in the 1930s; the country managed to turn a crisis into economic growth.[178]

Engineers who were part of Danish immigrant communities abroad are mentioned in some studies.[179] The importance of engineers trained in Denmark, albeit often native-born, for Icelandic technical development is a central topic in a book from the 1940s.[180] There is, as far as the author knows, no 'straightforward' Danish study on this topic, but it has been discussed in parts of other studies by Per Boje, Henrik Harnow, and Ole Hyldtoft who claim that Germany and the United States overtook Britain as a technical model for Denmark and gives a number of examples of returnee engineers and their contributions in his two accounts of technological changes in Danish industry from 1870 to 1930.[181] Hyldtoft has also, based on statistics compiled by Gunnar Viby Mogensen, revealed destinations of Danish university-trained engineers from 1860 to 1899. Germany was a marked destination and about every fourth journey went there. Britain was also common, but Hyldtoft notes that France and Belgium together received about the same number of Danish engineers. French technology had a prominent international position in the 1870s. Germany increased its share of the journeys in the 1890s, while the United States also became one of the most attractive destinations. At the same time, the shares to Britain and France-Belgium began to decrease. In addition, about every fifth engineer went to 'other countries' than the ones revealed above, and to Sweden and Norway. This implies, following Hyldtoft, transfer of technology from, rather than to, Denmark.[182]

Harnow includes a part on transnational mobility in his book on the history of Danish engineers, whose number abroad increased threefold from the 1890s to the 1910s. Some of them were recruited to good positions abroad. Others wanted to acquire experience in modern technologies such as concrete building. Some were adventurers, but many were simply 'forced' abroad as the period often implied high domestic unemployment. The United States was the

178 Lange, *Norske ingeniører*.

179 Christopher Bo Bramsen, *Open doors: Vilhelm Meyer and the establishment of General Electric in China* (Richmond, VA 2000); Tage Kaarsted, *Admiralen: Andreas de Richelieu: forretningsmand og politiker i Siam og Danmark* (Odense 1990).

180 Thorvald Krabbe, *Island og dets tekniske udvikling gennem tiderne* (København 1946).

181 Hyldtoft, 'Perioden 1896–1930', 41–51, 73, 98, 140, 148, 152, 156, 165, 172–180; Ole Hyldtoft, *Teknologiske forandringer i dansk industri 1870–1896* (Odense 1996) 47, 51–52, 92, 97, 118, 134, 167–172, 178–186, 204–206, 247–249.

182 Hyldtoft, *Teknologiske forandringer i dansk industri 1870–1896*, 264–265.

most common destination, followed by Germany and Britain, but Danish engineers were dispersed all over the world. France was an important destination for construction engineers specialising in concrete. Some were employed by the Dutch government in Java. South America, especially Argentina, attracted 'adventurers', but was to some extent also the 'last chance' to get a position. Many journeys were undertaken as an adventure, but there were also several Danish firms that offered more secure and formalised intermissions abroad. Because of working conditions and wages, potential travellers were often advised to apply for overseas jobs through European companies. Engineers from Denmark generally had a good reputation abroad; technical abilities were often viewed as eminent, but problems could occur with unaccustomedness to working conditions as well as language and mercantile skills.[183]

Boje has concluded that foreign industrial knowledge to a large extent was transferred to Denmark through industrial managers' intermissions abroad. Foreign experience among industrial managers increased from the 1870s to the 1930s, but decreased thereafter. The shares of engineers going abroad increased from the late 1880s to the outbreak of World War I. These intermissions acknowledged that countries such as Britain, Germany, the United States, and to some extent, France had reached further in industrial development. They derived from the journeyman tradition, but future industrial managers often had more specific goals with their journeys and utilised business contacts. Many spent time in several countries, because learning languages was one of the main purposes. Technical knowledge was also gained abroad: Future managers in the textile industry had, for example, often been in Germany. Many had been at the technical school in Mittweida in Saxony, which also became a model for the *Teknika* schools in Denmark. Other future managers gained experience through employment abroad for Danish companies. It is, following Boje, difficult to draw the line between shorter study trips and longer intermissions abroad; it was relatively common to go on study trips already in the 1870s as European communications were already developed. Public or private grants to go to fairs and so forth were also rather common. For the Paris fair in 1867, for example, grants were awarded for studies of preservation of food stuff and sewing and other machines. Study trips and other foreign intermissions had different implications, but some industrial managers could 'pick up' ideas and utilise them upon return. One example is N. J. Haustrup, who modernised a metal factory after he had studied conveyor belts in America.[184]

183 Harnow, *Den danske ingeniørs historie*, 232–235.

184 Per Boje, *Ledere, ledelse og organisation 1870–1972* (Odense 1997) 136–141.

In a more recent study, Boje focuses on engineers educated at lower-level technical schools and their importance for Danish economic growth. He concludes that these engineers have been important and that foreign impulses have been essential. New knowledge was brought to Denmark by foreigners working in the country, shorter or longer study trips as well as continuous reading of technical literature and catalogues from major foreign corporations. Danish technicians became increasingly more interested primarily in Germany and the United States going in to the interwar years. Study trips and periods of employment in these two countries confirm the pattern as do the import of machines, licenses, and collaboration agreements between Danish, German, and American companies. The earlier-mentioned studies at German technical universities were part of the pattern. Most of these engineers were employed by private companies upon return; some founded their own businesses. Around the turn of the century, Denmark's technical connections to Germany were strong, but American experiences became increasingly important in the light of the transatlantic mass emigration. Some of the returnees were people who had travelled to acquire technical knowledge and valuable experiences in America. The shift from British and French models to German and American ones implied an acknowledgement that companies in the latter two countries were technologically more advanced. The American engineering industry, for example, had become known for machines characterized by preciseness. These machines were a prerequisite for serial manufacture, which in turn was a prerequisite for the assembly line. The 1893 Chicago fair was, to a large extent, an 'eye-opener' when it came to American industrial capacity. The marvels of electricity were described in admiring terms. On to World War II, Danish engineers' ties to the United States were strengthened. The transfer of knowledge implied more than technology; organizational and marketing practices also were brought to Denmark and were important for the growth of some of the country's most successful businesses. It is however, following Boje, difficult to quantify the transfer of American know-how, partly because some of it also was mediated through Germany and Sweden. Nevertheless, Boje gives examples; Taylor's principles of scientific management are one. American-style rationalisation was often combated, but was dispersed in Denmark's iron and metal industry. Often, engineers from Teknika with experiences from America and Germany were driving forces. The founders of cable factories, electrical motor factories, refrigerator factories, and in the packing industry are some examples.[185]

185 Per Boje, 'Teknikumingeniører og dansk økonomisk vækst', in: Henrik Harnow, et. al (eds.), *Industriens Pionerer: Teknikumingeniørernes uddannelseskamp og betydning* (København 2011) 279–321.

In Sweden, some studies have described returnee technicians and their influence. Carlsson, Mats Fridlund, Grönberg, Staffan Hansson, and Lars O. Olsson are among the scholars who have dealt with the topic. Fridlund has written on the electrical industry, how young engineers went mostly to the United States and Germany to study rationalisation, mass production, and new technologies. Sweden's major electrotechnical company, ASEA, was reorganised after *American principles* under the management of returnee engineer J. Sigfrid Edström. Fridlund states that it was an often-used strategy to give engineers leave of absence systematically to go abroad, obtain employment at major foreign competitors, learn about their technologies, and return with this knowledge.[186] Fridlund and Hansson have described ASEA's involvement in water-power technology and how some of the major water-power stations had North American models.[187] Olsson has written on the shipbuilding industry. Despite British domination, Swedish naval architects more frequently went to the United States and Germany. In America, they learned the organisation of specialised shipyards, the use of labour-saving machines, template systems, and a more mechanised shipbuilding in general. These became solutions for an industry suffering from a shortage of skilled workers, and the introduction of the template system is described as one of the reasons behind early twentieth-century shipbuilding expansion.[188]

Some biographic studies of Swedish engineers in America exist, for example on radio-pioneer Ernst F. W. Alexanderson[189] and, of course, on John Ericsson.[190] Some studies have dealt more generally with Swedish contributions to American engineering: Chicago's Swedish Engineers' Society and the engineers' consciousness as Swedes and Americans.[191] Grönberg concludes that

186 Mats Fridlund, *Den gemensamma utvecklingen: staten, storföretaget och samarbetet kring den svenska elkrafttekniken* (Eslöv 1999); Jan Glete, ASEA *under hundra år: 1883–1983: en studie i ett storföretags organisatoriska, tekniska och ekonomiska utveckling* (Västerås 1983).

187 Fridlund, *Den gemensamma utvecklingen*, 63; Staffan Hansson, *Porjus: en vision för industriell utveckling i övre Norrland* (Luleå 1994) 182.

188 Lars O. Olsson, *Technology carriers: the role of engineers in the expanding Swedish shipbuilding system* (Göteborg 2000) 47–55; Lars O. Olsson, 'To See how things were done in a big way Swedish naval archtects in the United States, 1890–1915', *Technology and culture*, 39:4 (1998) 434–456; Lars O. Olsson, 'Amerikaemigrationen och återvändande svenska ingenjörer, 1890–1930', in: *Göteborgs-emigranten* 6 (Göteborg 1997) 229–243.

189 James E. Brittain, *Alexanderson: pioneer in American electrical engineering* (Baltimore, MD 1992).

190 Olav Thulesius, *The man who made the Monitor: a biography of John Ericsson, naval engineer* (Jefferson, NC 2007).

191 Carlsson, 'Swedish engineers in Chicago'; Per-Olof Grönberg, '"My kind of town?': ethnicity and class as determining factors for return migration or permanent settlement among

Swedish engineers constituted a geographically mobile group. Labour-market conditions and mass emigration contributed, but 'target migration' was the main pattern. More than two-thirds returned with valuable knowledge and 'symbolic capital' that could be used on the domestic labour market in a country where American models were important in the coveted rise to an industrial superpower. Return was connected to social, educational, and geographical backgrounds; engineers with higher social origin and education returned more frequently and had smoother career paths. Returnee engineers were numerous, at least in the top management, at four studied Swedish companies: ASEA, one steel and ironworks, one traditional engineering workshop, and one mining company. Their most frequently implemented ideas concerned rational organisation in Taylorist spirits, often combined with American-inspired welfare practises. Not all practises came from the United States, but America was most important and dominated the steel and iron industry more than other branches. The fact that returnee engineers filled many key positions in leading companies suggests that their influence was considerable. The fact mentioned above points in the direction that returnee engineers were a source of technical development during Sweden's second industrial breakthrough.[192]

5 Methodology

The methodology will be discussed below. First, there will be a description of the database and its construction followed by a discussion on quantitative statistical as well as comparative methods used.

5.1 *The Construction, Content, and Implications of the Database*

The biographical notes have been collected from a variety of sources which will be discussed below. The original work was conducted in Excel, but elaborations with the database have also been conducted in Access.

Information connected to the graduates' background contains the following variables: last name, first name, year of birth and in some cases also month and date, sex, father's occupation, social class, place of birth plus location in

Swedish engineers in Chicago 1910–1930', in: Daniel Lindmark (ed.), *Swedishness reconsidered: three centuries of Swedish-American identities* (Umeå 1999) 121–142; Byron J. Nordstrom, 'Trasdockan: the yearbook of the Swedish Engineers' Society of Chicago', in: Philip J. Anderson and Dag Blanck (eds.), *Swedish-American life in Chicago: cultural and urban aspects of an immigrant people, 1850–1930* (Uppsala 1991) 193–212.

192 Grönberg, *Learning and Returning*.

country and region and the character of it. Variables connected to education include country and city of education, whether it was a capital, whether the graduate studied in his or her country of birth, specialisation, the year of graduation, and age at time of graduation. The database also contains columns connected to transnational mobility after graduation. Columns connected to occupational mobility contain information on positions of graduates, mobile or not, in 1900, 1910, and 1920. It includes the title of the position and whether the position implied upward occupational/social mobility compared to the titles 'engineer' and 'architect'. This information also contains data on the geographical location of the workplace, what type of area it was—larger or smaller city, rural, and so forth—and the relation between the graduate's place of birth and work place (Were they the same, located close to each other, or further away?). The employer is also included as well as information on age, the number of years since graduation at the time of observation and—of course—whether the graduate had been abroad or not. Below, we will present some information from the database in more detail and discuss some of its implications.

The names of 12,376 Nordic technicians have been collected in a database that represents four-fifths of the ones who graduated after at least three years of education in the four countries between 1880 and 1919. One striking thing in the database is the extremely unequal sex distribution: Sixty-nine graduates were women. The likelihood of women being more than two or three among the 'missing' graduates must be considered very low: This was a man's world. The view of technology as no place for women has, for example, been described by Berner.[193] Sweden was indeed very far from today's sought-after image as the world's most gender-equal nation. When Vera Sandberg graduated in chemical engineering at Chalmers in 1917, she became Sweden's first female engineer and she is also the only one in the cohort.[194] The Norwegian cohort includes ten women, whereas there are 27 women among the graduates in Helsinki. This gives Finland the highest percentage, even if 31 female graduates are in the Danish cohort. Finland was a 'spearhead country' when it came to women's positions and was among the first countries in the world to introduce universal suffrage in 1906. Signe Hornborg, who graduated in 1890, became one of the first female architects in the world.[195] The fact that the Finnish cohort consists of many architects is a major explanation of the pattern. As a matter of fact, the 1905 graduate in civil engineering, Jenny Markelin, was the

193 Berner, *Sakernas tillstånd.*
194 Indebetou and Hylander, *Svenska teknologföreningen,* 1044.
195 Riitta Nikula, 'Women in the history of Finnish art', in: Marja Manninen and Päivi Setäla (eds.), *The lady with the bow: the story of Finnish women* (Helsinki 1990) 88–89.

only female 'engineer'. The share of female graduates would thus have been higher in Denmark if we only had counted the engineers. All but one of the Danish female engineers were chemical engineers.[196] The overwhelming male domination must be taken into consideration when patterns of transnational mobility and return migration are discussed.

Occupations of the fathers have been classified with the help of the HISCO system, prepared to be internationally compatible, through cooperation between historical demographers in several European countries: including the Scandinavian ones.[197] The uppermost social group consists of children of 'higher managers and professionals' and represents 32 per cent of the database, 40 per cent in Finland, 38 per cent in Norway, 30 per cent in Sweden, and 26 per cent in Denmark. The second group is made up of graduates whose fathers were 'lower managers and professionals, clerical and sales personnel' and represents 31 per cent of the database, 38 per cent in Norway and around 30 per cent in the other countries. The third group were children of 'foremen and skilled workers, farmers and fishermen' and made up 18 per cent of the database, 21 per cent in Denmark, 19 per cent in Finland, 18 per cent in Norway, and 16 per cent in Sweden. The 'lowest' group, graduates originating from families of 'lower skilled and unskilled workers' made up 4 per cent of the entire Nordic database, between 6 and 7 per cent in Denmark and Finland and about 3 per cent in Sweden and Norway. It is striking that graduates from Norway and Finland seem to have had a higher social origin, but we need to consider a relatively high share whose fathers' occupation is not given: 15 per cent in the entire database. This deficiency is much more significant for Sweden (22 per cent) and Denmark (18 per cent) than for Norway (4 per cent) and Finland (3 per cent). Nevertheless, it is unlikely that all 'unknown' cases were in the two highest groups; they primarily come from the Danish and Swedish lower level technical schools, and there are reasons to assume that they provided a smoother path to enrolment for persons of lower social origin, compared to the technical universities.

Places of birth have been classified into capital cities and other major cities, that is, Aarhus, Odense, and Aalborg in Denmark; Tampere, Turku, and Viipuri in Finland; Bergen, Trondheim, and Stavanger in Norway; and Gothenburg, Malmö, and Norrköping in Sweden. Further categories are other cities and towns, rural-industrial and agricultural parishes and unknown places; the latter representing less than one percent in all countries. We have applied this

196 Harnow, *Den danske ingeniørs historie*, 167.

197 Marco H. D. van Leeuwen, Ineke Maas, and Andrew Miles, *HISCO: Historical International Standard Classification of Occupations* (Leuven 2002).

classification also for graduates who were born outside the country of education, that is, a Saint Petersburg- or London-born person was from a capital. A person born in Hamburg was from an 'other major city'. About 60 per cent of the graduates were, as figure 1 shows, urban-born—higher in Denmark and lower in Sweden.

Denmark notes a high share of capital-born graduates.[198] A higher share of the admission places was located in Copenhagen compared to Stockholm and Kristiania, but Denmark also clearly notes a higher share than Finland, although all Finnish graduates studied in Helsinki. Copenhagen's industrial dominance over the rest of the country was stronger than in most European nations.[199] Finland's industrial centre was Tampere, whereas Sweden's nineteenth-century industry tended to locate near natural resources rather than larger cities.[200] This is perhaps also one explanation of why the share born in the four largest cities is lower in Sweden. Also, this share would probably have been even lower if we had included the missing graduates of whom many studied at schools outside these four cities, even if Norrköping hosted one. Norwegian industrialisation was centred on the Oslofjord, but there was still a relatively low share of capital-born graduates. This could perhaps be explained by the fact that the capital itself did not stand in the forefront, but adjacent regions like Østfold and the neighbouring town of Drammen. Furthermore, the country's leading technical school was located in Trondheim rather than in the capital; a pattern that was 'formalised' when the Norwegian Institute of Technology was inaugurated in 1910. This fact also helps to explain why a comparably larger share of Norwegian graduates was born in 'other major cities'.

When it comes to country of education, Sweden represents about 45 per cent of the database, but Swedish schools really turned out 51 per cent of the Nordic technicians fulfilling the criteria. As we can see in table 1, there is still a higher share of missing cases from Sweden than from the other countries. Norway accounts for 24 per cent of the cohort but turned out 20 per cent. Corresponding shares for Denmark are 23 and 21 per cent. Finland accounts for 9 per cent of the cohort, and 8 per cent of the total graduation. Sweden is thus underrepresented rather than overrepresented, but we need to consider the

198 Copenhagen's share is probably somewhat overrated; many 'missing' graduates studied at the schools in Odense and Aarhus, but even if not a single missing graduate was born in the capital, the Danish capital share is still much higher than in the other countries.

199 Ole Hyldtoft, *Københavns industrialisering 1840–1914* (Herning 1984) 418.

200 Pertti Haapala, *Tehtaan valossa: teollistuminen ja työväestön muodostuminen Tampereella 1820–1920* (Tampere 1986) 393; Jan Glete, *Ägande och industriell omvandling: ägargrupper, skogsindustri, verkstadsindustri 1850–1950* (Stockholm 1987) 71–73.

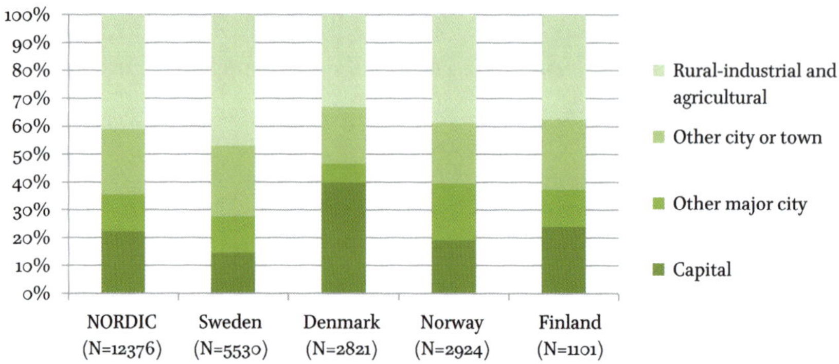

FIGURE 1 Birthplaces of Nordic technical school graduates, 1880–1919

SOURCES, DENMARK: *DANSKE INGENIØRER FRA TEKNIKA* (KØBENHAVN 1945); *DANSKE TEKNIKA OG DERES DIMITTENDER GENNEM 50 AAR, INGENIØR- OG KONSTRUKTØRSAMMENSLUTNINGEN GENNEM 35 AAR* (KØBENHAVN 1931); AAGE HANNOVER (ED.), *DANSK CIVILINGENIØRSTAT 1942, BIOGRAFISKE OPLYSNINGER OM POLYTEKNISKE KANDIDATER 1829–1941* (KØBENHAVN 1942); AAGE HANNOVER (ED.), *DANSK CIVILINGENIØRSTAT 1955: BIOGRAFISKE OPLYSNINGER OM POLYTEKNISKE KANDIDATER 1829–1955* (KØBENHAVN 1956); VOIGT, J., AND R. JESPERSEN (EDS.), *BIOGRAFISKE OPLYSNINGER ANGAAENDE DEN POLYTEKNISKE LÆREANSTALTS KANDIDATER 1829–1929, 3. UDG. AF J.J. VOIGTS BOG OM DE POLYTEKNISKE KANDIDATER, MED ET TILLÆG INDEHOLDENDE BIOGRAFISKE OPLYSNINGER OM MEDLEMMER AF DANSK INGENIØRFORENING, SOM IKKE ER UDGAAET FRA DEN POLYTEKNISKE, UDARB. AF R. JESPERSEN* (KØBENHAVN 1930). FINLAND: *MATRIKEL ÖFVER TEKNISKA REALSKOLANS OCH POLYTEKNISKA SKOLANS I HELSINGFORS SAMT POLYTEKNISKA INSTITUTETS I FINLAND LÄRARE OCH ELEVER 1849–1897* (HELSINGFORS 1899). SULO HEINIÖ (ED.), *MATRIKEL ÖFVER POLYTEKNISKA INSTITUTETS I FINLAND LÄRARE OCH ELEVER 1898–1908* (HÄMEENLINNA 1918), Y[RJÖ JUHO] TALVITIE, *SUOMALAISTEN TEKNIKKOJEN SEURAN NIMIKIRJA 1896–1936* (HELSINKI 1936). *SUOMEN INSINÖÖREJÄ JA ARKKITEHTEJÄ: INGENJÖRER OCH ARKITEKTER I FINLAND: 1948* (VAASA 1948). IISAKKI LAATI AND YRJÖ BLOMSTEDT, (EDS.), *KUKA KUKIN OLI: HENKILÖTIETOJA 1900-LUVULLA KUOLLEISTA JULKISUUDEN SUOMALAISISTA* (HELSINKI 1961). *PROFILES—PIONEERING WOMEN ARCHITECTS FROM FINLAND* (HELSINKI 1983); BRAGES PRESSARKIV HELSINKI (BP), BIOGRAFISKA AVDELNINGEN [BIOGRAPHICAL DEPARTMENT]. NORWAY: O. ALSTAD (ED.), *TRONDHJEMSTEKNIKERNES MATRIKEL: BIOGRAFISKE MEDDELELSER OM SAMTLIGE FASTE OG HOSPITERENDEELEVER AV TRONDHJEMS TEKNISKE LÆREANSTALT 1870–1915: MED CA.1300 UNGDOMSPORTRÆTTER* (TRONDHJEM 1916); O. ALSTAD (ED.), *TILLEGG TIL TRONDHJEMSTEKNIKERNES MATRIKKEL* (TRONDHJEM 1932); BJARNE BASSØE (ED.), *INGENIØRMATRIKKELEN: NORSKE SIVILINGENIØRER 1901–55 MED TILLEGG* (OSLO 1961), CHR. BRINCHMANN ET AL., *HVEM ER HVEM?: HAANDBOK OVER SAMTIDIGE NORSKE MÆNDT OG KVINDER* (KRISTIANIA 1912).GEORG BROCHMANN (ED.), *VI FRA NTH: DE FØRSTE TI KULL: 1910–1919* (STAVANGER 1934). LEIF ESKEDAL (ED.),

BTS-MATRIKKELEN: INGENIØRER UTEKSAMINERT VED BERGEN TEKNISKE SKOLE 1875–1975 (BERGEN 1975). EILIV FOUGNER (ED.), *NORSKE INGENIØRER OG ARKITEKTER: KORT OVERSIGT OVER DEN NORSKE INGENIØRFORENINGS OG NORSKEARKITEKTERS LANDSFORBUNDS HISTORIE, SAMT BIOGRAFISKE OPLYSNINGER OM DE TO ORGANISATIONERS NULEVENDE MEDLEMMER MED PORTRÆTTER* ([KRISTIANIA] 1916). *INGENIØRER FRA 1896 FRA KRISTIANIA TEKNISKE SKOLE: BIOGRAFISKE OPLYSNINGER SAMLET TIL 25 AARS JUBILEET 1921* (KRISTIANIA 1921); *INGENIØRENE FRA K.T.S 1897–1947* ([OSLO] 1947); *KTS: 50 ÅRSBERETNING OM INGENIØRKULLET FRA KRISTIANIA TEKNISKESKOLE 1896* ([OSLO] 1946); KRISTIANIA TEKNISKE SKOLE, *FESTSKRIFT I ANLEDNING AF KRISTIANIA TEKNISKE SKOLES 25-AARS JUBILÆUM I JUNI 1898* (KRISTIANIA 1898); EINAR JANSEN AND PAULUS SVENDSEN (EDS.), *NORSK BIOGRAFISK LEKSIKON* (OSLO 1954). HJ. STEENSTRUP, *HVEM ER HVEM?* (OSLO 1930); HJ. STEENSTRUP, *HVEM ER HVEM?* (OSLO 1937); *TIDSSKRIFT FOR KJEMI, BERGVESEN OG METALLURGI* (1950); OSLO CITY ARCHIVES: THE LIBRARY: BERETNING OM KRISTIANIA TEKNISKE SKOLE VIRKSOMHET, 1891, 1898, 1902, 1908; OSLO TEKNISKE SKOLES ARKIV, VOLUME G0001: BERETNING OM KRISTIANIA TEKNISKE SKOLE VIRKSOMHET, 1889–1890, 1893–1897, 1899–1901, 1903–1907, 1909–1910, VOLUME G0002, BERETNING OM KRISTIANIA TEKNISKE SKOLE VIRKSOMHET, 1911–1915. SWEDEN: GÖSTA BODMAN, *CHALMERS TEKNISKA INSTITUT: MATRIKEL 1829–1929* (GÖTEBORG 1929); *CIVILINGENJÖRER OCH ARKITEKTER UTEXAMINERADE FRÅN CHALMERS TEKNISKA HÖGSKOLA: JÄMTE REDOGÖRELSE FÖR CHALMERSSKA INGENJÖRSFÖRENINGENS OCH CHALMERSSKA FORSKNINGSFONDENS VERKSAMHET* (GÖTEBORG 1904–1930); EMIL FORSBERG AND EMIL ADLERS (EDS.), *TEKNISKA FÖRENINGEN I ÖREBRO 1875–1925: MINNESSKRIFT UTGIVEN MED ANLEDNING AV FÖRENINGENS FEMTIOÅRIGA VERKSAMHET* (STOCKHOLM 1925); GOVERT INDEBETOU AND ERIK HYLANDER (EDS.), *SVENSKA TEKNOLOGFÖRENINGEN 1861–1936: BIOGRAFIER* (STOCKHOLM 1937); *MALMÖ TEKNOLOGFÖRBUND: MINNESALBUM UTG. I ANLEDNING AV MALMÖ TEKNISKA LÄROVERKS 75-ÅRIGA VERKSAMHET: 1853–1928* (MALMÖ 1928); *TEKNISKA FÖRENINGEN I ÖREBRO. MEDLEMSFÖRTECKNING, M.M* (ÖREBRO 1926–1932).

domination when we are discussing pan-Nordic results. However, as we are comparing the four Nordic countries, it will be possible to shed light on specific traits for each country.

Sixty-one per cent of the graduates in the cohort studied in capitals. This is somewhat higher than the 'real' total, which was 54 per cent. The countries diverged significantly; all Finnish graduates in the cohort studied in Helsinki; inclusion of the Tampere graduates would still have given the capital 95 per cent. Graduates from the time before the transformation to a technical university in 1908 are overrepresented. In the Danish cohort, 97 per cent studied in Copenhagen. Inclusion of all graduates would have resulted in 92 per cent. The Polytechnic Institute is somewhat overrepresented in the cohort, while all other schools are underrepresented. Forty-nine per cent of the cohort's Swedish graduates attended school in Stockholm, the capital hosted in 'reality' 38 per cent. The Royal Institute of Technology, Chalmers and Malmö are overrepresented; Örebro is underrepresented, whereas Borås, Norrköping, and Härnösand are not included. In the Norwegian cohort, 35 per cent studied in the capital. Among all graduates, the two Kristiania schools hosted 38 per cent. The distribution of schools in the Norwegian cohort corresponds better with 'reality' than do the other countries; Bergen and the two Trondheim-based schools are somewhat overrepresented. The capital cities thus dominated technical education much more in Finland and Denmark, at least quantitatively speaking. Capital cities hosted a higher number of students and had the highest 'ranked' schools. Norway was an exception; Trondheim hosted—informally, and later also formally—the highest ranked schools and hosted more than 48 per cent of the students. This was somewhat higher than Kristiania, but we must consider that there is no capital-based school in Norway included during the last five years.

The total number of graduates born outside of the country of education in the database is 420 and represents slightly above 3 per cent. Sweden hosted 191 of them of whom 67 were Norwegians, and 45 were Finns. Others were, for example, born in Denmark, the United States, Russia, and Britain and often had Swedish parents. About two-thirds of the foreign-born students went to Gothenburg; they constituted about 8 per cent at Chalmers. This is a higher share than at any other school in the Nordic cohort. Denmark hosted 120 foreign-born students, a little above 4 per cent, of whom about one-fourth was born in the areas Denmark ceded in 1864. One-fifth was born in Sweden and one-tenth in Iceland. Norway and the United States were other 'larger' countries; some were also born in other dependencies as well as in East Asia, France, and Britain. Almost two-thirds studied at the Polytechnic Institute, but the other schools also note higher foreign-born percentages. Copenhagen's mechanical school, Odense, and Aarhus note about 6 per cent foreign-born students, but at least Aarhus's share is based on a small number. Norway hosted 62 (2 per cent) of whom 25 were children of returned Norwegian-Americans. The others were 'dispersed', for example, to Sweden, Denmark, Germany, Spain, and Britain. Kristiania's technical school hosted about a third, but three students at the university's mining department give this institution the highest foreign-born share. Finland hosted 47 (4 per cent) and about two-thirds of them were born in Russia by Finnish parents. The others were often born in Sweden and a few in German-speaking regions in Europe. There were more foreign-born students before the transformation to a technical university, also in shares.

The division of specialisations has been made to make possible national comparisons and to have graduates in the groups all through the period. This has implied merging of groups on national levels. These merges have shortcomings, but have been considered necessary. Danish, Finnish, and Swedish civil and construction engineers showed, for example, 'reciprocal' migration and return differences. Swedish naval architects went abroad to a much greater extent than mechanical and electrical engineers.[201] These three specialisations have nevertheless been merged to one group together with early graduates that followed a 'general' curriculum and a few whose specialisation has not been possible to identify. This group of 'mechanical and electrical engineers and naval architects' is the largest in all four countries. In Sweden, electrical engineering and naval architecture were independent subjects at an early stage. Norwegian and Finnish mechanical engineers often worked with electricity and at shipyards. Norwegian electrical engineers were, according to Astrid Wale, often men with domestic mechanical engineering degrees who deepened their knowledge of electrical engineering in Germany. The same pattern could be found in Finland. Many engineers in naval architecture also had mechanical engineering degrees.[202] We have merged these specialisations into one group to increase the comparability between the countries. Finland's graduates came from M (mechanical engineering) and the 'mechanical' sub-departments ME (electrical engineering), MF (manufacturing industry), and MM (mechanical engineering) after the transformation.[203] Norwegian graduates came from M (mechanical engineering), from the mechanical department's curriculum MS (naval architecture) as well as from E (electrical engineering) at the Norwegian Institute of Technology. Swedish technicians in this group graduated in M (mechanical engineering) at all the four schools. There were also graduates from the Royal Institute of Technology's 'mechanical' sub-departments: Mb (rock mechanics), Mm (heat engines and water turbines) and Mt (mechanical technology). Some also graduated in E (electrical engineering) at the Royal Institute of Technology, Chalmers, and in Örebro, S (naval architecture) at the Royal Institute of Technology and in Sk (naval architecture) at Chalmers. Their Danish colleagues graduated in M (mechanical engineering) at the Polytechnic

---

201 Grönberg, *Learning and Returning*, 97–98.

202 Timo Myllyntaus, *Electrifying Finland: the transfer of a new technology into a late industrialising economy* (London 1991) 151–155; Astrid Wale, *Nyhet, nytte, framskritt: introduksjonen av lokale elektrisitetssystem 1877–1900: Trondheim i et nasjonalt og internasjonalt perspektiv* (Trondheim 2004) 122–127, 131.

203 Bernhard Wuolle, *Den tekniska högskoleundervisningen i Finland 1849–1949* (Helsingfors 1949) 322.

Institute (called Mi from 1899 to 1902), Copenhagen's mechanical school, and in Odense. There were also graduates in E (electrical engineering) from the Polytechnic Institute and Aarhus.

Half the number of Nordic graduates belonged, as figure 2 shows, to this group; close to 60 per cent in Sweden and slightly above 40 per cent in the other countries. Despite the size of this group, the specialisation is underrepresented in all countries except Norway, whose distribution by specialisation in the cohort is almost the same as in 'reality'.

Civil and construction engineering was one specialisation in Norway; there were graduates from B (civil and construction engineer) at all four institutes. These groups are, therefore, merged in the other countries. In Denmark, this group includes graduates from B (construction engineer) at the Polytechnic Institute and Horsens as well as I (civil engineering) at the Polytechnic Institute. Finnish technicians graduated from I (civil engineering) before the transformation and from the sub-departments IL (agricultural engineering) and IV (civil engineering) thereafter. The total number also includes some construction engineers from Tampere. Swedish technicians in this group graduated from B (construction engineer) at all four institutes, V (civil engineering) at the Royal Institute of Technology and Chalmers, and H (house construction) at Chalmers. On the pan-Nordic level, civil and construction engineers represented about 30 per cent of the cohort, and this was a somewhat higher share than among all graduates. This group was stronger in Denmark and Norway than in Finland and Sweden. In the cohorts, civil and construction engineers were somewhat overrepresented in all countries except Norway.

Chemical engineers in Denmark include the specialisation AN (applied natural science, until 1884), K (chemical engineering, 1885–1889), and F (factory engineer, from 1888) at the Polytechnic Institute. All other technicians had graduated from K (chemical engineering), a specialisation existing in Helsinki and at all Norwegian and Swedish schools. Chemical engineers constitute

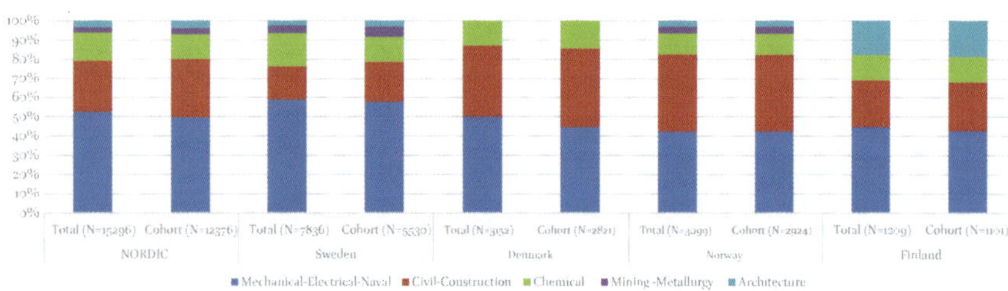

FIGURE 2 Distribution of specialisation among Nordic technical school graduates, 1880–1919

SOURCES: SEE FIGURE 1.

about 13 per cent of the Nordic cohort, which is somewhat lower compared to the reality. This is primarily because this specialisation is underrepresented in the Swedish cohort. In Denmark, this group was somewhat overrepresented.

Mining engineers and metallurgists did not graduate in Denmark and Finland. In Sweden, these graduates came from the Royal Institute of Technology's department of mining and had studied at the programmes G (mining, four years), g (mining, three years), H (metallurgy, four years) and h (metallurgy, three years). In Norway, these technicians graduated from the mining departments of the University of Kristiania until 1914 and from the Norwegian Institute of Technology until 1919. This group constituted about 3 per cent of both the cohorts and is somewhat overrepresented. This is also the case in Sweden.

Architects from Denmark are not included as they all studied at art schools, but graduated from A (architecture) in Helsinki, Trondheim, and Stockholm. Architects made up close to 4 percent of the Nordic cohort, and this is a somewhat higher share compared to all technical school graduates. A striking pattern is that this specialisation made up nearly 20 per cent of the Finnish graduates compared to about 3 percent of the Norwegian and Swedish counterparts. Architects were somewhat overrepresented in the national cohorts compared to the 'reality'.

Most graduates left school in the 1910s: 10 per cent graduated in the 1880s, 19 per cent in the 1890s, 32 per cent between 1900 and 1909, and 40 per cent in the 1910s. Figure 3 shows a minor pan-Nordic bias towards technicians graduating later. This is also evident in the Norwegian and Swedish cohorts, whereas the Finnish counterpart is biased towards technicians leaving school before 1909. The latter is not completely lucky, as we face the risk of overlooking some important technicians during Finland's earliest years of independence. The distribution over time in the Danish cohort corresponds very well to the distribution in general. We can note that the three first decades accounted for little in Denmark, while graduation 'exploded' in the 1910s.

Finland's second industrial breakthrough is placed significantly later than in Scandinavia, but a relatively large share of the Finnish technicians who left school between 1880 and 1919 did so in the 1880s. The same is true for Sweden, but this can be considered less surprising.

As for graduation age, we have created three categories: the group up to 24 years embraced 72 per cent on the pan-Nordic level, but the countries diverge: Norway notes 86 per cent, Sweden 79 per cent, Finland 56 per cent and Denmark 50 per cent. The group 25–29 years consisted of 25 per cent of the graduates on the pan-Nordic level and was larger in Denmark (44 per cent) and Finland (40 per cent) than in Sweden (19 per cent) and Norway (12 per cent).

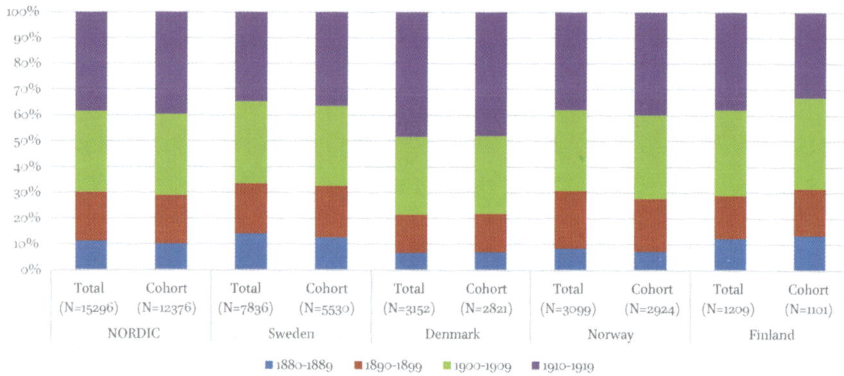

FIGURE 3 Distribution per decade of graduation for Nordic technicians, 1880–1919
SOURCES: SEE FIGURE 1.

The oldest group 30 years and older (including unknown ages) embraced 3 per cent in the Nordic countries: 6 per cent in Denmark; 3 per cent in Finland, 2 per cent in Sweden, and somewhat over 1 per cent in Norway. Norway turned out the youngest graduates and Denmark the oldest. Sweden's graduates generally left school at younger ages than their Finnish colleagues.

We have created dichotomous columns for transnational mobility, transnational migration, study trips, and return migration of migrants. Also, we have divided the destinations of the transnationally mobile graduates into eight categories based on geographical location and numbers of graduates going there; the German-speaking countries consists of Germany, Switzerland, and Austria. Acknowledging that French, Italian, and Rhaeto-Romanic were spoken in parts of Switzerland and that some of the Austrian parts of the Habsburg double monarchy spoke other languages, it was still a fact that graduates going to these countries almost always 'visited' the German-speaking parts. The German-speaking countries received the largest share of the transnational mobility. The second destination, both in this division and in terms of the number of graduates going there, is North America, that is, the United States and Canada. Greenland has been labelled a Nordic destination in this context, and Mexico a Latin American one. 'Europe' consists of all European countries except the Nordic and German-speaking ones, the British Isles and Russia. Turkey has been counted in this group, and so have countries that became independent from Russia, Germany, and Austria after World War I, for example, the Baltic States and Poland. 'Nordic countries' consists of mobility between Sweden, Denmark, Norway, and Finland as well as to Iceland and Greenland. The fifth destination is Britain or the British Isles. Ireland is included here, as it was a part of the United Kingdom through most of the studied period. Russia is, in this context, equal to the Russian Empire, except the Grand Duchy of Finland, and the subsequent Soviet Union after the October revolution and the Russian Civil War. Latin America and the Caribbean comprise the Americas south of the United States, and the final group is the rest of the world; Africa, Asia, and Oceania. In the database, we have dichotomous columns for these regions, that is, if the graduate was there or not, and we also have a column for what 'region' was the first destination. Information on the travellers' destinations is sometimes missing, as well as the time of departure. In some cases, we have only a note of a 'study trip abroad'.

There is also information on the length of the interval between graduation and migration and, consequently, departure year and decade. As for the returnees, the database has information on time spent abroad and permanent or temporary homecomings.

To look at occupational mobility, we have used the international HISCLASS system. The strengths of this system are its historical base and that it is based on data from different European countries. We have used the system to classify the graduates into social groups, but in this context, we need to refine our classifications. When engineers and architects leave school, they are, according to this scheme, in group two, that is, 'higher professionals'. There is only one group 'above', which is 'higher managers'. To classify by social origin, we have merged these two groups. In this context, we need to separate them. Upward occupational mobility is based on graduates stepping from group two to group one.

HISCLASS has, of course, shortcomings. Like most occupational classification systems, it is mainly constructed for studies of whole societies with arrays of different occupations and may be regarded as too coarse for a study like this. One example is the fact that executive engineers are still in group two. It is reasonable to assume that being an executive engineer at, for example, Burmeister & Wain or F. L. Smidth in Denmark or ASEA or Götaverken in Sweden was more prestigious than being president of a small company in a tiny town. Nevertheless, we will find a graduate in the latter category in group one, whereas one in the former is still in group two. This implies that we lose sight of some graduates who should be regarded as climbers, primarily people in intermediate leading positions. One alternative could have been to try to do our own classification into *top leaders* and *intermediate leaders,* but this implies a certain number of other problems. Using HISCLASS gives at least an indication of occupational mobility patterns, but it is, of course, a challenge for future researchers to develop more fine-meshed systems. Having stated this, we have registered the following number and shares of positions and upward mobility in table 1.

TABLE 1 Registered positions of Nordic technicians and upward mobility 1900, 1910, and 1920

	1900				1910				1920			
	Positions	Country %	Up-ward	Up %	Positions	Country %	Up-ward	Up %	Positions	Country %	Up-ward	Up %
Sweden	1234	49,8	102	8,3	2591	49,9	264	10,2	4331	47,0	526	12,1
Denmark	488	19,7	57	11,7	1068	20,6	121	11,3	2071	22,5	228	11,0
Norway	509	20,5	37	7,3	1006	19,4	86	8,5	2009	21,8	225	11,2
Finland	246	9,9	37	15,0	527	10,2	73	13,9	812	8,8	155	19,1
NORDIC	**2477**	**100,0**	**233**	**9,4**	**5192**	**100,0**	**544**	**10,5**	**9223**	**100,0**	**1134**	**12,3**

SOURCES: SEE FIGURE 1.

As we can see, almost half the investigated positions were in Sweden, about one-fifth each in Denmark and Norway, and around one-tenth in Finland. The share that was in group 'one' increased over time, probably because a higher number of technicians had been active for a longer time. In 1900, the technician was, on average 31 years old and had worked an average of almost nine years. The corresponding averages were 34 and 12 for 1910 and 37 and fourteen for 1920. Norwegian technicians had the lowest average ages at all three observation points, while Danes and Finns generally were somewhat older. Swedes had generally somewhat longer experience, while Norwegians and Danes were most 'inexperienced' when it came to time. Technicians in Finland 'climbed' more often than their Scandinavian colleagues. Intra-Scandinavian differences are small, at least in 1920.

To study the geographical dispersion of returnee technicians in a comparative perspective, we have also added the geographical locations of the workplaces in 1900, 1910, and 1920. They have been registered on the county level, the counties were called 'län'/'läänit' in Sweden and Finland, 'fylken' in Norway, and 'amt' in Denmark. The regional division in today's Norway is the same as in this study, whereas we use the Swedish division before 1997, separating today's Scania and Västra Götaland counties, into two and four counties respectively. We use the pre-1970 Danish regional division and the division in Finland between 1831 and 1938.[204] These were the regional divisions in force during the period we are studying and provide—as smaller units—a better picture of the dispersion of returnee technicians compared to today's larger regions, especially in Denmark and Finland.

We have also registered on the city/parish level. Somewhat over 40 per cent of the work places were located in the capital cities; Copenhagen diverged from the other capitals and hosted about two-thirds of the registered workplaces. Helsinki hosted between 40 and 50 per cent of the workplaces in Finland, whereas Kristiania and Stockholm hosted about one-third in Norway and Sweden respectively. Generally, we can conclude that Copenhagen's dominance contributed to a significantly lower Danish share born in other cities and towns as well as in rural areas.

5.2 *Quantitative, Qualitative, and Comparative Methods*

This study makes use of quantitative methods and is heavily dependent on descriptive statistics. The statistics used are rather basic: percentage calculations of how many migrated, went on a study trip, returned, and so

204 Petsamo County was added in 1921.

forth out of different criteria. We could have used, for example, logistic regressions to discuss the weight of different characteristics in the mobility processes as well as statistical significance, but the questions revolve more around numbers, shares and actions than weighting the importance of social origin and so on.

This study takes its point of departure in the mutual background of its subjects—they all graduated from a minimum three-year course at a technical school in the Nordic countries between 1880 and 1919. However, the group of over 12,000 graduates is too large to suit fully the prosopography, or collective biography, criterion. A prosopographical method usually takes its point of departure in a smaller group, maybe a couple of hundred, and follows them more in-depth compared to this study.[205] The study is an attempt to combine quantitative and qualitative research and to mix a different kind of data in the writing. It is up to the reader to judge whether this is an appropriate manner to write a manuscript on this topic.

The study also sets out to write comparative history, which has become increasingly common in many countries since around the mid-1980s, even if the method also had many proponents in the early twentieth century. The globalisation debate has spurred historians to make comparisons between different parts of the world. There are different kinds of comparisons. Generally, they are divided into individualising comparisons focusing on the demonstration of the uniqueness of one case through a comparison with other cases and universalising comparisons aiming to identify similarities between cases, often giving equal weight to all cases compared. Between these two ideal types, there are several hybrids, and the borders between different comparisons have become increasingly less visible. As Stefan Berger states, 'On a basic level, one could say that researchers in comparative history are always interested in establishing both differences and similarities between cases'.[206] This is also the case in this study.

The advantages of comparative history are manifold. We may, for example, improve our explanations of certain phenomena by comparing similar developments in other places. We know, for example, from previous studies that Norwegian technical education developed slower than in the other Nordic countries. This might explain the high propensity among Norwegian graduates to go abroad to complete their education. However, if we find similar

205 Lawrence Stone, *The past and the present* (London 1981).

206 Stefan Berger, 'Comparative history', in: Stefan Berger, et. al. (eds.), *Writing history: theory & practice* (London 2003) 160–163, quote from 163.

shares going abroad from the other countries, it is unlikely that the quality of the Norwegian technical education is the only reason. Comparisons may also draw attention to different causes and different patterns behind similar outcomes. If a rapidly industrialised country experiences as much out-migration of technicians as a slowly industrialising country, we might conclude that industrialisation level was important for the former country, but we need to find another explanation for the latter. Developments in one place can be explained better by comparing; we avoid studying countries, areas and places in isolation.

However, comparative history is, of course, not unproblematic. For example, a comparative historian needs to be familiar with different contexts. In this study, we have been fortunate to find relatively comparable archival material, that is, the biographical registers discussed below. The lack of comparable source material is often a problem in comparative historical studies, but this is less evident in this study. Secondary literature has also been a problem as it often diverges. Researchers have posed different questions in different geographical and time contexts, and a research topic might also be significantly more explored in one area compared to another. It appears, for example, that Danish studies of transnational mobility of technicians are less numerous than studies from Finland, Norway, and Sweden. This does not necessarily imply that the phenomenon was less common in Denmark. It might be a result of a different research tradition. Furthermore, it is difficult to gain knowledge of another societal context only through reading about it. First-hand experience is often a necessity. The author has already stated that the knowledge of his native Sweden is deeper and more thorough than his knowledge of the neighbouring countries.

In this study, we have been favoured by the mutually intelligible Scandinavian languages. Differences in languages are described by Berger as a 'veritable minefield' for comparative historians. Words that might appear as similar may on the one hand have different meanings in different languages.[207] On the other hand, even if the languages are mutually intelligible, some words, for example, occupational titles, may differ a lot. Language still ought to cause fewer problems in a comparative Nordic context than for studies comparing, for example, northern and southern Europe, but we should not go so far as to say that it is unproblematic. Finnish is, as mentioned, a completely different language. The author has as mentioned some working knowledge of Finnish, but his reading abilities in the language are significantly weaker compared to

207 Berger, 'Comparative history', 169.

the other languages that he can take in such as Scandinavian, English, and German. This needs to be considered, even if many Finnish documents, as mentioned, are also available in Swedish.

6 Sources, Selections, and Cohorts

This study is based on engineers and architects graduating from an at least three-year-long technical school education in Denmark, Finland, Norway, and Sweden between 1880 and 1919. The schools in table 1 offered the education. This selection leaves out several individuals who may be regarded as engineers and architects. First, there were several autodidacts performing what one may call engineer's or architect's work. Second, several individuals had only two years of studies, and they include graduates from the Norwegian mechanical schools in Horten and Skien as well as the schools in Bergen, Kristiania, and Trondheim after the educational reform of 1915. The Swedish mining schools in Falun and Filipstad also belong to this category. Biographies on some of these technicians exist but are largely incomplete. Even more important, especially in Finland and Norway, are the relatively large groups of technicians who went through their entire education abroad. They were, of course, important for modernisation and industrialisation. Norway's Sam Eyde, who studied in Germany and became one of his country's leading industrialists upon his return, is one example.[208] Acknowledging their importance, this group is not included in the statistical base of this study, although some of them are mentioned as examples in the text. One reason is that their numbers are difficult to estimate, especially in Denmark and Sweden. The biographies of the Swedish Technical Association, for example, include members in this society. Studies at the Royal Institute of Technology and later also at Chalmers implied membership. However, this society also accepted members who were viewed as skilled and experienced technicians, even if they had studied elsewhere.[209] Some technicians who had studied abroad were awarded membership; others were not members. Thus, it is difficult to get the appropriate numbers and we want to generate knowledge of the strength of transnational mobility.

A huge majority of the engineers—81 per cent—graduating from educational institutes providing at least three years of education between 1880 and 1919 are included in this study. It has not been possible to investigate all the engineers due to a

208 Ole Kristian Grimnes, *Sam Eyde: den grenseløse gründer* (Oslo 2001).

209 Indebetou and Hylander, *Svenska teknologföreningen*, VI.

lack of relevant source material. The choice of period to follow graduates—1880–1919 until 1930—requires some explanation. The stoppage year is easily explained; we do not have comparable source material after 1930; this is especially true for Sweden. We wanted to follow the careers for at least ten years after graduation; 1919 is the latest common denominator where we can do this with a reasonable work effort. As for the starting year, we wanted to study the technicians during the second industrial breakthrough, which started in Denmark in 1880. The three technical schools in Norway were inaugurated in the 1870s; Finland got its Polytechnic Institute one year earlier. As for Danish and Swedish technicians, we could have collected hundreds of them from earlier decades. This development, as well as judgements concerning work efforts, led us to opt for 1880 as the starting year. We might have considered 1870; then we would have had the entire existence of the Norwegian schools, and we would have had some Danish technicians graduating before the breakthrough mentioned above. All in all, as table 1.1 shows, we have collected biographies of many technicians. This allows us a high degree of generalisation, whereas we lose some of the in-depth perspectives that a 'smaller' cohort could have allowed for. This may, however, be a task for future researchers, possibly with this study as a point of departure.

6.1 *Biographical Sources*

The biographical registers have been a prerequisite for this study. Two criteria are especially important: at least three years of technical school education from Denmark, Finland, Norway, and Sweden, 1880–1919. The lack of technical education in Iceland makes it impossible to include the country in the study from the same points of departure as the other countries. Icelanders obtaining education in Denmark and Norway are, of course, included.

The biographical notes are collected from registers and source material mentioned below, and a database has been constructed. Notes on geographical, social, and educational background as well as the careers have formed the base for the discussion of these technicians and their transnational mobility. The biographical notes have primarily been combined with secondary sources.

In Denmark, very good directories issued in 1929,[210] 1942,[211] and 1956[212] cover the students at the Polytechnic Institute (PL) in Copenhagen. They are qualitatively the best ones in this study and contain detailed information on

---

210 R. Jespersen (ed.), *Biografiske oplysninger angaaende Den Polytekniske Læreanstalts kandidater 1829–1929* (København 1930).

211 Hannover, *Dansk Civilingeniørstat 1942*.

212 Aage Hannover (ed.), *Dansk civilingeniørstat 1955: biografiske Oplysninger om polytekniske Kandidater 1829–1955* (København 1956).

TABLE 2 Engineers and architects with at least three years of education at technical schools in the Nordic countries 1880–1919 in per cent per educational institute

COUNTRY INSTITUTE	All	Cohort	Coverage %
Denmark	**3152**	**2821**	**89**
Polytechnic Institute (PL), Copenhagen, 1880–1919	2111	2105	100
Copenhagen Mechanical School (KM), 1881–1919	798	618	77
Odense Mechanical School (OMT), 1881–1919	180	63	35
Aarhus Electrotechnical School (AE), 1913–1919	40	16	40
Horsens Construction School (HB), 1916–1919	23	19	83
Finland	**1209**	**1101**	**91**
Finnish Polytechnic Institute (SPO), Helsinki, 1880–1908	727	714	98
Finnish Institute of Technology (STK), Helsinki, 1908–1919	416	387	93
Tampere Technical Institute (TTO), 1915–1919	66	0	0
Norway	**3099**	**2924**	**94**
Kristiania (Oslo) Technical School, (KTS) 1880–1914	1100	952	87
Trondheim Technical Institute (TTL), 1880–1914	894	891	100
Norwegian Institute of Technology (NTH), Trondheim, 1915–1919	519	518	100
Bergen Technical School (BTS), 1880–1914	497	483	97
University of Kristiania (UiK), mining department, 1880–1914	89	80	90
Sweden	**7836**	**5530**	**71**
Royal Institute of Technology (KTH), Stockholm, 1880–1919	2993	2692	90
Chalmers Institute of Technology (CTI), Gothenburg, 1880–1919	1632	1566	96
Örebro Technical Upper-secondary School (TESÖ), 1880–1919	1141	583	51
Malmö Technical Upper-secondary School (TESM), 1880–1919	696	689	99
Norrköping Technical Upper-secondary School (TESN), 1880–1919	643	0	0
Borås Technical Upper-secondary School (TESB), 1880–1919	524	0	0

TABLE 2 Continued

COUNTRY INSTITUTE	All	Cohort	Coverage %
Härnösand Technical Upper-secondary School (TESH), 1904–1919	207	0	0
NORDIC COUNTRIES	**15296**	**12376**	**81**

SOURCES: SEE FIGURE 1.

background and careers. As we can see in table 1, they are almost complete. The directories covering students at lower technical schools, *Teknika,* that is, the mechanical schools in Copenhagen (KMT) and Odense (OMT), the electrotechnical school in Aarhus (AE), and the construction school in Horsens (HB), have far from the same high quality. One directory was issued in 1931[213] and a second in 1945.[214] Biographical notes exist for roughly 70 per cent of the graduates, but they are in many cases sparse and leave aside important information such as year of birth, social background, and above all, major parts of the careers. For the 1931 directory, questionnaires were sent a few years earlier. The board of the association of electrical and mechanical engineers requested that their members should not be included in the directory, something that the editors of the directory in most cases accepted. Therefore, many cases are missing.[215] Some of the missing persons may have been possible to investigate through extensive but time-consuming archival research in Denmark. Worth mentioning here is that architects are left out of the Danish cohort as they were educated at the Academy of Arts and not at the technical schools. The source material covering Danish architects is focused on those remaining in Denmark.[216]

Yearly lists of graduates from the Finnish Polytechnic Institute (SPO) and the subsequent Finnish Institute of Technology (STK) between 1880 and 1919 have been used for identification. There is one directory of students issued

213 *Danske teknika og deres Dimmitender gennem 50 Aar, Ingeniør- og Konstruktørsammenslutningen gennem 35 Aar* (København 1931).

214 *Danske Ingeniører fra Teknika* (København 1945).

215 Wilh. Hansen, 'Indledning', in: *Danske teknika og deres Dimmitender gennem 50 Aar, Ingeniør- og Konstruktørsammenslutningen gennem 35 Aar* (København 1931) 9.

216 *Dansk Arkitektstat 1941* (København 1941).

in 1899[217] and a second in 1918.[218] The latter includes students enrolling up to 1908; the year when the Polytechnic Institute was transformed. These directories are of varying quality, and many technicians are mentioned only by name. To make the study of Finland comparable, the directories issued by the Finnish technical associations in 1936[219] and 1948[220] have been used as well as a *Who's Who?* (*Kuka kukin oli*) from 1961,[221] a booklet on women architects,[222] and the Brage press cutting archive in Helsinki. This archive covers the lion's share of all persons the Swedish-language press in Finland has written about—including obituaries—arranged in alphabetical order. Its focus is on Swedish speakers, but several prominent Finnish-speaking persons are also included in the extensive material. It covers 65 per cent of the graduates from 1880 to 1919. We end up with a minor share of technicians followed only until 1900 or 1918. The lower level technical institute in Tampere is left out of this study.[223]

The Norwegian source material allows for inclusion of most technicians leaving the schools up to 1915. Qualitatively good directories exist for the schools in Bergen (BTS)[224] and Trondheim (TTL)[225] between 1880 and 1915 as well as for the Norwegian Institute of Technology (NTH) after 1915.[226] The technical school in Kristiania (KTS) did not issue a directory of its own, and the notes are mostly collected from Bjarne Bassøe's directory of Norwegian

217 *Matrikel öfver Tekniska realskolans och Polytekniska skolans i Helsingfors samt Polytekniska institutets i Finland lärare och elever 1849–1897* (Helsingfors, 1899).

218 Heiniö, *Matrikel öfver Polytekniska institutets i Finland lärare och elever*.

219 Y[rjö Juho] Talvitie, *Suomalaisten teknikkojen seuran nimikirja 1896–1936* (Helsinki 1936).

220 *Suomen Insinöörejä ja arkkitehtejä: Ingenjörer och arkitekter i Finland: 1948* (Vaasa 1948).

221 Iisakki Laati and Yrjö Blomstedt (eds.), *Kuka kukin oli: henkilötietoja 1900-luvulla kuolleista julkisuuden suomalaisista* (Helsinki 1961).

222 *Profiles—pioneering women architects from Finland* (Helsinki 1983).

223 *Ingenjörer: Insinöörit: Tampereen teknillinen opisto, Tekniska läroverket i Helsingfors, Helsingin teknillinen opisto, Turun teknillinen opisto* (Tampere 1957). The first candidates from Tampere left school in 1915 and years, numbers and type of education are available in a remembrance of the institute. However, the book does not reveal the names of the candidates and although it probably had been possible to identify them through other sources, this work has been evaluated as too time-consuming in relation to the numbers. In 1916, a Swedish-language lower technical institute was founded in Helsinki. As regards the latter, the first candidates did not graduate before 1920 and are, therefore, not within the time period of this study.

224 Leif Eskedal (ed.), *BTS-matrikkelen: ingeniører uteksaminert ved Bergen tekniske skole 1875–1975* (Bergen 1975).

225 Alstad, *Trondhjemsteknikernes matrikel*; Alstad, *Tillegg til Trondhjemsteknikernes Matrikkel*.

226 Georg Brochmann (ed.), *Vi fra NTH: de første ti kull: 1910–1919* (Stavanger 1934).

engineers[227] and from the press cutting archive at the National Library. Bassøe's directory has shortcomings; it does not cover graduates before 1900, nor does it cover architects. It was compiled long after the engineers' careers, and Bassøe bases some of his notes on earlier directories. Nevertheless, if his notes are compared with earlier directories, some information is left out, mainly visits to universities abroad.[228] Information based only on Bassøe, then, may under-report mobility rates. Materials from the technical school at Oslo City Archives have been used.[229] So has a variety of other biographical sources: a biographical collection issued by the Association of Norwegian Engineers and Architects in 1916,[230] booklets issued to various anniversaries by Kristiania's technical school and the students themselves,[231] some issues of Norway's *Who's Who?*,[232] and the country's biographical encyclopaedia.[233] We still end up with some missing technicians. Other sources must also be used for the mining engineers graduating at the University of Kristiania between 1880 and 1900.[234] All in all, 94 per cent of the engineers and architects fulfilling the criteria are included.

The two biographic volumes of the Swedish technical association (1936)[235] cover many of the country's late nineteenth- and early twentieth-century technicians. These biographies primarily include graduates of Stockholm's Royal Institute of Technology (KTH), but some are also from Gothenburg's

227 Bassøe, *Ingeniørmatrikkelen.*

228 The reason is probably that Bassøe covers a large number, around 10,000, Norwegian engineers (all graduates between 1901 and 1955) and, thereby, deliberately left out some information on the elderly graduates to give space to the more recent ones.

229 Oslo City Archives: the Library: Beretning om Kristiania Tekniske Skole virksomhet, 1891, 1898, 1902, 1908; Oslo Tekniske Skoles arkiv, volume G0001: Beretning om KristianiaTekniske Skole virksomhet, 1889–1890, 1893–1897, 1899–1901, 1903–1907, 1909–1910, volume G0002, Beretning om Kristiania Tekniske Skole virksomhet, 1911–1915.

230 Eiliv Fougner (ed.), *Norske ingeniører og arkitekter: kort oversigt over Den norske ingeniørforenings og Norskearkitekters landsforbunds historie, samt biografiske oplysninger om de to organisationers nulevende medlemmer med portrætter* (Kristiania 1916).

231 *Ingeniører fra 1896 fra Kristiania Tekniske Skole: biografiske Oplysninger samlet til 25 Aars Jubileet 1921* (Kristiania 1921); *Ingeniørene fra K.T.S 1897–1947* ([Oslo] 1947); *KTS: 50 årsberetning om Ingeniørkullet fra Kristiania tekniskeskole 1896* ([Oslo] 1946); Kristiania tekniske skole, *Festskrift i anledning af Kristiania tekniske skoles 25-aars jubilæum i juni 1898* (Kristiania 1898).

232 Hj. Steenstrup, *Hvem er hvem?* (Oslo 1930); Hj. Steenstrup, *Hvem er hvem?* (Oslo 1937); Chr. Brinchmann et al., *Hvem er hvem?: haandbok over samtidige norske mændt og kvinder* (Kristiania 1912).

233 Einar Jansen and Paulus Svendsen (eds.), *Norsk biografisk leksikon* (Oslo 1954).

234 *Tidsskrift for kjemi, bergvesen og metallurgi* (1950).

235 Indebetou and Hylander, *Svenska teknologföreningen.*

Chalmers Institute of Technology (CTI) and lower technical schools. The Gothenburg-based institute also has issued its own biographies.[236] Yearly lists of Chalmers graduates have, to some extent, been used to identify what specialisation these technicians had and to follow them up to 1930.[237] Biographies of the technical upper secondary schools in Malmö (TESM)[238] and Örebro (TESÖ)[239] are also used. These schools offered lower technical education. Örebro's register covers just under half of the graduates because it focused on candidates who were alive in 1925 and sent in biographies to the editor. We have used yearly lists to follow these graduates to the early 1930s.[240] Malmö's graduates have been followed only until 1928 because this was the directory's publishing year.

With the reservations mentioned above, there are good biographies of Swedish technicians. Sten Carlsson has used the same material for a study of Swedish engineers in Chicago and claims that the ones from the Swedish technical association and Chalmers are excellent, while he ascribes the ones from Malmö and Örebro shortcomings.[241] It seems, however, that there are shortcomings also in the directories of the 'big' institutes, whereas the notes in the intermediate schools' directories are almost as good qualitatively. There were also technical upper secondary schools in Borås, Härnösand, and Norrköping, but they are not included due to incomplete source material. Borås issued a directory in 1912, and yearly lists exist, but to work with them was not within the scope of this study. The source material from Härnösand and Norrköping is too sparse to allow for inclusion. It is possible to identify the candidates by name and education, but not to follow their careers.[242]

---

236 Gösta Bodman, *Chalmers tekniska institut: matrikel 1829–1929* (Göteborg 1929).

237 *Civilingenjörer och arkitekter utexaminerade från Chalmers tekniska högskola: jämte redogörelse för Chalmersska ingenjörsföreningens och Chalmersska forskningsfondens verksamhet* (Göteborg 1904).

238 *Malmö teknologförbund: Minnesalbum utg. i anledning av Malmö tekniska läroverks 75-åriga verksamhet: 1853–1928* (Malmö 1928).

239 Emil Forsberg and Emil Adlers (eds.), *Tekniska föreningen i Örebro 1875–1925: minnesskrift utgiven med anledning av föreningens femtioåriga verksamhet* (Stockholm 1925).

240 *Tekniska föreningen i Örebro. Medlemsförteckning, m.m* (Örebro 1926). 1926–1932.

241 Carlsson, 'Swedish engineers in Chicago', 182.

242 *Matrikel över ingenjörer, utexaminerade från Högre tekniska läroverket i Norrköping förutvarande Tekniska gymnasiet resp. Tekniska elementarskolan, från skolans början år 1857 t.o.m. vårterminen 1961* (Norrköping 1962); A. G. J. Eurenius, *Matrikel öfver eleverna i Tekniska elementarskolan i Norrköping från skolans början år 1857 till och m. v. till 1912* (Norrköping 1912).

6.2 *Secondary Sources and Technical Journals*

There are a variety of sources in the Nordic countries that are little used in this study; travel reports constitute one such source. There are numerous interesting travel reports in different archives throughout the Nordic area, and to survey them all is of course impossible. We have basically limited our qualitative parts to materials available in technical journals and earlier studies. Printed source materials such as remembrance books, newspapers, and technical journals are used. The problems with remembrance books issued by companies on different anniversaries and so forth are that people involved in company management or responsible for certain production often wrote and/or financed the accounts. Published works that aimed at celebrating the companies would hardly discuss sensitive technology imitation or the like. It is hardly likely that the companies chose 'critical' authors, at least not in those times. The same may be relevant for the technical journals used in this study: Denmark's *Ingeniøren*, Finland's Swedish-language *Tekniska Föreningens i Finland förhandlingar,* the Finnish *Teknillinen Aikakauslehti,* Norway's *Teknisk Ukeblad,* and Sweden's *Teknisk Tidskrift.* It was one thing to diffuse company secrets in internal messages within the board, another to spread news in accounts that could be read by others. This must be considered.

This could, of course, have been another type of study, less broad, embracing fewer individuals, and allowing for more in-depth studies of primary sources and the survey of documents from business and company archives and the like. It is up to the reader to judge, whether this largely quantitative approach is an appropriate way to deal with the subject. As stated earlier, this book can form a base for more specialised, in-depth studies of different 'subgroups' or individuals by future researchers. If this is realised, travel reports constitute one of the most, if not the most, important source material. These reports may, for example, be the series *Tiedonantoja* by the Finnish Board of Industries or the series available at Royal Institute of Technology's library in Stockholm.

So, we have packed our 'luggage' and are ready for departure. It is time to begin our journey in the footsteps of late nineteenth- and early twentieth-century Nordic technicians to 'shipyards and workshops in America, to railway building in Russia and laboratories in Zurich, to draftsman's offices all around the earth'.[243]

243 Cf. Berner, *Sakernas tillstånd,* 43. Original in Swedish: till skeppsvarf och verkstäder i Amerika, till järnvägsbyggnader i Ryssland och laboratorier i Zürich, till ritkontor jorden rundt.

CHAPTER 2

A Peregrine Profession

The quote ending the last chapter, from a 1906 Swedish student paper, illustrates the 'mobile mentality' among Nordic technical school graduates. We can note that the total number of transnationally mobile technicians in our cohort was 6,711 of whom 5,664 were migrants. Transnational mobility embraced 54 per cent of the graduates in our cohort and migration 46 per cent. The number of Swedish graduates going abroad was 2,745, the Norwegian number was 1,820, the Danish 1,322, whereas 824 technicians from Finland left the country for shorter or longer intermissions abroad. However, even if the real Finnish number was lowest, figure 4 shows that Finnish graduates going abroad comprised the highest percentage. A large share of the Finnish travellers made study trips; the 'migrant share' was higher from Norway.

Transnational mobility characterised technicians from several countries. It was not only a 'peripheral' trait but included technicians from industrial 'core' nations like France, Germany, and the United States. Braun has shown that the back-and-forth migration of German engineers to America also was important for technology transfer in both directions, and Nolan describes interwar study tours by German engineers and how they spurred debates on Taylorism and Fordism in Germany.[1]

Travel from 'peripheries' was a tradition with deep historical roots. Academic travel to renowned foreign universities was a trait in the Middle Ages, and pre-industrial and early industrial transnational crossings of specialists and journeymen have been a focus for many scholars. Myllyntaus calls the mobility of technicians a modern form of a medieval trait, where craftsmen trained under masters in foreign environments. Spanish, Portuguese, and Greek historians have shown that it was common among nineteenth-century technicians to go abroad for academic training and to improve their expertise.[2] Efmertova's studies conclude that technical studies abroad were important after the

1 Braun, 'Franz Reuleaux': 112–130; Braun, 'Technologietransfer im Maschinenbau': 238–252; Braun, 'A Technological Community in the United States': 447–463; Nolan, *Visions of modernity*; Also see: Day, 'The Making of Mechanical Engineers in France', 439–460; Higginson, 'Privileging the Machines', 1–34; Teisch, 'Home is not so far away', 139–160.

2 Cardoso de Matos, 'Asserting the Portuguese Civil Engineering Identity', 177–208; Cardoso de Matos and Diogo, 'Bringing it all back home', 155–181; Assimacopoulou et al., 'Elève en France', 25–41; Antoniou et al., 'Greek Engineers'; Anduaga, 'The engineer as a "linking agent"', 45–70.

 | DOI:10.1163/9789004385207_003

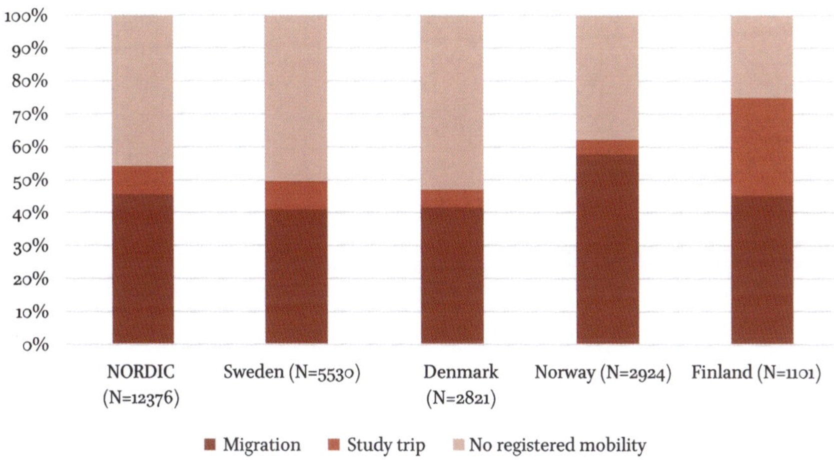

FIGURE 4 Transnational mobility before 1930 among Nordic engineers and architects leaving school 1880–1919
SOURCES: SEE FIGURE 1.

establishment of an independent Czechoslovakia in 1918, and their experiences were utilised in more or less every branch of domestic industry.[3] It parallels to some extent the Norwegian case and perhaps to an even greater extent the Finnish case: a newly independent country utilising travel and foreign knowledge to build up a domestic industry.

The phenomenon was common in many countries, but it is difficult to compare the 'strength' of these streams with the Nordic countries. As far as the author knows, there are no comprehensive statistical studies of other countries. The data in this study is based on technicians with a domestic pre-education, whereas a great deal of the students mentioned in earlier studies took their entire education abroad. This was also a trait among youngsters from the Nordic countries, especially from Norway and Finland because of the shortcomings in domestic technical education. The real numbers are difficult to estimate. Between 1901 and 1920, 388 engineers graduated abroad by Bassøe's count; of that number, 81 per cent returned to Norway.[4] The studies of Portugal, Greece, and the Basque regions emphasise enrolment at technical universities and travels to fairs rather than acquiring experience from working abroad, which is an important focus in this study.

3 Èfmertovà, 'Les professeurs èlectrotechniques', 513–523.
4 Bassøe, *Ingeniørmatrikkelen*, XII.

Our observations underline statements of engineers and architects as transnationally mobile professions, but this is not to say that other professions and academics were immobile. Hietala concludes that Finnish medical doctors often went to Germany. Scholars within humanities, forestry, natural science, law, theology, and political science also participated in these moves.[5] Transnational mobility was not limited to professionals and academics. Skilled workers also went abroad often. Myllyntaus has noted that there were almost 1,000 engineering shop workers with foreign work experience in Finland in 1911; Sixty per cent had been in Russia, mostly in nearby Saint Petersburg, and 25 per cent in the Scandinavian countries. Finnish engineering shop workers sought more foreign experience than their Swedish colleagues.[6] According to calculations by Fay Lundh Nilsson, around 6 per cent of Sweden's engineering shop workers had foreign work experience, often from other Nordic countries, but also from Germany and the United States. Modern and more specialised workshops often employed higher shares of foreign-experienced workers compared to older and traditional engineering companies.[7]

1 Aspects of Nordic Technicians' Transnational Mobility

There were many aspects of transnational mobility among technicians from the Nordic countries and below; we will discuss certain characteristics that made technicians prone to go abroad.

1.1 *Architects and Finns Study Travelled—Engineers and Norwegians Migrated*

Figure 5 shows the distribution between migration and study trips, revealing that migration accounted for about 80 to 90 per cent of the transnational mobility on the pan-Nordic level as well as from Sweden, Denmark, and Norway. Among Finnish technicians, there were more study travellers than migrants, and it was also much more common to combine the two 'mobility types'.

Thus, study travelling was a Finnish phenomenon in this context, whereas the Scandinavians migrated. The latter was especially true for the Norwegians. Graduating in any of the engineering specialisations also increased the likelihood that migration was the preferred type of mobility; the shares for

5 Hietala, 'Finnische Wissenschaftler', 373–394.

6 Myllyntaus, *The gatecrashing apprentice*, 39.

7 Fay Lundh Nilsson, *Lönande lärande: teknologisk förändring, yrkesskicklighet och lön i svensk verkstadsindustri omkring 1900* (Stockholm 2007) 95–96, 189.

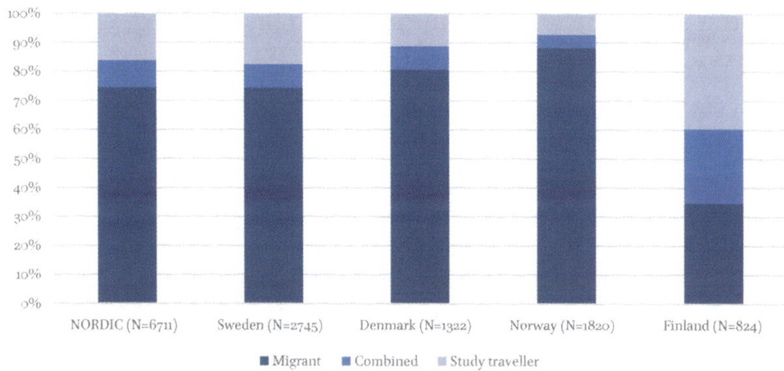

FIGURE 5 Distribution between 'migration' and 'study travel' before 1930 among transnationally mobile Nordic technical school students graduating 1880–1919
SOURCES: SEE FIGURE 1.

'migration only' fell between 70 and 80 per cent for the four engineering specialisations, compared to 30 per cent for architects.

1.2 *Extensive Study Travel Contributed to the Mobility from Finland*

Their extensive study travel distinguishes the Finnish graduates from their Scandinavian colleagues. Graduates stating, 'study trip' without further information on foreign work places and/or enrolment at educational institutes comprise about 30 per cent of the Finnish cohort, as figure 5 shows, whereas these shares are less than 10 per cent in the other countries. Two remarks can be made related to sources. First, we do not know how many unrecorded 'study trips' there were. Second, it is difficult to ascertain whether and to what extent Finnish graduates (alternatively editors) left out information on employment and/or enrolment and simply labelled relatively long-term foreign intermissions 'study trips'. It is a reasonable assumption that the Finnish 'migration' share was higher than the recorded 45 per cent, but that 'study travelling' still was a distinct Finnish pattern. Myllyntaus claims that organised study trips 'for collecting information from abroad became an institution in nineteenth-century Finland'.[8] New infrastructural investments were often preceded by several experts visiting advanced places in Europe, and these visits were crucial for the adoption of new technology. Also, they were often made with the purpose of purchasing new machines or recruiting foreign specialists. These visits had a stronger impact on technology transfer than 'ordinary' emigration,

8 Myllyntaus, 'The Best Way', 142.

and the reason was the comparably late starting and the peak point of emigration to America and that most emigrants never returned to use their skills.[9]

1.3 *Geographical Closeness and Political Ties to Russia Also Spurred Finnish Mobility*

Except for some coastal regions in the west and north, Finland was not struck by 'America fever' to the same extent as many areas in Scandinavia. Hypothetically, this could have had an effect also on transnational technical mobility from Finland. This was not case. We have touched upon the study travelling, but we should not forget that Russia functioned as a kind of 'Substitute America' in Finland.

The Russian Empire had, however, different implications on transnational mobility from the Grand Duchy. On the one hand, a nearby European metropolis like Saint Petersburg with developed artisan trades, major mechanical workshops, and so forth attracted people from all social classes in Finland. On the other hand, some tsarist politics had a repelling impact and implied westward migration. Myllyntaus writes that political conflicts and discrimination against ethnic minorities such as Jews drove many Russian-born students to universities primarily in Germany. He argues, nevertheless, that students normally did not leave Finland because of political oppression, but to acquire qualifications in their professions.[10] However, the Grand Duchy experienced intensified Russification in the early years of the twentieth century, and one manifestation was a new military service law which made it mandatory for young Finnish men to serve four years in the tsarist army, anywhere within the borders of the empire. This law was abandoned in 1905[11] but certainly implied that some recent graduates of this period might have considered spending a period abroad as a means of avoiding military service in some distant part of tsarist Russia.

1.4 *Slower Industrialisation Fuelled Mobility from Both Finland and Norway*

Nearby Saint Petersburg's possibilities can be characterised as 'pull factors' for technicians and many other groups in Finland, whereas the Russian military

9 Myllyntaus, 'The Best Way', 141–142.

10 Myllyntaus, *The gatecrashing apprentice*, 120–121.

11 J. E. O. Screen, 'The Finnish Army 1881–1901: A National Force in a Russian Context', *The Slavic and East European Review* 70:3 (1992) 453–476; J. E. O. Screen, 'The Finnish Cadet Corps, 1819–1903: A Reflection of Finno-Russian Relations and the Language Conflict in Finland', *The Slavic and East European Review* 81:2 (2003) 217–235.

service law can be viewed as a 'push factor'. However, other conditions also drew engineers and architects away from Finland. Pasi Tulkki estimates that about two-thirds of the mechanical engineers educated in Finland before 1900 had their first employment abroad, mainly because of the limited possibilities at home. The volume of late nineteenth-century industry was, as Michelsen points out, limited. Paper mills and sawmills were not normally in need of highly educated people during this period.[12] Large-scale industrialisation of Finland was still many years in the future; economic historians have, as mentioned, dated it to the 1920s and 1930s. Denmark's breakthrough has been dated to the 1880s and 1890s, and Sweden's to the 1890s and the first decade of the twentieth century.

Norway's counterpart can be placed in the first two decades of the twentieth century, but Stang argues that the country had relatively weak industrial development until 1930, at least if we look at the structure of employment.[13] The problems Tulkki and Michelsen identified in late nineteenth-century Finland were present also in Norway. In 1885, the editor of *Teknisk Ugeblad* received an anonymous letter in which he read the words: 'we might as well emigrate immediately'.[14] The background was an advert for a technical director position at a cellulose factory near Trondheim. The frustrated engineer had seen it in *Berliner Papierzeitung* but had been unable to find it in any Norwegian technical journal. This, he claimed, revealed a tragic lack of confidence in Norwegian ability. The factory owner responded that able countrymen were welcome to apply, but believed that the factory had to look to Germany and Sweden to find the required competence.[15] In 1940, Bjork sent a questionnaire to Norwegian-born engineers in the United States asking them: Why did you leave for America? We can quote some answers: 'No work to be had in Norway', 'prevailing salaries and particularly slow schedules of advancement', 'because it was next to impossible to secure technical employment in Norway', 'I was not born in my father's office', and 'America seemed the only way out'.[16] Norway was as mentioned only behind Ireland when it came to overseas emigration per capita, and this, of course, had an impact also on technicians. Problems making a decent living were also often forces behind transatlantic emigration in general.

---

12 Tulkki, *Valtion virka*, 149, 175; Michelsen, *Viides sääty*,166.

13 Stang, 'A measure of relative development?', 89–90.

14 X, 'Patriotisme', *Teknisk ugeblad* 9 (1885) 43. Norwegian original: 'vi kan ligeså godt gå til udlandet strax'.

15 X, 'Patriotisme', 43; Laur. Jensen, 'Patriotisme', *Teknisk ugeblad* 14 (1885) 67; Dixi X, 'Patriotisme', *Teknisk ugeblad* 17 (1885) 85.

16 Stang, 'A measure of relative development?', 96.

In 1903, *Teknisk Ugeblad* wrote that freshman engineers had to travel because they could not get employment upon graduation, but that the problem also included engineers with some years of experience who had lost their employment in public service or private companies.[17] All Nordic countries experienced periods of an oversupply of engineers. In both Denmark and Sweden, complaints were sometimes raised about the constantly increasing number of technical school graduates. This caused frustration and transnational migration sometimes became a necessity, as unemployment or very low-paid jobs were the only domestic alternatives.[18] Stang argues however that this case was clearer in Norway. Norway's 57 per cent migration is notably higher than the other countries.

1.5 *The Lack of Higher Technical Education 'Forced' Norwegians and Finns Abroad*

The 'basic immaturity of the Norwegian industrial economy' thus contributed to the patterns.[19] This can also be said about Finland. Slower industrialisation alone cannot, however, explain these mobility differences. For example, Denmark was also described as a technical laggard in a 1905 article written by a New Jersey-based Danish engineer and published in the Danish journal *Ingeniøren*.[20] Tomas Nilson has discussed how German technology was used as contrasting pictures in *Teknisk Tidskrift* to reveal technological shortages and show that Sweden also lagged behind.[21] However, Sweden possessed technical universities. Sweden's and Denmark's higher technical education can be dated to the 1820s. There were, de facto, two technical universities in Sweden around 1900. Chalmers received its 'formal' status only in 1937, but functioned in practice as a university much earlier.[22] The processes of establishing higher technical education were slower in Finland and Norway. Finnish industry continued to use elementary technology, and the need for highly educated technicians was viewed as limited.[23] Philosopher J. V. Snellman argued that

17 'N. I. A. F. i New York', *Teknisk ugeblad* 20 (1903) 208–209.

18 G. W., 'Teknikerna och utvandringen', *Teknisk tidskrift. Veckoupplagan* 51 (1910) 415; Sten Carlsson, *Swedes in North America: 1638–1988: technical, cultural and political achievements* (Stockholm 1988) 58; Harnow, *Den danske ingeniørs historie*, 235–236; Hugo Hammar, *Minnen 1. Från Ölands Alvar till livets* (Stockholm 1937) 230.

19 Stang, 'Ble det for mange ingeniører?', 33–42; Stang, 'A measure of relative development?', 94.

20 Harnow, *Den danske ingeniørs historie*, 235–236; Hannover, *Dansk Civilingeniørstat 1942*, 125; Søren Jensen Ellert, 'Overproduktion af Ingeniører', *Ingeniøren* 22 (1905) 144–145.

21 Nilson, 'Vacker, föredömlig, rationell', 43–67.

22 Sundin, *Den kupade handen*, 249.

23 Fellman, *Uppkomsten*, 92.

an establishment of higher technical education primarily would favour the Russian labour market. To award young Finns grants to go abroad for studies was still preferred.[24] Finland was not the only country where this idea had a foothold. Cardoso de Matos and Diogo have described a similar controversy in mid-nineteenth-century Portugal.[25] Nevertheless, the 1870s saw the establishment of a polytechnic institute in Helsinki. In a way, this was a 'higher institution of technology', but the curriculum was narrower, and the education never reached the same levels as in Denmark and Sweden. In Norway, strong agrarian interests saw no need for improved technical education, but urbanisation, industrialisation, and modernisation implied new demands and three new technical schools were inaugurated in the 1870s. However, none of them received university status. The Helsinki polytechnic attained formal university status in 1908, whereas the Norwegian Institute of Technology was inaugurated in 1910.

Thus, the lack of higher technical education constituted a difference between Norway and Finland on the one hand and Denmark and Sweden on the other. Some Norwegians and Finns obtained, as mentioned, their entire technical education abroad, but domestic graduates also viewed it as a necessity to complete their education with some semesters abroad. It was common, says Myllyntaus, to study first in Helsinki and then go abroad to specialise in a certain technological field.[26] He continues, 'Up to the 1910s, it was customary for Finnish students interested in electrical engineering to travel abroad to study it'.[27] Foreign studies were a 'must' for people who wanted to specialise in this field.[28] Germany was the major destination, but graduates also looked to schools in, for example, Zurich, Vienna, Paris, Prague, Liege, Glasgow, and Gothenburg. The longing for a more solid education and to experience advanced technological environments were reasons to go abroad. This was a common view in many peripheral regions, for example in the Basque area, where Anduaga has stated that Liege often was viewed as a better choice than a perceived insufficient technical education in Spain.[29] For Norwegians, Stang identifies another and perhaps the more important reason. Not only private but also public employers wanted prospective employees to possess a formally higher quality education than the one available in Norway. Many graduates were,

24 Engman, *Lejonet och dubbelörnen*, 152–157.

25 Cardoso de Matos and Diogo, 'Bringing it all back home', 162–163.

26 Myllyntaus, 'The Best Way', 142–143; Myllyntaus, *The gatecrashing apprentice*, 67.

27 Myllyntaus, *The gatecrashing apprentice*, 66.

28 Myllyntaus, *The gatecrashing apprentice*,

29 Anduaga, 'The engineer as a "linking agent"'.

therefore, 'forced' to study abroad to be competitive for engineering positions in their own country.[30]

1.6 *Going Abroad for Domestic Companies was Particularly a Danish Trait*

The observation that Norwegian and Finnish graduates were more transnationally mobile than their Danish and Swedish colleagues is further underlined when we look at the pattern of going abroad on the payroll of a domestic company. We will see that a few engineers served abroad for the Norwegian classification society *Det Norske Veritas* (DNV), and some Finnish technicians worked for domestic granite industries as well as for the Finnish national railways in Russia. However, only 6 per cent of the Finnish mobility and 1 per cent of the Norwegian were for domestic companies.

The pattern was stronger in the two higher industrialised countries. Harnow writes that companies such as F. L. Smidth, Møller & Jochumsen, Burmeister & Wain, Nielsen & Winther, and Christiani & Nielsen offered Danish engineers safer and more formalised stays abroad, compared to the ones who travelled without specific plans.[31] Danish companies seem to have offered this to a greater extent compared to companies in Sweden. Every fifth graduate from the schools in Copenhagen, Odense, Aarhus, and Horsens had a domestically based employer while working in a foreign country, and this was twice as high as for Sweden. Stang, nevertheless, calls it 'particularly a Swedish trait',[32] but this is probably a result of his focus on the Americas and perhaps also the inclusion of graduates from the 1920s. The only Swedish company that came near a Danish number was the electrical manufacturer ASEA with 43 engineers from the cohort abroad.

F. L. Smidth was quantitatively most important. As for the cement industry, Per Schybergson writes in his account of large enterprises in Denmark, Finland, and Sweden that Denmark 'has most strongly made its mark internationally' both in technological development and the delivery of machines and finished products.[33] The epoch-making coal-powder-fuelled rotating kiln and other innovations were exported to every corner of the globe.[34] Schybergson

30 Stang, 'Ble det for mange ingeniører?', 36.

31 Harnow, *Den danske ingeniørs historie*, 234.

32 Stang, 'A measure of relative development?', 96.

33 Per Schybergson, 'Large Enterprises in Small Countries', in: Hans Kryger Larsen (ed.), *Convergence?: aspects on the industrialisation of Denmark, Finland and Sweden 1870–1940* (Helsinki 2001) 126.

34 Georg Nørregaard, 'Ingeniørernes indsats. Danske ingeniører fra tekniske skoler og teknika', in: *Danske Ingeniører fra Teknika, Udgivet af Ingeniør-Sammenslutningen i Anledning*

claims that F. L. Smidth 'held a technologically superior position worldwide in its branch' after World War I.[35] This development implied that Danish engineers were sent out as experts. F. L. Smidth had three engineers in Chile in 1912. Portugal and Iceland were the only European countries where F. L. Smidth had no activities before 1920.[36] A total of 80 engineers in the Danish cohort had employment abroad for the company. Building contractors Christiani & Nielsen were the second largest in this context, but they had just over half as many engineers abroad as F. L. Smidth. The company opened branches in Russia, Scandinavia, Western Europe, South America, and New Zealand in the 1910s and 1920s.[37] Several engineers worked at the offices in Hamburg, London, and Paris and contributed to raising the Danish mobility share.

1.7 *Engineering and Transnational Mobility were Male Worlds*

Graduates travelling abroad for domestic companies were exclusively male, and—as we stated in chapter one—this book is dealing with a 'man's world'. The contemporary view implied that it was a man's task to construct cars, trains, ships, and to arrange electric light and care for other infrastructure. Sixty-nine women made the trailblazing step into these fields. This overwhelming male dominance contributed to the profession's transnational mobility. Berner writes:

> Only men could depart in this way, shift employment and place of residence, freely move over environments and employments. The women's world was narrower. Their education and the expectations led them back to their home, or to subordinate and static positions in the working-life.[38]

Women were less able than men to use geographical space for career promotion. A few mobile female graduates chose other paths in life. Randi Holwech[39]

---

af Foreningens 50 Aars Jubilæum (København, 1945) 88; Schybergson, 'Large Enterprises', 126–127.

35 Schybergson, 'Large Enterprises', 127.

36 Hyldtoft, 'Perioden 1896–1930', 75–76, 102, 190; 'F. L. Smidth & Co. 1882–1922', *Ingeniøren* 1(1922) 10–13.

37 Hyldtoft, *Københavns industrialisering,* 365; V. M., 'Christiani & Nielsens 25 aars jubilæum 8/9 1904—8/9 1929', *Ingeniøren* 7 (1929) 77–78.

38 Berner, *Sakernas tillstånd,* 43. Original in Swedish: Bara män kunde på detta sätt bryta upp, byta anställning och bostadsort, fritt röra sig över miljöer och anställningar. Kvinnornas värld var snävare. Deras utbildning och förväntningar på dem förde dem tillbaka till hemmet, eller till underordnade och statiska positioner i arbetslivet.

39 NTH, chemical, 1919.

went to continental Europe to become an artist in the 1920s. Bertha Enwald[40] was the first female in our cohort to go abroad when she left for Saint Petersburg in 1898. She supervised the construction of wooden houses but faced contemptuous views of female architects and had an employer with economic troubles. Because of these hardships, Enwald became a teacher upon her return.

Migration scholars have concluded that women move more often than men, but rarely move over longer distances. However, these scholars usually study internal mobility over significantly shorter distances than what we are dealing with in this study. The former is not reflected in our statistics: 25 female graduates in the cohort went abroad, making a transnational mobility share of 36 per cent. Sweden's only female graduate, Vera Sandberg,[41] went on a study trip in the 1920s. Fifteen mobile Finnish female graduates make a share of 56 per cent; four mobile Norwegian women make 40 per cent; while the same number of Danish female technicians going abroad makes 13 per cent

However, compared to men, the female short-distance pattern is reflected in the destination choices. Fifteen of the 25 mobile women went to the German-speaking countries, 12 to the Nordic neighbours, nine to 'other European' destinations, six to Britain and three to Russia. In 1906, a writer in *Teknisk Ugeblad* was amazed by the many female architects and draftswomen in America.[42] However, Nordic architects and female technicians rarely went to America before 1930.[43] Norwegian Gudrun Holst-Grubbe[44] was the only woman architect in our cohort to cross the Atlantic. She was employed by architect offices in New York, married late, and stayed in America for life. Aslaug Mølmen claims that Holst-Grubbe chose her career before establishing a family.[45] No woman went to Latin America and the Caribbean and Holst-Grubbe's Finnish colleague Elsi Borg[46] became the only woman in our cohort who went outside Europe and the Americas when she visited Morocco in the 1920s.[47]

40 SPO, architecture, 1894.

41 CTI, chemical, 1917.

42 'Kvindelige arkitekter og ingeniører i Amerika', *Teknisk ugeblad* (1906) 304.

43 Silja Laine, 'Americanism and Architecture in 1920s Helsinki' in: Christina Folke Ax (ed.), *Encountering foreign worlds: experiences at home and abroad: proceedings from the 26th Nordic Congress of Historians, Reykjavik 8–12 August 2007* (Reykjavík 2007) 27–28.

44 TTL, architect, 1906.

45 Aslaug Mølmen, Kvinnelige pionerer: en studie av 21 kvinnelige teknikere ved Trondhjem Tekniske Læreanstalt 1883–1915 (Trondheim 2008) 70–71. Master Thesis Norwegian University of Science and Technology.

46 STK, architect, 1919.

47 http://www.mfa.fi/arkkitehtiesittely?apid=2904 (29 October 2017).

Holst-Grubbe and Borg remained in their profession, and both got married. So, did chemical engineer Martha Skaanes,[48] who went to Nestle's laboratories in Switzerland and joined her husband in establishing a chemical laboratory upon return. The Finland-Swedish couples Eva Kuhlefeldt-Ekelund[49] and Hilding Ekelund[50] practised in Stockholm with some of the famous Swedish architects of the 1920s. The couples also made several study trips in Europe, Africa and Asia before they established a joint architecture office in Helsinki in 1927. Fellow architects and countrywomen Elna Kiljander[51] and Salme Setälä[52] went to Russia and London. Both got married. The former became known as a designer of kitchens; the latter received awards within furniture design.

Many female technicians who stayed in the profession remained unmarried. Sonja Gill,[53] for example, who was at the same Swiss place as Skaanes and became employed with Kristiania's Freia chocolate company upon return. The same was true for Berrit Fiehn,[54] who spent eight years in Germany and returned to Denmark to establish a chemical laboratory in Nykøbing Falster.

1.8 *Mobility of Young, Upper-Middle-Class Men from Non-Capital Cities and Towns*

Clearly, being unmarried and young without family duties gave the possibility to, to quote Berner, 'freely move over environments and employments'. Transnational mobility declined with increasing age as figure 6 shows. The only exceptions were the Norwegian mobility increase after from 25 to 28 years and the Finnish from 21 to 24 years.

A person who graduated at a younger age simply had a longer career ahead of him- or herself and more time to go abroad. Certainly, they were also less established and had only very rarely begun the creation of families.

Children from the upper social layers also went abroad more: Every second upper-class-born Nordic technical school graduate went abroad, and the share reached almost two-thirds in Norway. If we disregard the 'unknown' group, transnational mobility generally decreased stepwise from the uppermost group to the lowest in the four countries, even if group four was somewhat more mobile compared to group three in Denmark and Finland.

48 TTL, chemical, 1910.
49 STK, architect, 1916.
50 STK, architect, 1916.
51 STK, architect, 1915.
52 STK, architect, 1917.
53 BTS, chemical, 1913.
54 PL, chemical, 1915.

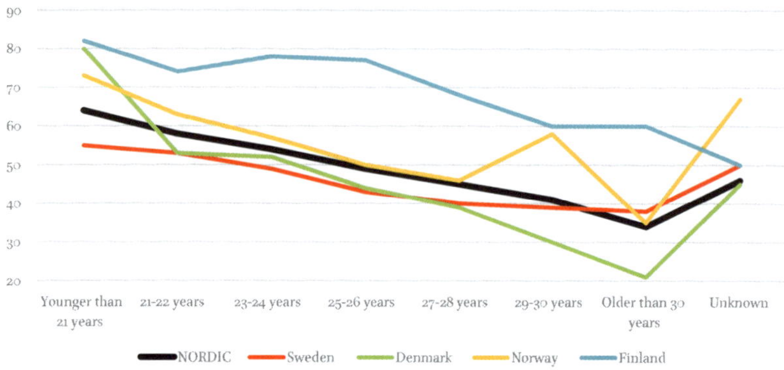

FIGURE 6 Transnational mobility before 1930 per graduation age among Nordic engineers and architects leaving school 1880–1919
SOURCES: SEE FIGURE 1.

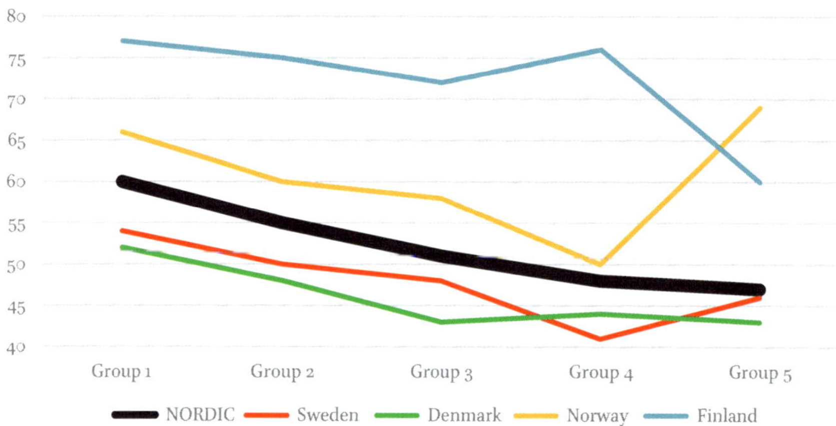

FIGURE 7 Transnational mobility before 1930 per social status among Nordic engineers and architects leaving school 1880–1919
SOURCES: SEE FIGURE 1.

It is reasonable to assume that graduates of a lower social origin lacked the contact networks that could facilitate employment abroad. According to Mats Lindqvist, young Swedish businessmen and industrialists used shorter or longer stays abroad as a part of the creation of an identity and a way to become more mature and responsible men.[55] To have a father who was managing

55 Mats Lindqvist, *Herrar i näringslivet: om kapitalistisk kultur och mentalitet* (Stockholm 1996) 46–47.

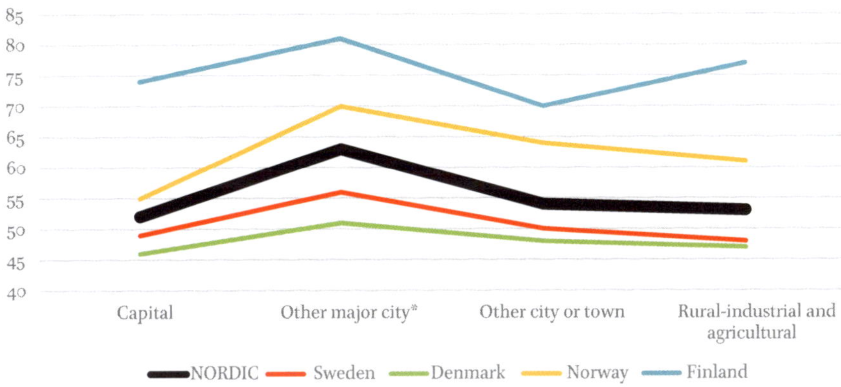

FIGURE 8 Transnational mobility before 1930 per birthplace among Nordic engineers and architects leaving school 1880–1919
SOURCES: SEE FIGURE 1.

director or on the board of a larger and well-known company was a valuable career asset, but it does not look like transnational mobility was a means to level out differences on the domestic labour market based on social origin. Rolf Torstendahl noted such a pattern about grades, but in this context, the general pattern was, as mentioned, the opposite.[56] Instead, transnational mobility widened the opportunity-gap based on family origin. Those well-off could increase the differences by accumulating experience abroad. From the Middle Ages and onwards, it was primarily wealthier elite groups who could go to universities, even if the journeyman patterns reveal a somewhat different picture. Upper-class children had, nevertheless, wealthier parents ready to finance journeys as well as board and lodging. The father of Ivar Magnusson[57] was managing director of the Sandviken ironworks in central Sweden. Magnusson studied mining in Austria and practised six years in the steel and iron industry around Philadelphia and Pittsburgh. Most of the time, he was financed by his father. Upon return, Magnusson headed and modernised Sandviken's cold rolling and wire drawing department.[58]

The main pattern was, as figure 8 shows, that graduates born in the second to fourth city more often performed transnational mobility.

---

56 Rolf Torstendahl, *Dispersion of engineers in a transitional society: Swedish technicians 1860–1940* (Uppsala 1975) 45.

57 CTI, mechanical-electrical, 1902.

58 Grönberg, *Learning and Returning*, 185–186.

One explanation is Bergen's, Trondheim's, and Stavanger's geographical locations in Norway and Gothenburg's and Malmö's in Sweden. Previous and contemporary international connections possibly also contributed. Gothenburg, Trondheim, and Bergen were major ports of departure for emigrants to America. Bergen also had traditions of European connections through its history as a major Hanseatic City. Malmö, on the southernmost tip of Sweden, neighboured Denmark and was close to continental Europe. In Finland, the cosmopolitan city of Viipuri lay close to Russia, and Turku was near Sweden.

The pattern for student city echoes the birthplace pattern. Nordic graduates who had studied in the capitals generally went abroad to a somewhat smaller extent compared to those who studied in other cities. This pattern was valid also for Norway and Sweden, whereas Denmark diverged.[59] One plausible explanation may be that graduates studying in capital cities were able to establish professional networks during their time as students. Copenhagen, however, was Denmark's major emigration port. Graduates who studied in other 'port-dominated' cities also went abroad more often: Bergen's 72 per cent is clearly above the Norwegian average, and Gothenburg's 59 per cent clearly above the Swedish average. However, an even more plausible explanation is that the non-capital Danish schools started late. In our cohort, the first graduate left the school in Odense in 1894, whereas departures from the counterparts in Aarhus and Horsens began in 1914 and 1916. As we will see, transnational mobility decreased during the 1910s.

1.9 *Foreign-Born Graduates Going Home*

Another group with high transnational mobility comprised, not surprisingly, foreign-born graduates, Sweden noted 83 per cent, Norway 73 per cent, Finland 66 per cent, and Denmark 59 per cent. Finland was the only country where native-born graduates were more transnationally mobile than graduates born outside the country. The 'foreigners' at the school mostly comprised Russian-born Finns and had not moved to Finland to undertake technical education. 'Foreigners' at the schools in Kristiania, Bergen, and Trondheim were often

59 52 per cent of those who had studied in the capitals went abroad, compared to 57 per cent of those who had studied in other cities. In Sweden, 53 per cent of the students from Gothenburg, Malmö, and Örebro went abroad compared to 46 per cent of the students from the Royal Institute of Technology in Stockholm. In Norway, 65 per cent of the students from Bergen and Trondheim went abroad compared to 58 per cent of the students at two schools in Kristiania. In Denmark, however, almost every second student from the two Copenhagen institutes went abroad compared to somewhat less than a third of those who had studied in Odense, Aarhus, and Horsens. All Finnish graduates studied in Helsinki.

born by Norwegian parents, but more transnationally mobile than 'Finns' born abroad. Many of them were born in the United States and, possibly, carried a positive picture of America.

Sweden attracted some students from Finland and Norway who later went back to their native countries. Since these students did not have a pre-education at home, they are in the Swedish cohort in this study. Thus, they are not defined as returnees in this context, but it was return migration. The number of Finnish technical students in Sweden rose towards the end of the nineteenth century, and Chalmers became an important institution, be it far from as populated as German and Swiss institutes.[60] Chalmers did not reach the same importance as schools in the German-speaking countries for nearby Norway either, but Norwegians studying in Gothenburg had deep historical roots, and we point to the 1869 graduate Gunnar Knudsen. He took over his father's shipping company in 1872. Later, he founded the Borgestad Company, but he is best known as Norway's prime minister in the 1910s.[61]

Denmark never assessed the same attraction on foreign technical students, but Icelanders found their way there. Icelanders interested in deeper knowledge of technology had to go abroad, and the stream went almost exclusively to the Danish capital.[62] All these graduates went back and were certainly important. There were strong ties between Danish and Icelandic technology and engineers from the Copenhagen Polytechnic participated in the build-up of an infrastructure from a minimal level.[63]

1.10 *Civil and Construction Engineers Were Less Mobile, Except among Those from Denmark and Norway*

Figure 9 shows that civil and construction engineers were less mobile than other specialisations on the pan-Nordic level. This is in line with Carlsson's observations in his study of Swedish engineers in Chicago.[64] However, the pattern is not reflected in Denmark and Norway. The four 'engineer specialisations' noted 54 per cent transnational mobility, whereas architects noted 56 per cent, but the mobility types differed. Engineers were migrants, while architects were study travellers. If we look at migration only, the engineer specialisations end up with 46 per cent, whereas architects note 25 per cent. It is,

60 Myllyntaus, 'The Best Way', 143.

61 Bodman, *Chalmers tekniska institut*, 77, 146, 249.

62 Krabbe, *Island og dets tekniske udvikling*, 354–355.

63 Magnússon, *Iceland in transition*, 59; Vilhelm Marstrand, 'Island. I anledning af Dansk Ingeniørforenings Islandstur Juni 1927', *Ingeniøren* 24 (1927) 300–301.

64 Carlsson, 'Swedish engineers in Chicago', 185–188.

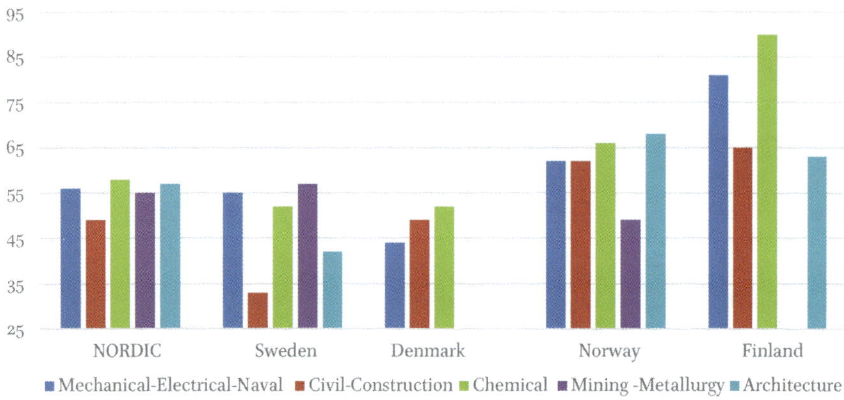

FIGURE 9 Transnational mobility before 1930 per specialisation among Nordic engineers and architects leaving school 1880–1919
SOURCES: SEE FIGURE 1.

however, worth mentioning that Norwegian architects diverged as they note 52 per cent migration. One plausible explanation is that they had more of a 'technician' identity than architects in Finland and Sweden.[65]

Chemical engineers were the most mobile specialisation from Denmark, Norway, and Finland. Swedish mining engineers and metallurgists, as well as mechanical, electrical, and naval architects, were, however, somewhat more mobile than chemical engineers.

Germany's leadership in chemical industry and closeness contributed to the high transnational mobility of chemical engineers. Mechanical and electrical engineers together with naval architects, chemical engineers as well as mining engineers and metallurgists also constituted more transnationally mobile specialisations. Many mechanical engineers went on to specialise in electrical engineering or naval architecture before these subjects were established as 'independent' ones at the technical schools in the Nordic countries.

1.11 *International Unrest and Establishments of Technical Universities Decreased Mobility*

Figure 10 shows that transnational mobility among the Nordic engineers and architects lay roughly between 60 and 65 per cent from the earliest graduates in the study to the ones leaving school around 1905. Thereafter, it decreased

65 Ingrid Pedersen, *Litt om tegne- og arkitektundervisning i Norge før høiskolens tid: bidrag til en historisk oversikt* (Trondheim 1935) 31–33.

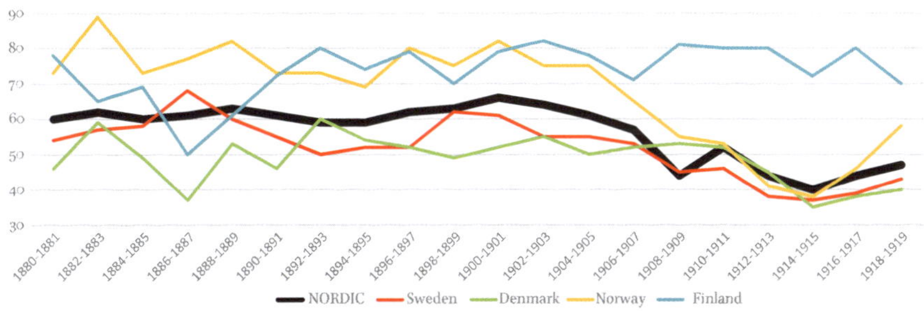

FIGURE 10 Transnational mobility before 1930 per graduation year among Nordic engineers and architects leaving school 1880–1919
SOURCES: SEE FIGURE 1.

gradually and was around 40 per cent for the graduates during the mid-1910s. The shares increased again among the latest graduates in our study, but were still considerably lower compared to the pre-1905 graduates.

One explanation, of course, is the international unrest, World War I and the Russian Revolution. Until about 1910, with a minor exception for the years 1886–1887, graduates from more sparsely industrialised Norway and Finland were more mobile than their Swedish and Danish colleagues. All through the 1910s, Finnish mobility continued at a higher level. Norwegian mobility, however, dropped, as figure 10 shows, to be on a par with the average Nordic level among the graduates from the early years of the decade, but it increased again among the graduates from 1916 to 1919. Mobility shares from Sweden and Denmark also dropped during these years, but the decrease was less dramatic compared to Norway.

The Finnish Polytechnic Institute was transformed into a technical university in 1908, whereas the Norwegian Institute of Technology was inaugurated two years later. Finnish and Norwegian youngsters now had access to technical education on the highest level within the borders of their respective countries. This is reflected in a decreased Norwegian mobility, but figure 10 shows no effect in Finland. However, as mobility for longer studies to foreign technical schools is labelled 'migration' and not 'study trip' in this study, the mobility statistics do not give the true picture. Figure 11 illustrates the difference between mobility and migration for Finland and Norway.

As we can see, the Finnish migration share dropped similarly to the Norwegian. Therefore, there are reasons to assume that the establishment of technical universities affected both Norway and Finland. Continued extensive study travelling kept the Finnish mobility share at a high level. Finland's need to build institutions after achieving full independence and the strong economic growth

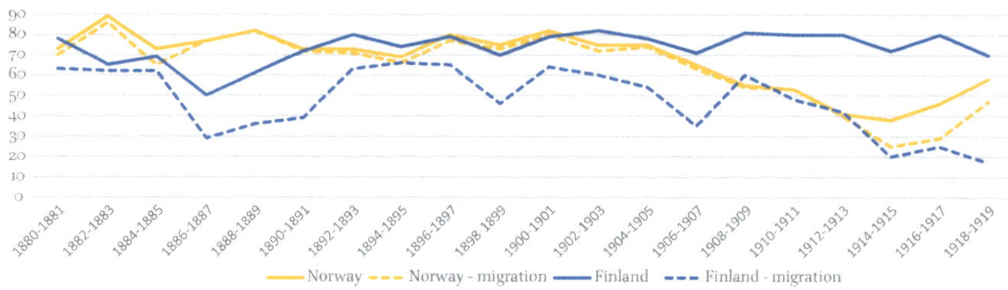

FIGURE 11 Migration and study travelling before 1930 per graduation year among Norwegian and Finnish engineers and architects leaving school 1880–1919
SOURCES: SEE FINNISH AND NORWEGIAN SOURCES, FIGURE 1.

in the interwar period[66] implied, on the one hand, a need to 'keep' technicians and other highly educated people in the country, but on the other hand, also a need to study technical development in advanced places, primarily in continental Europe. Shorter study trips were a solution that implied a possibility to keep up with technological development abroad without the loss of an educated workforce for significant periods.

2 Summary

Transnational mobility—for employment and/or longer studies abroad, 'migration', and 'study travelling' without employment—was a common feature among Nordic technicians; more than half of them went abroad. Also, there were engineers and architects who took their entire education abroad. Finnish technicians were the most mobile if we count total transnational mobility, whereas their Norwegian colleagues went abroad at a greater extent if we count only migration. The stronger Finnish and Norwegian transnational mobility should be viewed in the light of the pattern of going abroad on the payroll of a domestic employer. This characterised Denmark, with F. L. Smidth in the forefront, but was also more common among Swedish technicians. Nevertheless, transnational mobility was significantly stronger among Norwegian and Finnish technicians.

Earlier studies have shown that technicians from other European peripheries and Japan as well as from more industrialised societies such as Germany

66 Jari Eloranta, et. al., 'On the Road to Prosperity: An Introduction', in: Jari Ojala, Jari Eloranta and Jukka Jalava (eds.), *The Road to Prosperity. An Economic History of Finland* (Helsinki 2006) 20.

moved and travelled extensively over international borders. It was far from a specific Nordic trait, but it is difficult to compare different countries as comprehensive statistical studies of other countries are not found. We should underline that engineers and architects were extraordinarily mobile professions, but their mobility was not unique. Earlier studies revealed how other professions and skilled workers also travelled over national borders.

Some factors contributed to create the 'peregrine profession', including the extraordinary male domination among the graduates. While men moved freely, even technically educated women were often expected to return to housewifery once they had married. The age pattern and the domination of upper- and upper-middle-class sons are other factors. An overwhelming majority of the graduates left school at a young age, and young graduates were more mobile. Children of wealthier parents had other possibilities for economic support for longer foreign intermissions and extensive study trips. Spending time abroad was often a means for upper-class sons to create an identity as rich and affluent men. Children from the upper social layers could probably also utilise contacts to a greater extent compared to colleagues of lower social origin. Communication with other countries may also have been easily accessible as many graduates studied in port cities or cities located near an international border. We have observed that male graduates were more mobile than female, that the mobility decreased with decreasing social origin, and that graduates studying outside the capital cities also went abroad to a greater extent. The non-capital student cities—like Bergen and Trondheim in Norway and Gothenburg and Malmö in Sweden—were often the port cities and 'border cities' discussed above.

Other factors also influenced the decision to go abroad. There are, however, small differences between the specialisations, but civil and construction engineers were less mobile than the other groups. Generally, graduates leaving school before the international unrest of the 1910s went abroad more frequently. This was the case for all countries except Finland, but the decrease for the 1910s graduates was most dramatic in Norway. One reason was the 1910 establishment of a technical university in Trondheim; Norwegians no longer needed to go abroad to obtain—at least formally speaking—the highest degree of engineering. At a first glimpse, the establishment of higher technical education in Finland did not have the same mitigating effect on transnational mobility, but the reason is that study travelling continuously was at a high level, whereas migration fell in a way resembling Norway. In general, with a peculiar Danish exception, study travelling was a 'late' rather than an 'early' pattern. As we have seen, study travelling had a completely different foothold in Finland, compared to in Sweden, Norway, and Denmark.

Extensive study travelling partly explains the extraordinarily strong Finnish mobility. The organised nineteenth- and early twentieth-century Finnish strategy to collect information and adopt technology from abroad through study trips apparently had no counterpart in Sweden, Denmark, and Norway. This Finnish strategy continued into to the full independence and Finland's strong interwar economic growth—the beginning of her industrial breakthrough is placed in 1920—implied a dual and somewhat contradictory need; on the one hand, to keep technicians in the country but, on the other hand, to study development in advanced environments abroad. A continuation of the tradition of shorter study trips rather than 'longer foreign migration' intermissions became the solution.

Finland's proximity and political ties to Russia as a Grand Duchy under the Russian tsar until 1917, also contributed to transnational mobility. Finland's border with Russia proper ran just north of Saint Petersburg, and the imperial capital offered many opportunities in trade and industry. Whereas Finland was not struck by 'America fever' to the same extent as many other European countries, Russia functioned as a kind of substitute. However, Russia also had a different implication as early twentieth-century Russification politics—like the military service law—pushed some Finns to Western countries.

Finnish and Norwegian industrialisation was also slower than in Denmark and Sweden, which implied that many engineers had difficulties finding domestic employment. This also happened in Denmark and Sweden; Norway was the clearest Scandinavian case, and scholars have observed similar problems in Finland. Norway's and Finland's weaker industrial and economic development and the low level of technical education sometimes also entailed that domestic employers preferred foreigners in major engineering positions. Graduates were simply 'forced' abroad to obtain higher technical education as some specialisations, such as electro-technology, were absent for a long time in Norway and Finland. Germany led the world in technical education; it was a major place to go for graduates who wanted to complete their domestic education, and as such, it also was the most frequently selected destination on a pan-Nordic level. However, Germany was far from the only destination. The Nordic technicians' mobility and migration patterns give, instead, the impression of a worldwide labour market.

CHAPTER 3

The Choice of Destinations

Stang states that engineers 'travel because the market for their competence has since the mid-nineteenth century been world wide [sic!]'.[1] Our observations underline Stang's statement. Nordic technical school graduates went, as the appendix shows, to at least 93 countries around the world. They included at least 21 countries or dependencies in Latin America and the Caribbean and at least 39 in Africa, Asia, and Oceania. Costa Rica, Honduras, the Dominican Republic, Haiti, Mozambique, Senegal, Niger, Nigeria, Namibia, Madagascar, Mongolia, Burma, and Korea are among the countries that received Nordic technicians. Employment was even registered in Pacific and south Atlantic island nations and dependencies such as Fiji, Guam, the Solomon Islands, New Caledonia, and South Georgia.

1 Domination of the German-Speaking Countries and North America

In table 3, we can still identify two dominating destinations: the German-speaking countries of Europe—Germany, Switzerland and Austria—and North America, that is, the United States and Canada. On the pan-Nordic level as well as in Norway, the German-speaking countries attracted more than North America. North America was the fifth destination in Finland, according to this division. As we can see, Norwegian graduates went more often to North America than their Danish colleagues, even if it was the first Danish and the second Norwegian destination. Relatively speaking, the importance of North America as a destination was strongest among the Swedish technicians.

The reason for this main pattern is simple; the graduates viewed the German-speaking countries and the United States as the most interesting destinations. Boje, Harnow, and Hyldtoft have revealed how the interest in these countries grew among engineers in Denmark around the turn of the century.[2] The United States had, argues Thomas P. Hughes, developed into the world's

1 Stang, 'A measure of relative development?', 93.

2 Boje, 'Teknikumingeniører og dansk økonomisk vækst', 279–321; Hyldtoft, 'Perioden 1896–1930', 15; Hyldtoft, *Teknologiske forandringer i dansk industri 1870–1896*, 264–265; Harnow, *Den danske ingeniørs historie*, 232–235.

 | DOI:10.1163/9789004385207_004

TABLE 3 Distribution on destination (numbers-per cent) before 1930 among transnationally mobile Nordic technical school graduates 1880–1919.

DESTINATION	NORDIC	Sweden	Denmark	Norway	Finland
German-speaking Europe	3015–45	1153–42	433–33	910–50	519–63
North America	2628–39	1230–45	450–34	763–42	185–22
Nordic countries	1349–20	470–17	231–18	215–12	432–52
Europe*	1199–18	462–17	241–18	210–12	286–35
British Isles	994–15	409–15	242–18	176–10	167–20
Russia	521–8	187–7	94–7	28–2	212–26
Latin America & the Caribbean	411–6	124–5	142–11	137–8	8–1
Africa, Asia, & Oceania	401–6	127–5	171–13	89–5	14–2
Unknown destinations	139–2	43–2	39–3	37–2	20–2

*Except for the Nordic and German-speaking countries, the British Isles and Russia.

SOURCES: SEE FIGURE 1.

most innovative country and a leading industrial nation by 1900.[3] Norway's commissar at the 1893 Chicago fair viewed America as the most advanced country in manufacturing. Returnees told about the marvels of electricity and engineers reported about modern American tools. In 1902, America was described as technologically superior to Europe in a lecture in Kristiania.[4] The 'discovery' of America, was of course not limited to the Nordic countries. Interwar transatlantic journeys from Germany were extensive, and technicians were, as Nolan has revealed, one of the most important groups.[5] Braun, too, has shown how late nineteenth- and early twentieth-century German engineers went back and forth over the Atlantic and contributed to both the American and German industrial economies.[6]

3 Thomas Parke Hughes, *American genesis: a century of invention and technological enthusiasm, 1870–1970* (New York, NY 1989) 15.

4 Skard, *USA i norsk historie,* 180; Hyldtoft, 'Perioden 1896–1930'; Grönberg, *Learning and Returning.*

5 Nolan, *Visions of modernity.*

6 Braun, 'Franz Reuleaux'; Braun, 'Technologietransfer im Maschinenbau'; Braun, 'A Technological Community in the United States.'

However, the German-speaking countries were also admired in many parts of the world. Switzerland was looked upon with a certain amount of admiration. Myllyntaus emphasises the nineteenth-century transport revolution, which made it possible for Nordic tourists to enjoy Swiss scenery, which also often was described in journals back home. Switzerland stood for neutrality and independence, which was especially relevant in semi-independent Norway and Finland. The Finns, under the pressure of the tsarist Russian government, also admired press freedom and the freedom to form associations. Furthermore, in the light of an increasingly tense situation between Finnish and Swedish speakers in late nineteenth-century Finland, it was a country where people of different native tongues lived together. However, not only their beauty and democracy were described. Articles also started to emphasise 'Switzerland as a land of technology, manufacture and trade'. At the outbreak of World War I, the Swiss GDP was significantly ahead of all Nordic countries except Denmark.[7] The Swiss Polytechnic in Zurich has been described as continental Europe's leading scientific university and many of its most remarkable achievements took place during the period of our study; the late nineteenth and early twentieth centuries have been described as the 'golden age of Swiss engineering'. To study in Zurich was interesting, and many Swiss companies also offered employment afterwards.[8]

Germany still attracted more, partly because it was a larger country in both population and area. Germany offered more possibilities. As Kim notes, German research centres achieved worldwide recognition in a number fields in the late nineteenth century and attracted scholars from around Europe as well as Japan and the United States.[9] The attraction was also high among Nordic citizens wanting to study technology. Håkon With Andersen writes, for example, 'During the second half of the nineteenth century the German institutes (*Hochschulen*) were the ultimate centres for higher technological education in Norway'. Germany educated many Norwegian engineers; it was the model for education, provided textbooks and, later, professors.[10] Björck has described how the German educational system and the country's industrially oriented research policy was admired among technicians in Sweden.[11] However, studies of Greece and Portugal, for example, reveal that late nineteenth- and early twentieth-century engineering students from these countries more frequently

7 Myllyntaus, 'Discovering Switzerland', 301–303, quote from 303.
8 Myllyntaus, 'Discovering Switzerland.'
9 Kim, 'Shifting patterns', 390.
10 Andersen. 'Germany and the education of Norwegian engineers', 100–114.
11 Björck, *Staten, Chalmers och vetenskapen.*

went to France, even as Germany, Switzerland, and Belgium became more important over time.[12] Germany's attraction lay, of course, also in the fact that the country developed into Europe's leading industrial nation and, thereby, a major source of cutting-edge technology. The chemical and electrical industries were especially important. Electrical manufacturing equalled America's and was twice the size of Britain's in 1914. The synthetic dyestuff discoveries were behind the rise in the chemical industry. Germany was also a prime source of knowledge in wood pulping and water turbines.[13] Descriptions of German technology in *Teknisk Tidskrift* have been discussed in an article by Tomas Nilson who concludes that Germany's development often worked as a positive contrast to Sweden's. By applying German examples of technology and industrial organisation, Sweden could develop as an industrial nation. At the 1909 annual meeting of the Swedish technical association, German technical development was celebrated and described as a threat to aging industrial nations. In the early 1910s, an engineer described German machines and workers as superior to their British counterparts.[14]

A 'placement abroad' in one of the leading industrial countries was often viewed as more valuable for future learning than experiences from most other countries. The 1904 response from an electrical engineer to an employment offer from Sweden's leading electrotechnical company ASEA is one example of how experiences in the two leading industrial countries were evaluated:

> I must deeply regret, Mister Director, that it is impossible for me to accept your kind offer in the nearest future, because I think I am too young and inexperienced to be able to fulfil such a place you are offering me in a worthy manner. My intention is however, before I settle in Sweden, to complete my two years of valuable American practise with 3 or 4 years of equally valuable practise in Germany, and after that I would most

12 Assimacopoulou et al., 'Elève en France'; Antoniou et al., 'Greek Engineers'; Cardoso de Matos, 'Asserting the Portuguese Civil Engineering Identity'; Cardoso de Matos and Diogo, 'Bringing it all back home.'

13 William Carr, *A History of Germany 1815–1990. Fourth Edition.* (London 1991) 160; Thomas Nipperdey, *Deutsche Geschichte: 1866–1918. Bd 1, Arbeitswelt und Bürgergeist* (München 1990) 234, 237; Grimnes, *Sam Eyde,* 50, 54; Mikael Hård and Andrew Jamison, *Hubris and hybrids: a cultural history of technology and science* (New York, NY 2005) 87; Nerheim, 'Patterns of Technological Development in Norway', 61; Nilson, "Vacker, föredömlig, rationell", 56, 58.

14 Nilson, "Vacker, föredömlig, rationell", 43–67.

> certainly be more mature for successful work in my own native country, where the railway problem still awaits its 'electrical solution'.[15]

The statement that 'placements abroad' in the German-speaking countries or North America were of greater learning value is, of course, an overarching generalisation: It was, to a large extent, dependent on educational background and personal intentions. Also, as mentioned earlier, far from every move was made primarily for learning purposes. Some of the technicians left because of limited domestic possibilities, some certainly travelled to make their living and to be able to practise their profession, rather than acquiring experiences for a future career at home. However, a recession in the native country could very well be used as an opportunity to acquire valuable experiences for a future return. Thereby, these two aspects were combined. Technicians who moved abroad to work for a domestic company primarily brought knowledge from the Nordic area to another country or region, but this was not necessarily a one-sided pattern. They could also have learned something valuable to use back home.

2 Characteristics of Destination Choices

All these aspects implied that technicians moved not only to study at renowned technical universities or to work at technological spearhead companies in Germany and the United States. They sometimes moved to America to settle forever; they sometimes moved to other destinations. Finns, Norwegians, and Icelanders studying in Sweden and Denmark were not the only ones moving to a neighbouring Nordic country; such short-distance transnational crossings attracted every ninth graduate. Britain, the early and mid-nineteenth-century industrial model, still was attractive. Every twelfth Nordic technician found a reason to cross the North Sea. The Russian Empire, by some described as 'Finland's America', could also count in Nordic technicians; about four per cent of the graduates took an eastward path. Of course, there also were interesting things to see as well as employment and university enrolment possibilities in countries like France, the Netherlands, Belgium, Italy, and other regions around the Mediterranean and Eastern Europe. About one-tenth of the Nordic technicians went to these areas. Of course, some technicians also saw prospects in other parts of the world. Latin America and the Caribbean were the goals for about every fifteenth transnational move

15 Cf. Grönberg, *Learning and Returning*, 74.

and overseas destinations outside the Americas (including unknown destinations) for every sixteenth.

2.1 *Different Attractions of the German-Speaking Countries and North America*

Table 3 reveals differences on the country level when it came to the choices of destination. Swedish graduates were more inclined to choose North America than colleagues from the neighbouring countries, notably Denmark and Finland. Two plausible explanations can be identified. Sweden had, as mentioned, one of the highest rates of transatlantic emigration in Europe, and Lars. O. Olsson states that engineer journeys to the United States partly can be viewed in this light.[16] Sweden also hosted more large-scale and mass-production-oriented companies than the other Nordic countries, which implied that knowledge and experience acquired in North America had higher relevance upon a return compared to the other Nordic countries. Norway had an even stronger general transatlantic emigration per capita, but the objects of technician migrants to America from the two neighbouring countries diverged somewhat. Norwegian technicians more often settled in America and more often chose areas such as Chicago and Minnesota where Scandinavian immigrants constituted major shares of the total population.[17]

Swedish and Danish technicians were less inclined to go to Germany, Switzerland, and Austria than their Finnish and Norwegian colleagues. This can be connected to smaller needs in Sweden and Denmark to go abroad to complete technical education. The explanation of the Norwegian and Finnish inclination for the German-speaking countries is the reversed. While the Norwegians and Finns had the 'German pattern' in common they differed in another respect. The Norwegians seldom went to all 'non-German language' European destinations, especially Russia. Finnish technicians generally travelled shorter distances than their Scandinavian colleagues and concentrated largely on Europe; they show the highest shares for Russia, Europe, and the neighbouring Nordic countries, but the lowest for all overseas destinations. Russia was—hardly surprising considering its geographical closeness and political ties—a very typical 'Finnish' destination.

2.2 *The 'Payroll Migration' Dispersed the Danes All Over the World*

The mobility patterns of Danish technicians also reflect a somewhat different orientation. Denmark's ties to East Asia—especially Siam—clearly shines

16 Olsson, 'To See how things were done in a big way'.

17 Grönberg, 'Journeymen or Traditional Emigrants?', 197–218.

through in the shares and probabilities for going to Africa, Asia, and Oceania, and the Danish West Indies contribute to a high likelihood and percentage also for Latin America and the Caribbean. Britain was also, in a comparative Nordic perspective, a 'Danish' destination. London was a major node for the Danish companies that, at an early stage, established subsidiaries abroad. If the Finnish mobility was most concentrated to a limited number of destinations, the Danish counterpart was most dispersed. Harnow writes that potential travellers abroad were advised to look for employment through Danish or European companies.[18] This was, as mentioned, more of a trait in Denmark than elsewhere in the Nordic area, and a company like F. L. Smidth, with its net of subsidiaries around the world, contributed to this dispersion. At the same time, there was—as mentioned—a relatively large Danish community in a country like Siam. On a general level, the Danish migration system probably linked the country to more destinations compared to the systems in Finland, Norway, and Sweden. This exerted an impact also on technicians.

2.3 *The Upper Class Looked to Germany and the Middle-Class to America*

Social origin seems to have had no or minor impact on mobility to overseas destinations outside North America. European countries were more often destinations for children from the upper and the lower societal layers, while North America was a more likely target for those who came from the middle class. Figure 12 shows that the shares going to German-speaking Europe decreased from group one to group three, whereas the pattern was the opposite for North America. Children of lower-skilled and unskilled workers, however, went more often to Germany, Switzerland, and Austria than across the Atlantic. We can note this increase from group three to group four also for Europe, Britain, and the neighbouring Nordic countries.

Jürgen Kocka has emphasised that nineteenth- and early twentieth-century European middle-class stressed achievement, education, self-reliance, and work and carried visions of modernity, secularism, self-regulation, and enlightenment; they were against the autocratic privilege-society.[19] It is reasonable to assume that many of these ideas were associated with the United States and that children growing up in such environments carried a positive picture of America. Hans Lindblad, a past member of the Swedish Parliament, describes, as mentioned, transatlantic emigration as the 'people's project', while the privileged

18 Harnow, *Den danske ingeniørs historie*, 232–235.

19 Jürgen Kocka, 'The Middle Classes in Europe', *The Journal of Modern History* 67, no. 4 (1995) 785–786.

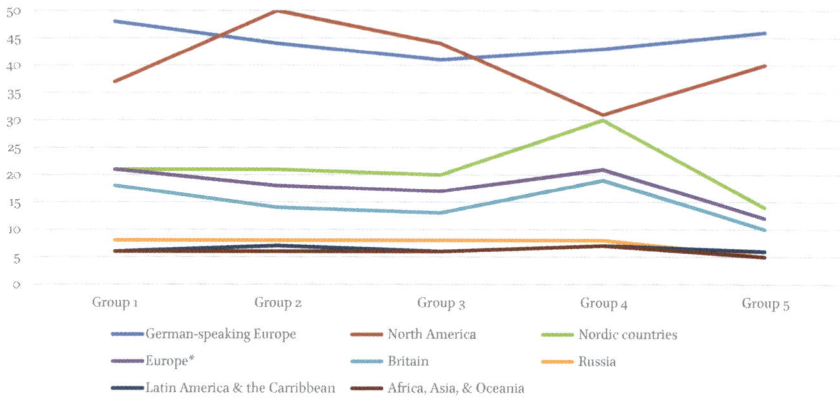

FIGURE 12 Distribution on destination (in per cent) before 1930 per social status among transnationally mobile Nordic technical school graduates 1880–1919
SOURCES: SEE FIGURE 1.

classes looked more to continental Europe for inspiration.[20] In our case, however, we can observe that graduates of working-class origin also were less prone to go to North America. One explanation may be connected to more limited financial resources, that is, these groups may not have had parents who could afford to finance long transatlantic journeys, nor may they have had—to use Bourdieuan expressions—enough social and symbolical capital to be awarded grants and other financial means. Some scholars have described transatlantic emigration as a kind of 'middle-class project'; wealthier people did not need to leave for America, and poorer people could not afford it. Simone A. Wegge has noted similar patterns in a study of mid-nineteenth-century transatlantic migration from Hesse-Cassel.[21] The North American social-class pattern was, of course, influenced by the fact that some graduates can be characterised as 'ordinary' emigrants who aimed at settling for good. It diverged from the general mobility pattern, with decreasing probability with lower social origin.

2.4 *Urban-Born Went to Europe, Rural-Born to America, and Foreign-Born Back Home*

Did it matter for the choice of destinations where the graduates were born? The patterns are ambiguous. It definitively mattered if a graduate was born in the

20 Hans Lindblad, 'Emigrationen—folkets projekt', in: Olov Isaksson (ed.), *Utvandrare och invandrare i Sveriges historia 1846–1996* (Stockholm 1997) 26–38.

21 Simone A. Wegge, 'Occupational self-selection of European emigrants: Evidence from nineteenth-century Hesse-Cassel', *European Review of Economic History* 6 (2002) 378–389.

educational country or not: 42 per cent of the foreign-born went to the neighbouring Nordic countries, compared to 19 per cent of the native-born, while the shares of native-born going to the two major destinations were higher.[22] This was simply return migration of, primarily, Finnish and Norwegian students from Sweden and some Icelandic students from Denmark. However, foreign-born graduates also went to Britain, Russia, and to countries outside Europe and the Americas to a somewhat greater extent.

Britain appears, together with other European countries as urban destinations: the former a destination where a lesser share of rural-born graduates went, the latter one where a greater share of capital-born graduates went.[23] Architects were inclined to travel to other European countries, and Swedish architects were certainly born in the capital to a much greater extent than other graduates, but this was not reflected in Finland and Norway. Rural origin increased the likelihood to choose North America and the neighbouring Nordic countries.[24] The North American pattern can, to some extent, be related to the rural character of transatlantic emigration in general. Reino Kero claims that Finnish emigration had a more rural character than Scandinavian.[25] However, Nils Olav Østrem also points to a rural emphasis in Norwegian emigration, even if urban emigration relatively speaking was stronger in the late nineteenth century, the transatlantic movement once again took a rural turn after 1900.[26] Donald Harman Akenson states that Swedish emigration was primarily rural on to World War I, also when accepting the broadest definition of urbanity.[27] The rural propensity to choose the neighbouring countries may be connected to the fact that many Nordic borderland regions were rural, although the Copenhagen-Malmö district constituted an exception. Finally, 54 per cent of the

---

22 45 per cent of the native-born went to the German-speaking countries, compared to 36 per cent of the foreign-born; 39 per cent of the native-born went to North America, compared to 32 per cent of the foreign-born.

23 Twelve per cent of the transnationally mobile rural-born graduates went to Britain, compared to fifteen to eighteen percent of the urban-born; 22 per cent of the transnationally mobile capital-born graduates went to 'other European countries' compared to 16 to 17 per cent of the ones born in other cities and towns and rural areas.

24 42 per cent of the transnationally mobile rural-born graduates went to North America, compared to 36–40 per cent of the urban-born; 22 per cent of the transnationally mobile rural-born graduates went to North America, compared to sixteen to 20 per cent of the urban-born.

25 Reino Kero, *Migration from Finland to North America in the years between the United States Civil War and the First World War* (Turku 1974) 90.

26 Nils Olav Østrem, *Norsk utvandringshistorie* (Oslo 2014) 36.

27 Donald Harman Akenson, *Ireland, Sweden and the great European migration, 1815–1914* (Liverpool 2012) 193, 198 n21.

transnationally mobile graduates who were born in the second to fourth cities went to the German-speaking countries, while the shares for the other categories lay somewhat above 40 per cent. Plausible explanations relate to geographical locations in Sweden. Malmö, but also Gothenburg, were relatively close to Germany. In Norway, the second city of Bergen had historical connections to Germany, beginning in the Middle Ages, when the city was the northernmost node in the Hanseatic system. Almost two-thirds of the transnationally mobile graduates born in Malmö and Bergen went to the German-speaking countries.

2.5 *Anglo-Saxon Attractions of Mechanical and Mining Engineers*

Hypothetically, specialisation mattered for the choice of destination. Different countries had leadership in different branches, and chemical engineers, for example, can be expected to choose Germany more than other specialisations. The patterns can be viewed in appendix 3 (mechanical, electrical, and naval), appendix 4 (civil and construction), appendix 5 (chemical), appendix 6 (mining and metallurgy), and appendix 7 (architecture).

The mechanical, electrical, and naval group shows, together with mining engineers and metallurgists, higher shares and propensities to choose North America. The United States was often viewed as the best place to study rational and efficient organisations of mechanical and electrical industries, steel and ironworks as well as shipyards; industries in America were often organised in a Taylorist or Fordist spirit.[28] The two specialisations mentioned above were educated for these industries. The three largest electrical manufacturers—General Electric in Schenectady in upstate New York, Westinghouse in Pittsburgh, and *Allis-Chalmers* in Milwaukee—had a lot to offer graduates who wanted to learn more about rationality, as had the large shipyards, locomotive and other workshops located in and around cities like Boston, New York, Chicago, and Philadelphia. *Carnegie's* steel and ironworks in the Pittsburgh district, more rational and efficient than anywhere else in the world, also attracted mechanical engineers. Pittsburgh and its surroundings were, of course, an attractive environment also for mining engineers and metallurgists. American mining offered, to a certain extent, the same kind of experience of rational production, and North America was an attractive place to go to also for this group.

To some extent, we can say the same about Britain for these two groups of engineers. Mining environments like the ones in northeast England drew some attention in Sweden and Norway, but the mechanical, electrical, and naval group had an even stronger inclination for the United Kingdom. There were highly automated—partly Americanised—workshops, like *Bruce, Peebles*

28 Grönberg, *Learning and Returning*, 145.

& Company in Edinburgh. The major American electrical manufacturers had British subsidiaries, where engineers often stopped by on their way to America or on their way back home. The most relevant explanation, however, is related to shipbuilding, where the British world position remained strong into the twentieth century. The shipyards on the Lagan in Belfast; on the Tyne, Tees and Wear around Newcastle, Sunderland and Middlesbrough; and on the Clyde in Glasgow belonged to the technologically most advanced in the world around 1900. Also, the University of Glasgow was one of the most renowned institutes for technicians who wanted to deepen their knowledge of naval architecture.[29]

2.6 *Mechanical and Mining Engineers and Sub-Groups to Different Places in Europe*

Whereas these two groups of technicians shared an inclination for the Anglo-Saxon world, they acted significantly differently when it came to the choices of other European destinations and especially overseas countries outside North America. They also acted rather differently when it came to the German-speaking countries. Germany, Switzerland, and Austria were more attractive to the mechanical, electrical, and naval group who had several interesting environments to go to such as Berlin's electrical industry, the shipyards along the Baltic Sea coast, as well as technical universities in cities like Berlin, Darmstadt, Munich, and Zurich; all of them pioneers in the teaching of electro-technology. Mining engineers and metallurgists had, of course, interesting objects too, Silesia, for example, and the mining universities in Freiberg in Saxony and Leoben in Styria in Austria. Nevertheless, there were less to choose from for mining engineers and metallurgists. They were—compared to other engineering specialisations in general and the mechanical, electrical, and naval group—more likely to set out for areas on the other side of Germany's western border, that is, France and Belgium, but Spain was also a relatively common destination. These countries attracted with prominent technical universities with an emphasis on mining as well as valuable practise in technologically advanced mining districts like the one around Liege. Intra-Nordic mobility was another likely trait among mining engineers and metallurgists. Mining engineers from Sweden had, for shorter or longer periods, leading positions in Norwegian mining and metallurgic ventures such as the electro-steel-works in Stavanger, the nickel plant in Kristiansand,[30] *Orkla* in Løkken Verk near Trondheim—a company

29 Olsson, 'To See how things were done in a big way Swedish naval archtects in the United States, 1890–1915.', 434–435; Sidney Pollard and Paul Robertson, *The British shipbuilding industry, 1870–1914* (Cambridge MA 1979) 49–51, 108–129, 231.

30 Sandvik, *Falconbridge nikkelverk.*

that also awarded study tour grants to Sweden for Norwegian engineers—and the *Skandia* copper mining company, based in the northern town of Narvik. Also in the north, some Swedish engineers worked for their fellow countryman and consul Nils Persson who owned the mines in Sulitjelma. It is reasonable to assume that Swedish knowledge was utilised to a large extent in Norwegian mining. In Finland, too, Swedish mining engineers had leading positions; the mines in Pitkäranta by Lake Ladoga and Orijärvi in the southwest are two examples.

2.7 *Mining and Civil Engineering in Overseas Countries Outside North America*

However, compared to other specialisations, the strongest positive effects for mining engineers and metallurgists are noted for overseas areas outside of North America. Together with civil and construction engineers, they went to Latin America and the Caribbean and Africa, Asia, and Oceania more often than other specialisations. Chile, Brazil, Argentina, and Bolivia received the most mining engineers and metallurgists in Latin America, whereas some also were involved in mining in Africa. South Africa, Congo, and Ethiopia, and Madagascar received a few Nordic mining engineers. These patterns are connected to the exploitation of natural resources; these engineers often travelled for European or American mining companies. The gold mines in Kimberley in South Africa employed Nordic mining engineers, as did Belgian and French ventures in Congo and Madagascar, respectively.[31] The American Guggenheim family formed an exploration company in the late 1880s to search for and buy profitable mines. The pit copper mine in Chuquicamata in northern Chile was one environment where Nordic technicians served the Guggenheim family.

Thus, the pattern for civil and construction engineers resembles the one for mining engineers and metallurgists. Infrastructure building in overseas countries outside North America created a significantly strong inclination. Argentina was the overseas country outside North America that received most Nordic technicians, and an overwhelming majority were civil and construction engineers. From the mid-nineteenth century, there were several Argentine civil engineering projects such as railway construction, water supply, and drainage in Buenos Aires. Many British engineers were also involved; only the colonies of India and Australia hosted greater numbers of civil engineers from Britain and Ireland in 1890.[32] Similar types of civil engineering projects took

31 Indebetou and Hylander, *Svenska teknologföreningen*, 953.

32 Michael M. Chrimes, 'British and Irish Civil Engineers in the Development of Argentina in the Nineteenth Century. Queens' College. Cambridge University 29th March – 2nd April

place in other overseas countries. Danish engineers in Siam primarily worked in civil engineering projects, and so did Nordic technicians in countries such as Brazil, South Africa, China, and Dutch East India, where the government of the Netherlands had problems recruiting the country's own engineers. Among the Nordic nationalities, Danish engineers, in particular, functioned as substitutes. Civil and construction engineers were also inclined to go to other European destinations. Today's Indonesia was not the only Dutch territory they frequented; the motherland offered interesting insights in construction of barrages and sea walls. France was another interesting country to go to: The Paris-based company founded by engineer Francois Hennibique was a pioneer in reinforced concrete. Italy, too, was interesting, primarily for study travelling, since some construction engineers wanted to specialise in architecture. It was a common feature to develop the studies of architecture at German technical universities and through study tours in the classical countries.

2.8 *Architects Found the 'Eternal Laws of Beauty' in Europe*

Almost 60 per cent of the transnationally mobile architects visited other European countries, which is between three and four times higher than for the other specialisations. We can also note a high share to Africa, Asia, and Oceania. The former is not surprising since the region includes their classical destination Italy, but also France. Many architects were, of course, attracted by Paris. Spain and Greece were also relatively speaking common destinations for architects if we compare with other specialisations. The high probability for Africa, Asia, and Oceania is perhaps more unexpected but can be explained by study tours in the Mediterranean which sometimes also were extended to Egypt, Tunisia, Algeria, and Morocco.

Architects were also inspired by the German-speaking countries, and they went to a somewhat greater extent than chemical engineers. Many studied at technical universities such as the ones in Dresden, Charlottenburg, and Munich that had an emphasis on neo-renaissance and Art Nouveau architecture, but inspiration from these architectural styles could also be acquired through employment at architect offices. Austrian architect and city planner Otto Wagner and his modern Art Nouveau style, emerging in Vienna, was another source of inspiration. Architects also show a strong inclination for intra-Nordic mobility. Some made study trips and worked in Norway and Finland. Most important, however, was travel to Denmark and Sweden. Danish national romanticism and brick architecture, most prominently represented by Martin Nyrop's

---

2006', in: Malcolm Dunkeld, et. al. (eds.), *Proceedings of the Second International Congress on Construction History*, vol. 1 (Cambridge 2006) 677, 681–692.

Copenhagen Town Hall, was one force that attracted Swedish, Norwegian, and to an even greater extent Finnish architects to Denmark. Contemporary Swedish architects in the so-called Nordic classicism—Hakon Ahlberg, Gunnar Asplund, Sigfried Lewerentz, and Ivar Tengbom—attracted study travellers, more often from Finland than from Norway. Architects show a much lower share to North America. Martin Borgstedt[33] was, however, employed in New York and San Francisco around 1890.[34] His 1892 travel report reveals a divided picture of American architecture. Efficiency, innovativeness, and creativity impressed, but the 'eternal laws of beauty' were set aside. Borgstedt concluded that an architect's priority should be esthetical studies in Europe. After that, if resources were available, he or she could cross the Atlantic and have a look around in North America.[35]

2.9 *Chemistry with Germany*

Chemical engineers constituted another specialisation with a comparably low share and propensity to choose North America. Britain was another destination that chemical engineers 'avoided', and the country's development in the chemical industry was following Alfred D. Chandler and Takashi Hikino very slowly.[36] Chemical engineers shared this low interest for these two Anglo-Saxon destinations with civil and construction engineers, who also showed relatively less interest for the German-speaking countries. Chemical engineers looked more often than the mechanical, electrical, and naval group for overseas employment outside North America, but not as inclined as civil and construction engineers and mining engineers and metallurgists. The Danish West Indies were one area that attracted comparably many chemical engineers from Denmark to the sugar business. However, chemical engineers were, of course, most attracted by the German-speaking countries, which is easily explained by Germany's prominence within this industrial branch. Germany took an early lead in new fields such as pharmaceuticals and dyestuff, and many industries along the Rhine, many near Frankfurt-am-Main, attracted many chemical engineers from the Nordic countries. Studies of brewing, often in Bavaria, was another trait and so was practice in areas like sugar-refining, pulp and paper, and electrochemistry.

---

33 KTH, architect, 1884.

34 Indebetou and Hylander, *Svenska teknologföreningen*, 239.

35 Martin Borgstedt, 'Reseberättelse af Arkitekten Martin Borgstedt öfver nyare byggnadskonstruktions-sätt, byggnadsmaterialer, m. m. i Nordamerikas Förenta Stater (1892)', 1892, 2, Reseberättelser 269, Royal Institute of Technology Library, Stockholm.

36 Alfred D. Chandler and Takashi Hikino, *Scale and scope: the dynamics of industrial capitalism* (Cambridge, MA 1994) 485.

2.10 *Booms, Depressions, and Germany's and America's Emergence as Industrial Leaders*

Figure 13 shows that the German-speaking countries and North America were the most common destinations from the 1880s to the 1920s. However, it was only among the graduates of the 1890s that the difference between the two destinations was large: The number going to the German-speaking countries was twice as large as the number to North America.

It is hardly surprising that technicians leaving school in the 1880s more frequently went to Britain compared to the ones graduating later. This can be interpreted as a reminiscence of Britain's early industrial dominance.

North America's low share among the 1890s graduates can be viewed in light of American and Canadian depressions in the mid-1890s. Latin America also experienced problems in the 1890s. The so-called Baring crisis led to an economic meltdown for the whole continent[37] and is reflected in a low share and a propensity for Latin America and the Caribbean among the graduates of that decade. Instead, the German-speaking countries were more popular for graduates of the 1890s compared to those who graduated both before and after that decade. German industry grew rapidly, and the same was true for the number of students accepted at *Technische Hochschulen*. Wolfgang König notes an oversupply of engineers entering the German labour market, which led to fierce competition for jobs in the early years of the new century.[38]

The highest share to North America is however noted for graduates from the following decade, the twentieth century's first. The years around 1900 were a period of recovery and prosperity in the United States and Canada.[39] Nordic technical school graduates were beginning to look across the Atlantic again to a country that now was described as technologically superior to the rest of the world. After the turn of the century, Nordic graduates were also more likely to go to overseas destinations outside North America and especially to Africa, Asia, and Oceania from 1910 onwards. Technical improvements in communication contributed. The latest graduates in this study were also more attracted by other European countries. The turbulent times contributed, as Myllyntaus has described, to a decrease in attraction for the German-speaking countries;

---

37 Kris James Mitchener and Marc D. Weidenmier, 'The Baring Crisis and the Great Latin American Meltdown of the 1890s', *The Journal of Economic History* 68:2 (2008) 462–500.

38 König, 'Technical education', 80.

39 James Arthur Estey, *Business Cycles: Their Nature, Cause and Control* (New York, NY 1941) 53. Statistics also downloaded from Angus Maddison's homepage, http://www.ggdc.net/MADDISON/oriindex.htm, (14 November 2017).

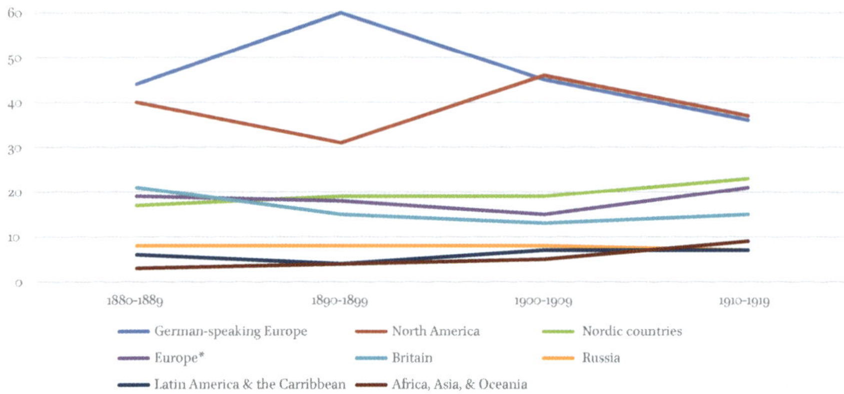

FIGURE 13 Distribution on destination in per cent per decade of graduation among transnationally mobile Nordic technical school graduates 1880–1919 going abroad 1880–1930

SOURCES: SEE FIGURE 1.

no other graduates were as unlikely to choose to go there as the ones leaving school in the 1910s. France became a popular destination for interwar technical studies.

2.11 *Freshman Mobility to Learn—Experienced Mobility to Teach*

As for graduation age, figure 14 reveals four tendencies. The German-speaking countries and North America were young destinations: The shares going there decreased with increasing age of graduation. Mobility to the Nordic countries, Britain, and other European countries was most common in the group that graduated between the ages of 25 and 29, while the shares going to Russia increased with higher age. Overseas destinations outside North America, finally, lay on the same levels regardless of graduation age.

Germany, Switzerland, and Austria as learning destinations are reflected in figures 14 and 15. The younger a technician was when he or she left school, and the shorter the time that had passed between graduation and departure, the likelier the graduate was to choose to go to German-speaking Europe. Britain shows a somewhat similar pattern, but less marked. This is in part also echoed for North America, but whereas German shares continued to fall on to the oldest groups, the American ones levelled out at around age 25. In figure 15, we can see that the share choosing to go to North America remained the same for departures within a ten-year period after graduation. The difference between the two major destinations is probably reflected here; whereas the German-speaking countries—as mentioned—only were learning destinations, there

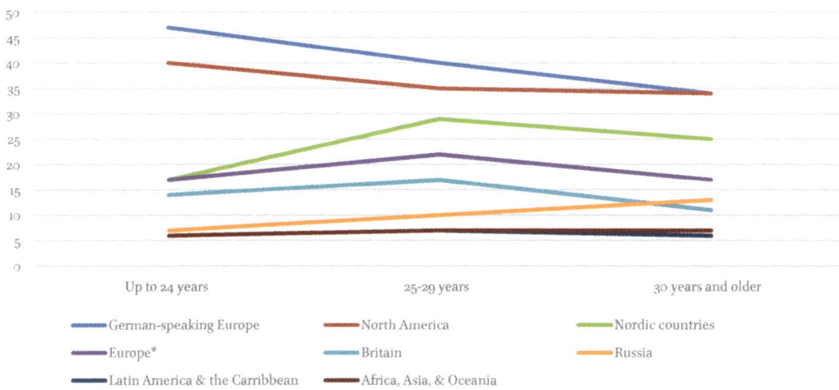

FIGURE 14 Distribution of destination in per cent per age at graduation among transnationally mobile Nordic technical school graduates 1880–1919 and going abroad 1880–1930
SOURCES: SEE FIGURE 1.

was a duality in the mobility across the Atlantic. North America was primarily a learning destination, but mobility there was a combination of technicians on placements abroad and traditional emigrants.

Russia was a destination many travelled to as specialists; technicians were often supposed to contribute to Russian development by bringing their experience there, rather than acquire experiences to return. Therefore, it is not surprising that somewhat older and more experienced technicians went to Russia. Other Nordic countries were also more experienced destinations, suggesting frequent moves as technical experts, often for domestic companies. Africa, Asia, and Oceania show in part a similar pattern; the shares going there increased among graduates who postponed their departures for at least five years after graduation. Journeys to these parts of the world hardly had the learning purposes characterising many of the other destinations.

2.12 *Migrants Travelled Longer—Study Travellers Shorter*

Figure 16 shows that there were more migrants than study travellers to all destinations except other European countries, that is, Europe outside the Nordic and German-speaking countries, Britain, and Russia. However, study travellers were overrepresented compared to the average in streams to all European destinations except Russia. It was very unlikely to go to Latin America and the Caribbean and Africa, Asia, or Oceania as a study traveller, even if as mentioned, there were some north African journeys related to visits to Italy and other countries in the European part of the Mediterranean.

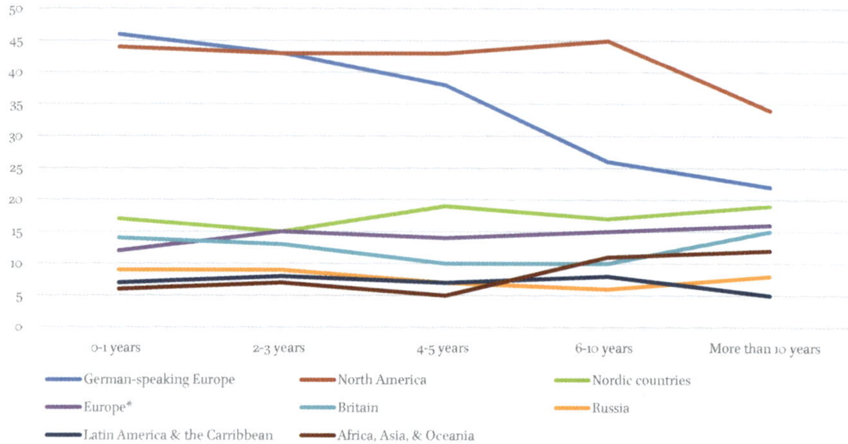

FIGURE 15 Distribution on destination in per cent per time interval between graduation and migration among Nordic technical school graduates 1880–1919 migrating abroad 1880–1930
SOURCES: SEE FIGURE 1.

Architects were as mentioned significantly more prone to study travel than other specialisations. The European region included classical architect destinations such as Italy and France, with their magnificent capitals, as well as Spain and Greece. These were destinations where graduates travelled to see things rather to work; this pattern was relevant not only for architects but also for construction engineers who wanted to become and work as architects. Another area where study travellers were overrepresented was German-speaking Europe. Lisa Brunnström has stated that architects often were more inspired by Germany than America,[40] and this held true also for construction engineers. The British Isles and unknown destinations were also study trip rather than migration destinations; this implies that the latter were nearby, rather than remote, countries. The neighbouring Nordic countries constituted a pronounced study trip area, but only for Finns and Norwegians. Two explanations are plausible. One relates to geography; Sweden and Denmark lay on the way to other destinations for technicians from these two countries. We may, however, also relate to the degree of industrialisation. Danish and Swedish technicians migrated, on the one hand, to other Nordic countries as technology carriers and set up, for example, cement factories for F. L. Smidth in Finland and water power stations for ASEA in Norway. On the other hand, Norwegian

40 Brunnström, *Den rationella fabriken*, 202.

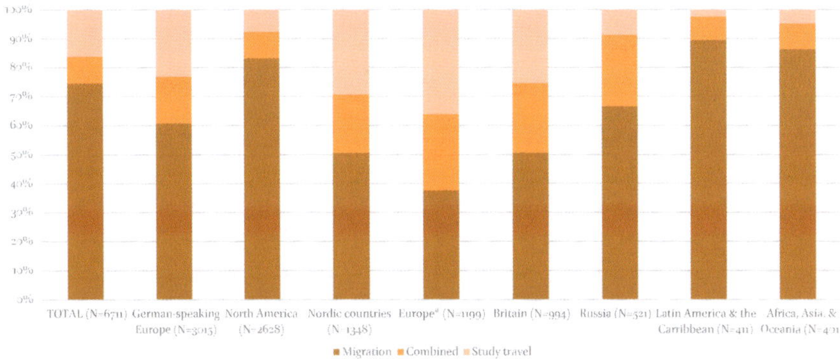

FIGURE 16 Distribution between 'migration' and 'study travel' before 1930 per destination among transnationally mobile Nordic technical school graduates graduating 1880–1919
SOURCES: SEE FIGURE 1.

and Finnish technicians probably had more to see in Sweden and Denmark as these countries had reached further in industrialisation.

Travelling with a smaller grant or at their own expense and without the purpose of obtaining employment to make a living was, of course, both cheaper and less time-consuming to nearby countries such as Germany or Nordic neighbours. With Russia as an exception, it was the European areas that were study trip destinations, whereas the migration countries were located in the Americas, Africa, Asia, and Oceania.

3 Summary

Stang's suggestion of a worldwide market for engineers' competence is confirmed by the fact that Nordic technicians went to almost 100 countries and dependencies all around the earth, some of them located as far away as the South Pacific. However, two destinations dominated: the German-speaking countries in Europe, that is, Germany, Switzerland, and Austria, and North America, that is, the United States and Canada. They were number one and two on the pan-Nordic level and in Norway, whereas the order was reversed in Denmark and Sweden. Finland diverged more: German-speaking Europe was number one, but North America was only the fifth destination.

The attractions of these areas were manifold, but both were interesting places to go for a placement abroad. Germany and Switzerland hosted some of the world's most renowned technical universities and were often destinations for students who wanted to complete their domestic education. However, Germany and her linguistic neighbours were also countries of both technology and architecture and, thereby, of interest both for the other type of placements abroad, that is, employment consciously organised for learning purposes and study trips. Developments in the chemical industry, especially along the Rhine, as well as Berlin's electrotechnical industry, drew Nordic technicians to Germany. There were also exciting architects, like Otto Wagner in Vienna, who contributed to the attraction. One difference between the two main destinations was that North America offered the employment path almost exclusively when it came to a placement abroad. At least before the 1910s, it was uncommon to cross the Atlantic for technical university studies, whereas Nordic technicians often took employment at, for example, mechanised and rationalised electrical manufacturers, steel and ironworks around Pittsburgh, mines, and shipyards along the American East Coast. At these industries, Nordic technicians could learn a lot about rationality and organisations in Taylorist or Fordist spirits. Another major difference between North America and the German-speaking countries was the former's minor share of study travellers. According to the definition used in this study, study travellers did not earn their own money while they were abroad. Therefore, study travelling was a much more common trait in short-distance mobility. Study travellers were not only overrepresented in the streams to German-speaking countries but to all European destinations except Russia, whereas they—as mentioned—constituted small shares to North America and even smaller to other overseas destinations. However, it was not only the distance that directed study travellers to European destinations. Architects were, as mentioned, study travellers rather than migrants, and some of their classical destinations—Germany, France, Italy, and Greece—lay in Europe. Some architects undertook a Mediterranean study trip and also visited north African destinations like Morocco, Tunisia, and Egypt, but Nordic architects were rare in other parts of the world. It was in Europe, not the United States or elsewhere, that the architects found what Martin Borgstedt called the 'eternal laws of beauty'.

Mechanical, electrical, and naval engineers together with mining engineers and metallurgists were not necessarily looking for these laws but were the specialisations that were supposed to work with rationalisations of workshops and the like. They had more to learn, and more to gain, from going to America. Specialisation clearly mattered more for the choice of destination than for the very decision of going abroad. The mechanical, electrical, and naval group was, for example, relatively more attracted also by Britain, a destination that also, because of its early industrial prominence, was more popular among earlier than later graduates. However, automated mechanical workshops and advanced shipyards in northern England and Scotland attracted into the twentieth century. British subsidiaries of larger American corporations were often visited on the way to or back from America.

Berlin and its electrical industry were of interest for mechanical, electrical, and naval engineers, and so were the mechanised shipyards along Germany's Baltic Sea coast. Several German technical universities were world-leading in teaching electro-technology. Chemical engineers showed, however, a higher likelihood to go to German-speaking Europe, and architects also went there to a greater extent. Chemical engineers were attracted by the development along the Rhine but also by brewing in Bavaria, and, for example—sugar refining. Architects could study Art Nouveau and other interesting architectural styles at universities and also through employment.

Mining engineers and metallurgists also had interesting regions in German-speaking Europe, such as Silesia, but were also likely to find their ways to areas inside the Nordic countries. Swedish mining engineers held, for example, prominent positions in Norwegian and Finnish mining. Mining districts in Belgium and France also attracted. The latter also interested civil and construction engineers because of, among other things, her development in reinforced concrete. So was Italy because many construction engineers wanted to develop as architects. However, mining engineers and metallurgists and civil and construction engineers also went overseas relatively often to areas outside North America. Exploitation was one reason for mining engineers; natural resources in countries like Congo, Madagascar, South Africa, Brazil, and—not least—the Guggenheim family's explorative activities in Chile engaged Nordic technicians. Civil and construction engineers often went to build infrastructure. There were many civil engineering projects in Argentina in the late nineteenth century. Siam was another country with similar projects, and Danish engineers were involved in many of them. Danes also functioned as substitutes when the Dutch government had problems finding engineers for infrastructural building in Dutch East India, that is, today's Indonesia. Countries in Africa, Asia, and Oceania were also likelier choices for the latest graduates in this study, perhaps because communications had developed.

Nevertheless, country of education, departing as a migrant or study traveller, and specialisation were not the only traits that mattered when it came to the Nordic technicians' choice of destination. At least to some extent, class and geographical origin played roles. Upper-class origin implied stronger mobility to German-speaking Europe, but this pattern is relevant for most European destinations. Middle-class origin implied stronger mobility to North America. This may be connected to different orientations and inspirational sources on class-level, but graduates originating in the working class often shared the short-distance pattern with the upper class. One explanation might be lack of traditions and resources for longer journeys and foreign intermissions; previous studies have described transatlantic emigration as a 'middle-class project', the upper classes did not need to go, the lower classes could not afford it. Of course, another difference between North America and German-speaking Europe was the fact that the United States and Canada also were destinations for graduates who can be characterised as ordinary emigrants. This is reflected in the Norwegian and Swedish shares to North America being higher than the Danish and Finnish (see below), but also in the rural touch of the Nordic technicians' transatlantic crossings, and the urban character of Nordic technical mobility to German-speaking Europe and other European destinations, be they not the neighbouring Nordic countries. Nordic emigration to America also had a rural character in general. All in all, North America can be described as both a learning destination and destination for permanent settlement, whereas the German-speaking countries basically constituted only the former.

The timing of the mobility also reflected this. The 1890's growth in German industry and the extensions of its technical education system, contemporaneous with transatlantic depressions, gave a share for the German-speaking countries that was twice as high as North America's for the graduates who left school between 1890 and 1899. The 1890's graduates make the difference between the destinations; the shares to German-speaking Europe and North America were almost the same among both earlier and later graduates. A pattern reflecting the difference between the two destinations is the fact that North America remained a popular destination also among those who departed at an older age and after a longer interval from graduation. The shares for German-speaking Europe gradually decreased with both increasing age and increasing time between graduation and departure. This reflects the domination of learning mobility in the streams to German-speaking Europe, as a younger freshman, at least hypothetically, has more to learn than an older and more experienced technician. This also has implications for other destinations. The shares, and the likelihoods, to choose Russia, other Nordic countries, and overseas destinations outside North America increased with age and interval. This implies more the mobility of experts, possibly with the aim to contribute to technical and industrial development in the receiving countries. Younger freshmen travelled to learn, whereas older and more experienced technicians travelled to teach.

Finnish technical mobility was considerably more short-distance; the main destinations were the German-speaking countries, Scandinavia, Russia, and other parts of Europe. The lower quality of domestic technical education and the extensive study travel contributed to shaping the system. Study travel was, in part, a result of the comparably high share of architects, which indirectly contributed to shaping the mobility. The extensive Finnish study travelling contributed to giving German-speaking Europe an exceptional position as a destination; almost two-thirds of the mobile Finnish graduates visited the countries. It is obvious that acquiring technology and knowledge through study travel was a distinct Finnish pattern in the Nordic context. Finland's position as a Grand Duchy under the Russian tsar and the geographical proximity also contributed: The border with Russia proper ran just outside the imperial capital of Saint Petersburg. Russia functioned as a kind of substitute-America when the transatlantic fever never really struck Finland. Finland in general can, at least until the latter part of the nineteenth century, be viewed as part of the eastward migration system, even if Finnish technicians flocked to western core nations like Germany and Switzerland. This contributed to the United States' and Canada's more minor role in Finland than in the other countries.

As opposed to the Finnish mobility system, the Norwegian counterpart was migration- rather than study-trip-based. This fact contributed to giving it more of a long-distance character compared to Finland. Relatively large contingents went to overseas destinations outside North America. Nevertheless, Norway's long lack of a domestic technical university gave Germany and Switzerland—like in Finland—prominent positions in the system. Norwegian technicians did not have as much to gain as their Swedish colleagues when it came to learning about large-scale production in the United States, but the country still shared North America with Sweden. Technicians were, of course, also influenced by Norway's significant contemporaneous transatlantic migration. Only Ireland lost a greater share of its population to North America.

Sweden's transatlantic migration followed only after Norway's, so Swedish technicians were definitively also influenced by general emigration patterns. The country also possessed more large-scale and mass-production-minded industry than the other Nordic countries, which implied relevance for learning in the United States. This combination gave North America its strongest position as a destination among the Swedish technicians. The Swedish system was migration-based, even if study trips played a somewhat stronger role than in Norway and Denmark. Access to a domestic technical university hampered to some extent the mobility to Germany, Switzerland, and Austria, even though the area constituted the second largest destination.

This pattern was also reflected in Denmark, by 1880 the most industrialised Nordic country with a higher technical institute established already in the 1820s. North America was the largest destination in Denmark also, but not as dominant as in Sweden, and transatlantic crossings also attracted relatively fewer than in Norway. Danish emigration to America was lower than in Norway and Sweden. Denmark's system was still migration-based, and study travel played a minor role. Certain traits contributed to shaping the Danish technical mobility somewhat differently. The existence of a colony such as the Danish West Indies is one, and a relatively large Danish population in Siam is another.

The tendency to go abroad on payrolls of worldwide Danish companies was, however, the main contributor to this shape. This mobility dispersed Danish technicians and gave relatively large streams to countries where technicians from the other Nordic countries almost were absent.

The German-speaking countries noted—perhaps somewhat surprisingly considering the geographical proximity—their lowest share from Denmark. The area was, nevertheless, the second Danish destination. On the pan-Nordic level, the German-speaking countries constituted the largest destination. We will now join our technicians on a journey there.

CHAPTER 4

To Study and to Practise in German-Speaking Europe

On the evening of 3 March 1908, flags were flying in front of the principal hotel in Sweden's southernmost town Trelleborg. Inside, there were lively goings-on among local politicians and other potentates. In the afternoon, a telegram had arrived with the happy news: the two chambers in the Parliament had voted in favour of a steam-powered train ferry connection from the town to Sassnitz on the German island of Rügen. King Gustaf V telegraphed his satisfaction over the decision to the chairman of the town council the following morning. He was certainly satisfied that new, modernised, and technologically improved train ferries were about to bring Sweden and large parts of the Nordic area somewhat closer to one of the major model countries in many different fields.[1]

Technology and industry were no exceptions. On the contrary, Germany was viewed as a high-tech country whose industrial organisation and rationality, production technology, and—not least—technical education system were admired. When the Swedish Technical Association gathered for its annual meeting in 1909, one of the speakers emphasised modern industry's dependency on cutting-edge technology. Countries that realised this had overtaken ageing industrialised countries.[2] The speaker continued: 'I am of course in the first place referring to Germany'.[3]

German was engineering's language of the time, and it was also spoken in parts of Switzerland and Austria. Switzerland's German-language technical university in Zurich was, as Myllyntaus has revealed, also an interesting destination. Switzerland was also viewed as being in the technological forefront.[4] High-voltage engineering and transformers were fields where the Swiss were leading. Germany, Switzerland, and Austria have been collected as 'German-speaking Europe' in this study, which is somewhat misleading as some

1 See for example: Simensen and Grimnes (eds.), *Tyskland—Norge*; Nilson, 'Vacker, föredömlig, rationell', 43–67; Henningsen and Högvall, (eds.), *Skandinavien och Tyskland: 1800–1914*; Hösch (ed.), *Deutschland und Finnland im 20. Jahrhundert*.

2 Runeby, 'Tyskland som teknisk förebild.', 389; Nilson, "Vacker, föredömlig, rationell'.'

3 Cf. Nils Runeby, 'Tyskland som teknisk förebild', 389. Swedish original: Jag syftar naturligtvis närmast på Tyskland.

4 Myllyntaus, 'Discovering Switzerland', 299–328.

 | DOI:10.1163/9789004385207_005

technicians also went to the French and Italian speaking parts of Switzerland.[5] Nevertheless, many of them took the route via the Swiss Polytechnic, and a majority went after all to the German-speaking parts. Development in, for example, architecture also drew students to Austria and primarily to Vienna. The mining university in Leoben in Styria was also attractive, not least since studies could be combined with relevant nearby industries such as charcoal-based ironworks, steelworks, and the open-cast mines.

The German-speaking countries were visited by every fourth graduate in the Nordic countries: almost every second from Finland, every third from Norway, every fifth from Sweden, and every sixth from Denmark. The ideal type was a young Finnish or Norwegian chemical engineer, a reflection of Germany's prominence in the chemical industry and the shortcomings in domestic technical education in the two less industrialised Nordic countries. Graduates going to Germany, Switzerland, or Austria were likely to depart shortly upon graduation, in many cases the same year or the year after, but rarely after more than five years. These countries were not places to go for permanent settlement, but learning destinations. Of course, this made them into freshman targets, more attractive destinations during a period when the technicians were, to paraphrase Stang, unfinished products. The technician was also likely to be a graduate of the 1890s. German-speaking Europe was always an alternative to the United States but appeared as a very good one during the late nineteenth-century American recession. The graduates were also likely to be of upper-class origin and from major cities in the countries of education, but not the capitals. Possibly, the upper social layers of society looked more to Germany for inspiration, whereas the major non-capital cities were located relatively to the south or had historical connections to the country.

Nils Jørgen Tokheim[6] projected water power stations in the Alpine regions in the early 1910s and was our cohort's only visitor to Liechtenstein. Table 4 reveals that more than 90 per cent of the graduates going to this region found their way to Germany and that Switzerland, hardly surprising considering the importance of the school in Zurich, was a more common destination than Austria.

Mobility between the German-speaking countries was, as table 4 reveals, stronger among Finnish technicians. The extensive Finnish study travel contributed to a pattern where Finnish technicians travelled around to a greater extent than their Scandinavian colleagues.

---

5 The Austrian territory also included some areas where German was not the main language, for example Bohemia, Dalmatia, Galicia, Bucovina, Carniola, Moravia and Küstenland.

6 BTS, mechanical, 1896.

TABLE 4 Number and percentage of Nordic technical school graduates, 1880–1919, migrating or study travelling to Germany, Switzerland, and Austria before 1930

	Germany		Switzerland		Austria		TOTAL
	Number	%	Number	%	Number	%	Number
Sweden	1067	93	158	14	129	11	1153
Denmark	406	94	53	12	66	15	433
Norway	822	90	143	16	46	5	910
Finland	502	97	123	24	105	20	519
NORDIC	**2797**	**93**	**477**	**16**	**346**	**11**	**3015**

SOURCES: SEE FIGURE 1.

Germany attracted, relatively speaking, most architects and chemical engineers, which can be related to the country's prominence in these fields. Civil and construction engineers and mining engineers and metallurgists were underrepresented. Every third visit was made by a study traveller, compared to every fourth foreign visit in total. Finnish graduates were most overrepresented in relation to all Nordic mobile technicians. Norwegians were somewhat overrepresented, Swedes somewhat underrepresented, and Danes most underrepresented. This is explained by the student migration.

Switzerland had an overrepresentation of civil and construction engineers as well as chemical engineers, which can be connected to the interest in power stations and the like. The mechanical, electrical, and naval group as well as mining engineers and metallurgists were underrepresented. Almost half of the visits in Switzerland were made by study travellers. Finns and Norwegians were overrepresented, and this is once again explained by the moves to complete domestic technical education, that is, to study at the Swiss Polytechnic. In the Norwegian case, we may also relate to the topographical similarities. Danish technicians were more underrepresented than Swedish. Both countries possessed higher technical education, but a tradition of continuing the studies in Zurich was nevertheless established at Chalmers in Gothenburg. Denmark was also the Nordic country that was most different from Switzerland when it came to nature and landscape.

Austria was to an even greater extent a study trip destination; more than two-thirds of the visits were study trips. Chemical engineers, mining engineers, and metallurgists were overrepresented, which can be connected in part to the mining school in Leoben. Architects were strongly overrepresented, which can be explained by interest in the works of persons such as Otto Wagner and Camillo Sitte. Finland, too, was strongly represented. This is connected to the higher percentage of architects among the graduates in Helsinki.

1 Different Types of Learning Mobility

Except for a small share of work for domestic employers, German-speaking Europe was only a target mobility area; graduates went for placements abroad and study trips, and almost no one settled permanently. Migrants who did not return to their Nordic native countries usually continued to North America. One important difference compared to North America was that Germany, Switzerland, and Austria offered technical university studies besides employment and interesting objects for study travellers.

We may speak of three types of learning mobility to German-speaking Europe. Study travelling is one. The migration pattern can be interpreted as 'dual-target migration' either for enrolment at a technical university, and to some extent at lower technical schools, or for one, usually self-organised, employment at an industry and, to some extent, municipal bodies. In most cases, graduates realised only one of these options. About 35 per cent were only enrolled; around 20 per cent were both enrolled and employed, whereas 45 per cent were only employed. However, there were distinct national patterns in this context. Norwegians noted 65 per cent enrolment only and 15 per cent employment only and Finns noted 45 per cent enrolment only and 35 per cent employment only. Swedes had 15 per cent enrolment only and 65 per cent employment only, and Danes 10 per cent enrolment only and 80 per cent employment only.[7] Almost needless to say, these diverging patterns can be traced to the national differences in domestic technical education.

1.1 *Studying at Technical Universities and Other Schools*

At least before the interwar years, the German-speaking countries constituted the by far most frequented among Nordic students of technology, be it students who took their entire education abroad or the ones in focus in this study: those who had a pre-education from one of the Nordic countries. For students from all over the world and from all over academia, Germany was, as mentioned,

7 Grönberg, 'To study or to work?', 77–78.

a major destination around the turn of the nineteenth century. Hietala has noted, for example, that not only technicians, but also medical doctors and scholars in many fields left Finland to take up studies in Germany.[8] As for engineering studies, Pauer has as identified Germany as the most popular destination for those who left Japan in the decades before 1900.[9] Nevertheless, cultural and linguistic proximity made engineering students from southern Europe primarily choose other destinations. Cardoso de Matos and Diogo write, for example, that only 15 per cent of the pre-1900 Portuguese engineering students abroad studied in Germany, whereas 70 per cent chose France.[10] Greek engineering students had similar preferences for France as Antoniou, Assimacopoulou, Chatzis, and Mahera have revealed.[11]

For the Nordic area, however, German-speaking Europe was geographically, culturally, and linguistically closer than France, and German was also engineering's language. Academic travel from the Nordic countries to Germany had, as mentioned, a long history. During this period, there were a few engineering students at the technical universities in Vienna and Prague, but an overwhelming majority studied in Germany and in Zurich. The Swiss Polytechnic was a model for many of Germany's *Technische Hochschulen,* although many of the latter did not have an equally broad curriculum. German technical education has traditionally been looked upon as a powerful system producing scientists and engineers for German industry. However, Wolfgang König, who claims that the system's strength was its heterogeneity and its ability to produce a large number of engineers has questioned that view.[12] Nordic admiration for German technical education was, nonetheless, strong; modern equipment and the broader German education, adjusted to industrial and business demands for subjects like industrial economy and statistics besides technology, were two areas.[13] German *Technische Hochschulen* were especially important in Norway and Finland, considering the lack of their own technical universities. Germany was the model for Norwegian technical education, providing textbooks and later professors.[14] Panu Nykänen states that Helsinki Polytechnic recruited many well-educated teachers from the German-speaking countries in the late nineteenth century and onwards. The institute became a 'German-style

8 Hietala, 'Finnische Wissenschaftler in Deutschland', 373–394.
9 Pauer, 'Technologietransfer und industrielle Revolution in Japan', 34–54.
10 Cardoso de Matos and Diogo, 'Bringing it all back home', 155–181.
11 Yannis Antoniou et al., 'Greek engineers'; Fotini Assimacopoulou et al., 'Elève en France'.
12 König, 'Technical education', 65.
13 Nilson, "Vacker, föredömlig, rationell", 61–64; Runeby, 'Tyskland som teknisk förebild.'
14 Andersen, 'Germany and the education of Norwegian engineers', 100–101.

Technische Hochschule in practise', but followed also many paths from Zurich towards the 1908 technical university status.[15]

The most frequented German-language technical university was *Technische-Hochschule Berlin-Charlottenburg*. The 285 graduates in the cohort that went there equalled every fourth student migrant to the German-speaking countries. The numbers and shares of the student-migrants on country level were: Finland 42 (31 per cent), Sweden 72 (28 per cent), Norway 161 (24 per cent), and Denmark ten (18 per cent). Students were attracted by a technical university that was described as a model and a pioneer, establishing one of the first electrotechnical laboratories in 1884 and attracting famous scientists. These scholars attracted new students. Heinrich Müller-Breslau taught statistical calculation of building materials and is described as the founder of scientific building statistics. The real number of Nordic students was higher than the ones revealed above, since many arrived without Nordic pre-education. One well-known example is Sam Eyde, later a leading Norwegian industrialist, who arrived in the 1880s.[16] Ole Kristian Grimnes writes on his choice:

> Once he had chosen technical progress, an academic engineer-education and prospects of an occupation connected to technical modernisation, he could not have chosen a more central, advanced and up to date educational institution.[17]

Eyde acquired experience and knowledge during seven years of practise as an engineer in a Germany undergoing a very rapid infrastructural development. The challenges lay in the construction of railways and canals into the city centres, and to adjust ports to constantly increasing steamship traffic. He specialised in the shaping of urban railway stations, canals and port establishments and managed, with this expertise in his luggage, to establish an engineering office in Kristiania upon return in 1898. He paid interest in water power, but he developed an electric arc method to tie up nitrogen in the air, manufacture

15 Panu Nykänen, *Kortteli sataman laidalla. Suomen Teknillinen Korkeakoulu 1908–1941* (Porvoo 2007); Panu Nykänen, *Turning the wheel. The history of Helsinki University of Technology TKK* (Porvoo 2008) 15–23, quote from 20.

16 Eydetook his entire education in Germany and is, therefore, not in our cohort, see chapter 1.

17 Grimnes, *Sam Eyde*, 50. Norwegian original: Når han først hadde valgt tekniske fremskritt, en akademisert ingeniørutdannelse og utsikter til et yrke knyttet til teknisk modernisering, kunde han ikke ha valgt en mer sentral, avansert og tidsriktig utdannelseinstitusjon.

nitric acid, and—most important—produce fertilizer. Later in life, Eyde was known as a businessman rather than an engineer.

Eyde took his entire education at Charlottenburg and is not included in our cohort. We can, nevertheless, identify important Norwegian returnees from the school, whose main building neighboured one of Wilhelmine Germany's most fashionable strolling districts. Haakon Hauan (see page 1) has already been mentioned. Thorvald Lindeman[18] went to Charlottenburg but continued his studies at the university in Jena where he also earned a doctoral degree. After some practise in the German chemical industry, and various positions in Norway, Lindeman became manager for Borregaard's chemical factories in Sarpsborg. He was appointed professor of organic chemistry at the Norwegian University of Technology in 1917. Mechanical engineer Christoffer Kahrs Kielland[19] completed his education in electro-technology at Charlottenburg. Upon return, he headed the *Norsk Hydro* project for the factories at Rjukan. He was honoured for his important contribution to Norwegian industry at his retirement in 1953.[20] Axel Guldahl[21] and Johan Osness[22] became practising architects in Trondheim upon return and brought German impulses but were also inspired by other ideas. Osness's E. C. Dahl foundation building from 1908 is one example of Art Nouveau. We can trace Art Nouveau and German influences in some of Guldahl's works, but also Dutch, English, and other influences.[23]

Finland's Albert Petrelius[24] is another example of an architect inspired by Art Nouveau and Neo-Renaissance.[25] Many returnees from Charlottenburg made important marks in the Finnish society, not least as professors at Helsinki's technical university. Bernhard Wuolle[26] was appointed in applied mechanics and industrial economy in 1922 after having headed Helsinki's electricity works as well as serving as director-general for the national railway board and minister of communications. He used his experiences from Germany to

18 TTL, chemical, 1887.

19 BTS, mechanical, 1901.

20 For more on Kielland, see: Ketil Gjølme Andersen, *Flaggskip i fremmed eie: Hydro 1905–1945* (Oslo 2005).

21 TTL, construction, 1885.

22 TTL, architect, 1898.

23 Alstad, *Trondhjemsteknikernes matrikel*, 141; 'Arkitekt i tredje generasjon. Arkitektprofesjonen har gått i arv i familien Guldahl i Trondheim. Lars Axel Guldahl utøver faget i tredje generasjon, godt over 100 år etter at bestefaren Axel Guldahl senior tegnet sine første hus i Trondheim', *Adresseavisa*, 17 December 2002.

24 SPO, architect, 1890.

25 Patrick Eriksson, 'Petrelius, Albert—Uppslagsverket Finland', http://www.uppslagsverket.fi/sv/sok/view-103684-PetreliusAlbert, 14 November 2017.

26 SPO, mechanical, 1900.

promote electrification of Finnish railways.[27] Ossian Aschan[28] was appointed in chemistry in 1908 and initiated a research institute for the pulp, paper, and cardboard industry.[29] Leonard Hjelmman[30] promoted technological-scientific research as a professor of geometrics and was the university's principal from 1919 to 1937.[31] Harald Kyrklund,[32] appointed in mechanical engineering in 1917, contributed to the development of combustion engines and surrogate fuel that became important in the war years.[33]

Gottfried Strömberg[34] is perhaps Finland's clearest example of German influence. He first served as a teacher of electro-technology and spurred students to deepen their knowledge abroad. So, did, for instance, Finland's two first electrotechnical professors. Strömberg also performed the first large-scale Finnish electrical machines. After he had founded his own company in 1888, Strömberg brought German manufactured lamps and dynamos to Finland. Strömberg's modern factory in Helsinki grew over time and was described as Finland's most superior electrotechnical company.[35] Emil Holmberg[36] continued to the 1893 Chicago fair where he got the idea to compare German and American bridge building. Rationality characterised both countries and German bridges were reliable, whereas the American ones were more remarkable. Holmberg was also engaged as a teacher and later a professor at the Polytechnic Institute and was involved in several railway projects. The railway between Loviisa and Lahti has been described as based on his extended expert knowledge acquired abroad. He also performed the Finnish pioneer works in precision-leveling, a practise developed in Germany around 1890.[37]

---

27 'Wuolle Bernhard—Uppslagsverket Finland', http://www.uppslagsverket.fi/sv/sok/view-103684-WuolleBernhard, 14 November 2017.

28 SPO, chemical, 1881.

29 'Aschan, Ossian—Uppslagsverket Finland', http://www.uppslagsverket.fi/sv/sok/view-103684-AschanOssian, 14 November 2017.

30 SPO, civil, 1890.

31 'Hjelmman, Leonard—Uppslagsverket Finland', http://www.uppslagsverket.fi/sv/sok/view-103684-HjelmmanLeonard, 14 November 2017.

32 SPO, chemical, 1903.

33 'Kyrklund, Harald—Uppslagsverket Finland', http://www.uppslagsverket.fi/sv/sok/view-103684-KyrklundHarald, 14 November 2017.

34 SPO, mechanical, 1885.

35 Michelsen, *Viides sääty*; '60 år', *Hufvudstadsbladet*, 16 December 1923; '100 år sedan Gottfr. Strömbergs födelse', *Björneborgs Tidning*, 30 December 1963; Johan Stén, in: 'Elektroteknikern som blev ett varumärke. Om Axel Gottfrid Strömberg (1863–1938)', in: Henry M. Ericsson, et. al. (eds.), *Finlandssvenska tekniker. Bd 4* (Helsingfors 2003) 44–61.

36 SPO, construction, 1880.

37 *Suomen Insinöörejä ja arkkitehtejä*, 631; 'Professor C. E. Holmberg 75 år', *Västra Nyland*, 8 June 1935.

Johannes Christensen Granholm[38] is one of few Danish students at Charlottenburg. In the 1920s, he started a coachwork factory in Elsinore and was also manager of some motor companies in northern Zealand.[39] The Swedes were more numerous. Frans Fredriksson[40] studied at Charlottenburg and *Akademie zur Künste der Berlin.* Fredriksson designed several buildings in Germany before he returned to become a practising architect in Malmö. He was inspired by German 'brick castle' architecture and also designed several Art Nouveau houses in Malmö and the neighbouring town of Lund.

One of his class mates in Malmö, Årad Svensson Nobell[41] worked for *Skånska Cementgjuteriet* and was assigned to plan a power station, sawmill, and cellulose factory in Jämtland in northern Sweden. He later became executive director of this establishment. Edy Velander[42] also studied at the *Massachusetts Institute of Technology* before he returned to Sweden in 1920, He was employed by the National Water Power Board, and later also was chairman of the Royal Swedish Academy of Engineering Science where he initiated the expansion of research and development.[43]

The *Technische Hochschule* in Dresden was another attractive German-language educational institute, and the second most frequented by the graduates in our cohort. Erland Thaulow[44] became a professor of mechanical technology at Copenhagen's Polytechnic Institute as well as manager of the technological laboratory at this university upon his return. Fredrik Carlson[45] went to Dresden to investigate and elaborate how calcium cyanamide could be produced cheaply. He concluded that a slight adding of fluorite to the carbide prior to adding the nitrogen was efficient and patented this method upon return to Sweden. He later headed the construction work of a carbide-factory near Sundsvall and became its manager.[46]

Architects also found their way to Dresden. Lambert Petterson[47] became municipal architect in Tampere and county architect in Hämeenlinna upon

---

38 KM, electrical, 1912.

39 *Danske teknika*, 120. Granholm was also a succcesful marathon runner; four times Danish champion, and a participant in the marathon at the Olympic Games in Stockholm in 1912.

40 TESM, construction, 1885.

41 TESM, construction, 1885. Årad Svensson Nobell's name is spelled with two l, whereas the Nobel brothers' names are spelled with one l.

42 KTH, electrical, 1916.

43 Fridlund, *Den gemensamma utvecklingen.*

44 PTI, mechanical, 1903.

45 KTH, mining, 1902.

46 Oscar Carlson, 'Om kalkkväfvetillverkningen. Föredrag vid Ljungaverkens invigning den 24 september 1912', *Svensk Kemisk Tidskrift* 24 (1912) 136.

47 SPO, architecture, 1885.

his return to Finland. During Tampere's 'building boom' he became important as the 'city's only technically educated civil servant'. Petterson was behind one of Tampere's most beautiful buildings, the Ruuskanen house by the Hämeen Bridge.[48] The building has connections to German neo-renaissance, taught in Dresden by, for instance, Carl Weissbach. Architect Gottfried Semper was active in the city before 1850 and in the 1870s. Dresden developed into a centre for an architectural course inspired by renaissance and baroque

Norwegians made up the second largest contingent of foreign students in Dresden, both at the civil engineering and the mechanical departments in the 1880s; only Swiss students were more numerous. One explanation for the Norwegian and Swiss dominance at the civil engineering department is that bridge and tunnel construction was important in those two countries.[49] Many of the Norwegian Dresden students worked in national or regional road administration as well as at the national railway company upon return. Some, like Olav Heggstad, made other careers.[50] He studied hydraulic engineering under the distinguished professor Hubert Engels and returned to participate in the building of Europe's then largest power station at Svelgfoss. Later, he served as professor of hydraulic engineering at the Norwegian Institute of Technology. Johan Kinck[51] returned from Dresden and started one of Norway's most prominent consulting firms in hydraulic engineering. Ragnvald Lie[52] was another important engineer in the same field. He specialised in power stations and was involved in most waterfall projections in Norway in the 1910s. The 1916 power station in Glomfjord in northern Norway is often seen as his 'masterpiece'.[53] Fredrik Selmer[54] practised in Germany and Austria before he continued to work one year in New York. Upon his return in 1906, he started a construction company and became a Norwegian pioneer for the use of reinforced concrete.[55]

48 Rolf Sonnemann, *Geschichte der Technischen Universität Dresden: 1828–1978* (Berlin 1978) 70, 75–76; Olaf Devik, *N.T.H. femti år: Norges tekniske høgskoles virksomhet 1910–1960* (Oslo 1960) 69; 'Dödsfall', *Tammerfors aftonblad*, 1 February 1928.

49 Sonnemann, *Geschichte der Technischen Universität Dresden*, 70, 75–76; Reiner Pommerin, *Geschichte der TU Dresden* (Köln 2003) 113.

50 TTL, construction, 1896.

51 TTL, construction, 1895.

52 KTS, mechanical, 1897.

53 Ivar Sekne, *Hundre års oppfinnsomhet. Historien om Multiconsult 1908–2008* (Oslo 2008) 5–9.

54 TTL, construction, 1896.

55 Jon Skeie, 'Fredrik Selmer—utdypning (Norsk biografisk leksikon)', https://nbl.snl.no/Fredrik_Selmer (15 November 2017).

The Swiss Polytechnic in Zurich was the third university in German-speaking Europe for the graduates of our cohort and the most popular German-language university for Swedish and Danish graduates. The school accounted for 30 per cent of the Swedish graduates who studied in Germany, Switzerland, and Austria, 21 per cent of the Danish, 11 per cent of the Norwegian, and 10 per cent of the Finnish. We shall, however, point out that these percentages do not give a true picture of the distribution of Nordic students at these institutes since students taking their entire education abroad are not included. Norwegians constituted, for example, half of the Nordic students in Zurich and were the largest foreign nationality. Swedes made up one-fourth of Zurich's Nordic students.[56]

The Swiss Polytechnic's emphasis was on the practical sides of the engineering profession, alongside with its broad curriculum and ambitious goals. Once it was established in the 1850s, 'word spread throughout the Nordic countries that this new technical institution possessed some outstanding qualities'.[57] The ambitious international orientation was reflected in a bulletin written for the Paris fair in 1889. The Zurich-based Polytechnic was not only to be an institute for Switzerland, but 'it should and will provide a place of education for the whole world'.[58] From the beginning, the school attracted foreigners and there were certain periods when Swiss students even constituted minorities. From the late 1880s, however, the number of Swiss students and teachers grew faster than the foreign numbers. In terms of enrolment-per-inhabitant, the Nordic countries were strongly represented at the school, even if real numbers were higher from neighbouring countries such as Germany and Austria and at times also from Russia. The Nordic peak year was, following Myllyntaus, 1904, when 53 Scandinavian and Finnish students enrolled.[59]

According to Myllyntaus, there were some differences in subjects studied between students from the Nordic countries, primarily in the nineteenth century. Norwegians focused on civil engineering, Swedes concentrated on mechanical engineering, Finns were divided between these two disciplines. Chemistry was studied more by Norwegians and Swedes than Danes and Finns. The latter, constituting the smallest Nordic student group, were nevertheless more dispersed between various disciplines compared to the Scandinavian nationalities.[60] These patterns diverge from the observations made here, which

56 Myllyntaus, 'Discovering Switzerland', 317.
57 Myllyntaus, 'Discovering Switzerland', 305.
58 Myllyntaus, 'Discovering Switzerland', 306.
59 Myllyntaus, 'Discovering Switzerland', 306–307, 314.
60 Myllyntaus, 'Discovering Switzerland', 318–319.

can be explained by this study's sole focus on graduates with a degree from the Nordic countries and that Myllyntaus's study starts at an earlier date. The observations in this study reveal that civil and construction engineers often came from Denmark and Norway, whereas chemical engineers arrived from Finland and Sweden.

The Swiss Polytechnic was a pioneer in teaching electro-technology, and Aage Rørbye Angelo[61] specialised in this subject, worked in Berlin, and upon return, became central in Denmark's switch to alternating current.[62] Kolbjørn Heje[63] became executive engineer of the Norwegian state railways, was appointed Norway's professor of road and railway building, and served as principal of the Norwegian Institute of Technology between 1917 and 1920.[64] Gustav Jebsen[65] studied at four universities in continental Europe before he earned a doctoral degree in Zurich in 1905. Upon his return in 1906, he became a pioneer in the Norwegian electro-chemical and electro-metallurgic industry and made use of some electro-chemical innovations of the time.[66] Guss Mattson[67] also studied chemistry. He returned to teach chemical technology and electrochemistry at Helsinki Polytechnic but became most known as one of the best publicists in the Nordic countries.[68]

Among the Swedish students in Zurich, some became connected to J. Sigfrid Edström's[69] ASEA. Edström was spurred to specialise in electrotechnology while studying mechanical engineering at Chalmers. Myllyntaus states that Edström 'made an outstanding career in the electrical engineering industry, rising to the post of executive manager of ASEA, the leading Swedish manufacturer of electrical equipment'.[70] He followed a common Chalmers pattern and continued his studies in Zurich before he continued to the United States. Danish-born Jens Lassen la Cour had his entire education from Zurich and graduated in 1899, but had been employed in the electrical

61 PTI, construction, 1899.

62 Harnow, *Den danske ingeniørs historie*, 219.

63 TTL, construction, 1891.

64 'Kolbjørn Heje—Store norske leksikon.', https://snl.no/Kolbj%C3%B8rn_Heje, (17 November 2017); Ingrid J. Brissach, 'NTH s første veiproffesor. Kolbjørn Heje var som sin forgjenger Sem Sæland opptatt av «kunnskapens praktiske anvendelse» og hadde selv mye praktisk erfaring', *Adresseavisa*, 13 November 2007.

65 BTS, mechanical, 1901.

66 Gunnar Nerheim, 'G Jebsen—utdypning (Norsk biografisk leksikon).', https://nbl.snl.no/G_Jebsen (17 November 2017).

67 SPO, chemical, 1895.

68 http://www.uppslagsverket.fi/sv/sok/view-103684-MattssonGuss, (17 November 2017).

69 CTI, construction, 1891.

70 Myllyntaus, 'Discovering Switzerland', 316.

industry in Switzerland and Germany as well as executive engineer at the heavily mechanised electrical workshop Bruce, Peebles & Co., in Edinburgh. Knowing la Cour from Zurich, Edström engaged him as technical director in 1907. La Cour rationalised and standardised a lot of ASEA's work and reconstructed several machine series in accordance with the latest technical development.[71]

The most famous Swedish student in Zurich is probably Gustaf Dalèn.[72] He studied in Zurich in the latter part of the 1890s, had various positions in Sweden before he became executive engineer for the industrial gas company AGA in 1906 and executive manager in 1909. He was awarded the Nobel Prize in Physics in 1912 for his invention of the sun valve that automatically switched the so-called Dalèn light on and off in a lighthouse, depending on the daylight; the light was produced by burning carbide gas. The innovation gave AGA a leading position in the world market.

Nordic students were, of course, also present at other German-language schools; Hannover (8 per cent), Darmstadt (7 per cent), Munich (5 per cent), Karlsruhe (5 per cent), Aachen (2 per cent), and Freiberg (1 per cent) were the most common ones in our cohort. Hannover and Munich were important for graduates from Norway and Finland. Darmstadt had a strong appeal on graduates from Norway, but they shared this school with colleagues from Sweden rather than technicians from Finland. As for Norway, our results correspond fairly well with Nerheim's statistics on 458 members of the Norwegian Society for Civil Engineers in 1916; the calculation is dominated by Dresden, Charlottenburg, Hannover, Zurich, Munich, and Darmstadt.[73]

The technical university in Munich held an early position in electrotechnology but was also central for graduates who wanted to deepen in Art Nouveau architecture. Albin Brag[74] returned from Munich to design several buildings, mainly around Stockholm.[75] In Darmstadt, Erasmus Kittler's professorship in electro-technology was the world's first and marked a position the university would keep for a long period.[76] Several Swedish and Norwegian students went there until the outbreak of World War I, and most of them worked in the electrotechnical field upon their return; some in municipal authorities.

71 Grönberg, *Learning and Returning*, 129–130.

72 CTI, mechanical, 1896.

73 Nerheim, 'Patterns of Technological Development in Norway', 81–82.

74 KTH, architect, 1902.

75 Malin Hollberg, 'Storängen—samhällsplanering under sekelskiftet', in *Nackaboken*, vol. 36 (Nacka 2000) 69–96.

76 Nerheim, 'Tysklands rolle', 120.

However, far from every graduate found their way to the Technische Hochschulen. The intermediate school in Mittweida in Saxony attracted many engineers from Denmark and functioned as a model for the technical school in Odense. Ordinary universities also attracted some engineers, like Alfred Lehmann,[77] who studied psychophysics with Professor Wundt in Leipzig in the 1880s and returned to start a laboratory with support from the government. Lehmann also became a long-time manager of the psychophysical laboratory at the University of Copenhagen.[78] Spinning and weaving schools like the renowned one in Reutlingen were other frequented institutions, especially among technicians from Finland. Magnus Lavonius[79] studied in Reutlingen and became weaving manager for the Tampere linen and iron manufacturer, later manager for the cotton industry Tampeeren Puuvillateollisuus and an important person in the early twentieth-century Finnish textile industry. Suomen Trikootehdas became one of largest tricot factories in northern Europe in the early twentieth century. The manager, Harald Jensen,[80] studied German and British textile industries.[81] Graduates who wanted a deepened knowledge of brewing often went to the Weihenstephan brewery in Bavaria and the appurtenant agricultural school. This school became known for educating some of the world's best brewers and attracted interest in many countries. Studies of centrifuges and fermentative processes were common experiences for Nordic brewery owners, and master brewer like Erik Olson,[82] who was, for example, appointed executive engineer at one of Stockholm's larger breweries and manager of J. A. Pripp in Gothenburg, later growing into Sweden's largest brewery. *Carlsberg* in Copenhagen also constantly employed engineers who had studied in Bavaria.[83]

77 PL, chemical, 1882.

78 Jespersen, *Biografiske Oplysninger,* 96.

79 SPO, mechanical, 1893.

80 SPO, mechanical, 1905.

81 Elias Lodenius, *Tammerfors linne och jern-manufaktur aktie-bolag 1856–1906: historik öfver Tammerfors linnefabrik och bolagets öfriga värk* (S.l., 1908) 86; 'Magnus Lavonius 60 år', *Svenska pressen,* 4 November 1930; 'Harald Jensen 60 år', *Tammerfors aftonblad,* 1 July 1942.

82 KTH, chemical, 1910.

83 Other examples include: Axel Bergh, founder of a labortory in Stockholm and manager of a brewery in Malmö; Napoleon Orell, manager of breweries in southern Finland; Einar Vendelbo, manager *Nora* brewery, Kristiania, see *Malmö teknologförbund,* 122; Indebetou and Hylander, *Svenska teknologföreningen*, 814; Voigt and Jespersen, *Biografiske Oplysninger 1829–1929*, 129–130; Gårdlund, *Industrialismens samhälle*, 254–256.

1.2 *Employment and Practise in the German-Speaking Countries*

Germany's chemical and electrical industries were especially important for Nordic engineer employment in the country. Just as for the university students, Berlin was the most popular destination for those seeking to learn from a work environment. More than every third employment in Germany, Austria, and Switzerland was in Berlin. Germany's capital had developed into an international centre for electro-technology with large corporations such as Siemens, AEG, and Union, whose rational factories attracted much of interest. The manufacture of large machines and electrical coil windings could, for example, be studied in the city.[84] This pattern was most pronounced for graduates from Sweden, perhaps because of its advanced electrical industry. However, Berlin dominated among technicians from all Nordic countries.

The two most popular German work regions outside of Berlin were—based on today's regional division—North Rhine-Westphalia and Hesse. They were part of the industrial heartland along the Rhine and in the Ruhr. As opposed to Berlin, this area attracted somewhat higher shares of graduates from Denmark, Finland, and Norway. Chemistry is behind the search for employment along the Rhine. Höchst, close to Frankfurt-am-Main, was one centre for the chemical industry. Several engineers also went to the heavily industrialised Silesia and Hamburg.

Rational and large-scale production and social relations between employers and workers were, of course, fields that to a large extent were connected to the United States and Frederick Winslow Taylor and Henry Ford. Corporate welfare programmes were widely dispersed in large American industries after the turn of the century. Edström was clear in his 1907 letter to a fellow Swedish engineer; America was the best place in the world to learn how to run a workshop cheaply and efficiently.

However, rational workshop organisation and more or less Taylorist and Fordist methods could be experienced not only in the United States, but also in Germany and a number of other countries. Bosse Sundin has pointed out that most ideas on rational and large-scale production in Sweden originated from Germany, at least before 1910.[85] We should nevertheless underline, in accordance with Ulrich Wengenroth, that late nineteenth- and early twentieth-century German industry was inspired by America. However, it was not simply

---

84 For example, Edström urged engineer Oscar Hellman, a man he was thinking of as head of an ASEA workshop, to study this in Berlin, see Grönberg, *Learning and Returning*, 158.

85 Bosse Sundin, *Ingenjörsvetenskapens tidevarv: Ingenjörsvetenskapsakademin, Pappersmassekontoret, Metallografiska institutet och den teknologiska forskningen i början av 1900-talet* (Umeå 1981) 86–87; Nilson, "Vacker, föredömlig, rationell", 56.

an imitation of American models: These ideas 'did change notably to become an integral part of something clearly different—the German manufacturing enterprise'.[86] Wengenroth refers to Joachim Radkan who claims:

> After 1870 a good part of the German history of technology can be delineated as a succession of thrusts of Americanization, but also of justification of German tendencies by American models, adaptation of American technology to German conditions and counteractions against this 'Americanization'.[87]

This is partly reflected in the travel report by a British engineer who was engaged by ASEA in Sweden after he had been working in the United States and managed the large 'American inspired' Bruce, Peebles & Co. workshop in Edinburgh. When he wrote to Edström from his study trip in Germany, he was impressed with some workshops' rational organisation and work environment, but he viewed American workshops as well as his 'own' Scottish one as more rational.[88] There are nevertheless examples of returned engineers from Germany modernising and rationalising production at mechanical workshops in, for example, Copenhagen, Kristiania, and Pori in Finland.[89] One interesting place to look for employment was the heavily mechanised Blohm & Voss shipyard in Hamburg. It had for example the world's largest shipbuilding area and floating dry dock. Sven-Olof Ahlstedt[90] became head of all state-owned shipyards in Finland after he had worked there and at Vulkan in Stettin. In Stettin and at the same company's shipyard near Bremerhaven, electric cranes were used for transportation, and the carpentry and metal work were remarkable. The company's success was based on its speed of production.[91]

---

86 Ulrich Wengenroth, 'Germany: Competition abroad—cooperation at home, 1870–1990', in: Alfred D. Chandler et al. (eds.), *Big business and the wealth of nations* (Cambridge 1997) 139–155.

87 Cf. Wengenroth, 'Germany', 140.

88 Grönberg, *Learning and Returning*, 145–148.

89 Even Jonassen at *Kværner Brug* in Kristiania and Viggo Bro at Smith, Mygind & Hüttemeier in Copenhagen, see Hugo Lagus, *W. Rosenlew & Co. Aktiebolag: 1853–1928* (Pori 1928) 224–225; 'Dir. John M. Gylphe 50 år', *Hangö*, 19300821; 'Femtioåringar', *Åbo underrättelser*, 22 August 1930; Alstad, *Trondhjemsteknikernes matrikel*, 165; Georg Brochmann, 'Kværner brug', *Teknisk ugeblad* 46 (1923) 303–305; Ole Hyldtoft, *Københavns industrialisering*, 316.

90 SPO, mechanical engineer, 1905.

91 Other examples of returnees from Vulkan include Sigurd Pauss, shipyard manager, Trondheim and Kristiania; Nils Ljungzell, teacher and professor of naval architecture, Royal Institute of Technology, editor for shipbuilding in *Teknisk Tidskrift*; Johan Löfgren head of

German factories and workshops were generally described in more positive terms in the Nordic countries than in the accounts of the British engineer writing to Edström. They were well-built, practically organised with well-timed machines, and primarily aimed at improving the quality of their products. The workers were often more well-educated, mainly because specialisation was not carried as far in Germany as in America. It was really the picture of American factories that was more divided; they were viewed from time to time as somewhat standardised and stereotyped and directed only towards labour-saving and automation. One of ASEA's workshops in Västerås was inspired by the turbine factory in Berlin, drawn by Allgemeine Elektrizitäts-Gesellschaft's (AEG) head architect Peter Behrens. AEG and other German engineering companies were, however, inspired by American workshops, since AEG's manager Emil Rathenau had travelled in the United States. However, the workshops were, to a large extent, 'Germanised' by Behrens. Västerås's city architect Erik Hahr was involved in the erection of the workshop and had visited Berlin. Another participant was Carl Silvander.[92] He had worked in Berlin before Behrens was engaged.[93]

European companies were also interested in Taylor's and Ford's ideas. AEG employed a former co-worker of Taylor to introduce scientific management in 1914.[94] AEG and other German workshops also had American models earlier that were inspirational sources for several engineers.[95] A Norwegian engineer noted, for instance, many American machines and an American style of mass production in several German workshops and factories in 1897. The machines were either German-made after American patterns or imported. American machines had made a breakthrough, and Germany had acknowledged America's leadership in this field. Several factories, Ludwig Loewe, for example, had introduced American work methods. Loewe is described by Wengenroth

---

draftsman's office at the Karlskrona navy yard in southeast Sweden; Åge Lønberg-Holm manager of a shipyard near Copenhagen, see Bodman, *Chalmers tekniska institut,* 229, 231; Indebetou and Hylander, *Svenska teknologföreningen*, 395, 569, 586, 630, 660, 706; Voigt and Jespersen, *Biografiske Oplysninger 1829–1929*, 252; *Malmö teknologförbund*, 189; *Suomen Insinöörejä ja arkkitehtejä*, 6; Eskedal, *BTS-matrikkelen*, 36; 'Sven-Olof Ahlstedt', *Åbo underrättelser*, 21 March 1951; H. E. Mohn, 'Notiser fra tyske verfter', *Teknisk ugeblad*, no. 39 (1905) 400–401; H. E. Mohn, 'Notiser fra tyske verfter. Bremer Vulkan, skibsbygning og maskinfabrik, Vegesack ved Bremen', *Teknisk ugeblad* 51 (1905) 529–530; 'Dödsfall.'

92 CTI, electrical, 1898.

93 Brunnström, *Den rationella fabriken*, 77, 86, 167–168, 173; Grönberg, *Learning and Returning*, 149; Nilson, 'Vacker, förebildlig, rationell', 59.

94 H. I. H., 'Taylor's System indført i Tyskland', *Ingeniøren* 40 (1914) 356; Robert Kanigel, *The one best way: Frederick Winslow Taylor and the enigma of efficiency* (New York, NY 1997).

95 Grönberg, *Learning and Returning*, 148–149.

as a 'translator' of American metalworking practise to German.[96] This Berlin workshop was also described as 'one from a technically, hygienically and social point of view modern mechanical workshop' in a 1902 Swedish lecture. It had washbasins, lockers, and several showers. The relation between employers and workers was characterised by respect and trust, and the workers were offered free education.[97]

Corporate social welfare has also been described as a characteristic of the Krupp company.[98] Specialising in steel, ammunition, and armament manufacturing, the Essen-based company grew to one of the largest corporations in early twentieth-century Europe. Krupp can be described as a German equivalent to Carnegie Steel, and its plant expanded enormously; from a workforce of 1,800 men in 1860 to about 41,500 employees in Essen alone by the outbreak of World War I.[99] Thomas J. Misa has revealed that Krupp attracted interest among representatives of Carnegie Steel and other larger American corporations as well as among French steel and iron makers.[100] Krupp was also described in Nordic technical journals. In 1908, for example, a Norwegian engineer told the readers of *Teknisk Ugeblad* about Krupp's new Friedrich Alfred Hütte near Duisburg. This 'enormous' new plant was described as magnificent. Seven blast furnaces guaranteed the production. Communications were excellent with several railways, its own harbour, different cranes, and modern transport facilities.[101]

Germany's engineering industry was, together with its chemical industry, leading the nation's transformation to an industrial superpower around 1900. Manufacture of machine tools was one area. American influenced, but 'Germanised' standardisation systems were dispersed in the early years of the twentieth century and created an industrial advantage. One trait was the manufacture of multi-purpose components that could be used as parts of different

---

96 Brunnström, *Den rationella fabriken*, 77; Karl E. Sundt, 'Af en stipendieindberetning fra tekniker Karl E. Sundt, vedrørende verktøimaskiner. 1897–98', *Teknisk ugeblad* 31 (1899) 347–350; W. Hoffstedt, 'Anordning och drift af en ur teknisk, hygiensk och social synpunkt modern mekanisk verkstad', *Teknisk tidskrift. Afdelningen för mekanik och elektroteknik* 5 (1902) 83; Wengenroth, 'Germany', 146.

97 Hoffstedt, 'Anordning och drift af en ur teknisk, hygiensk och social synpunkt modern mekanisk verkstad', 84–87.

98 Eugene C. McCreary, 'Social Welfare and Business: The Krupp Welfare Program, 1860–1914', *The Business History Review* 42, no. 1 (1968) 24–49.

99 McCreary, 'Social Welfare and Business', 25.

100 Thomas J. Misa, *A nation of steel: the making of modern America 1865–1925* (Baltimore, MD 1995) 109, 123.

101 V. B., 'Krupps nye Jern- og Staalverk ved Rhinen: Friedrich Alfred Hütte', *Teknisk ugeblad*, no. 14 (1908) 83–84.

products. German tools were flexible and yet capable of mass production, and machines were redesigned with the help of engineers from Germany's technical universities. Companies like Berlin's pioneer locomotive producer A. J. Borsig as well as Gebrüder Körting and Schwartzkopff belonged to the most advanced in Europe within heavy engineering, manufacturing a diverse range of products that, according to Wengenroth, were more or less unique. They attracted some engineers from the Nordic countries. Borsig was Europe's largest locomotive manufacturer and second in the world to Baldwin in Philadelphia. The company also became known for many other products and also for its welfare practises.[102] Some engineers, mostly in Sweden, had worked at Borsig. Joel Stern[103] served as a builder of locomotives between 1901 and 1904. Upon his return, he became a builder at a factory in Falun, manufacturing locomotives and railway equipment and headed the railway department at an engineering firm in Stockholm.[104] John Tuneld[105] returned from Borsig to head the technical department at one of Stockholm's largest mechanical workshops, Ludvigsberg.[106] Union was another Berlin-based company. It specialised in tramways and had made several around the world and was thus a valuable experience for tramway engineers. All Nordic capitals, for example, employed tramway engineers in responsible positions with experience from Union.[107] The Swiss Locomotive and Machine Works in Wintherthur was also a successful manufacturer. A few Finnish and Norwegian engineers returned to responsible positions from this company.[108]

Development in electricity was another reason to travel to Germany. There were, for example, positive reports in a 1903 Finnish lecture about experiments with high-speed electric railways near Berlin. Poul Gerlow[109] wrote admiringly

---

102 Chandler and Hikino, *Scale and scope*, 459.

103 CTI, mechanical-electrical, 1898.

104 Indebetou & Hylander, *Svenska Teknologföreningen*, 461.

105 TESM, mechanical, 1891.

106 *Malmö teknologförbund*, 150; Tuneld is, however, most known as industrialist in Saint Petersburg, and for his work for the Swedish community in that city, see Bengt Jangfeldt, *Svenska vägar till S:t Petersburg: kapitel ur historien om svenskarna vid Nevans stränder* (Stockholm 1998).

107 For example, Carl Åström, head of machines and traffic in Stockholm; tramway engineers Jacob Worm in Copenhagen, Gabriel Idström in Helsinki, and Jacob de Rytter Kielland in Kristiania, see Indebetou and Hylander, *Svenska teknologföreningen*, 400; Brochmann, *Vi fra NTH*, 264; Bassøe, *Ingeniørmatrikkelen*, 478; *Suomen Insinöörejä ja arkkitehtejä*, 127; Talvitie, *Suomalaisten teknikkojen*, 75; Hannover, *Dansk Civilingeniørstat 1942*, 227.

108 Karl Lilius, assistant manager, Finnish stone and paper industry; Hans Sørbye, mechanical workshops in Kristiania and later *Norsk Hydro*. *Malmö teknologförbund*, 120, 155; 'Ingeniör Karl Lilius död', *Hufvudstadsbladet*, 23 December 1935; Bassøe, *Ingeniørmatrikkelen*, 505..

109 PL, mechanical, 1905.

in *Ingeniøren* about the electricity works in the German capital in 1907. He later brought this experience to Copenhagen's counterpart. The same is true for Mikko Heikinheimo.[110] He returned to serve at Helsinki's municipal electricity works and became a professor of electro-technology.[111] Most of all, Nordic engineers admired German production of electrical machines. In 1914, it was twice as large as Britain's and almost equalled the American.[112] Unlike the chemical industry, machine building and other engineering industries were basically well-established and mature already in the late nineteenth century. Wengenroth underlines the American influences: 'If there was one industry in Imperial Germany where the adoption of American technology went hand in hand with the adoption of an American way of management, it was the electric power industry'.[113] Rolf Staubo[114] experienced production at Siemens and took over as manager of an electric factory in Skien upon his return to Norway. He turned a threating bankruptcy into profitability. In the 1930s, he started Staubo Apparatfabrikk in Oslo.[115] Manufacture of large machines and ways of winding were following Edström practices that an engineer should study in Berlin and at the larger American corporations.[116]

Germany's chemical success story was, of course, noticed in the Nordic countries. Chemical engineers were most likely to go to Germany, which is not surprising considering the country's leading position. Wengenroth claims that the worldwide success of the German chemical industry depended partly on the ability to turn academic research into marketable products quickly. Germany's early lead in new fields like pharmaceuticals and dyestuff was also important. In the early years of the twentieth century, Germany produced around 80 per cent of the synthetic dyestuff in the world. According to Wengenroth, the chemical industry in Germany was a unique branch in Germany as it did not apply American models; it saw itself as the model and was regarded as the model abroad. The chemical industry could utilise Germany's higher technical

110 SPO, mechanical, 1905.

111 Paul Gerlow, 'Berliner Elektricitäts-Werke. Af en Rejseberetning', *Ingeniøren*, no. 40 (1907) 279–287; Hugo Mäklin, 'Om de nyaste försöken i Tyskland röramde elektrisk drift å banor med stor hastighet', *Tekniska föreningens i Finland förhandlingar* (1904) 132–137. Heikinheimo also became professor of electro-technology, published extensively and his knowledge was utilised in several public committees.

112 Nipperdey, *Deutsche Geschichte*, 235–237.

113 Wengenroth, 'Germany', 149.

114 NTH, electrical, 1919.

115 Brochmann, *Vi fra NTH*, 263–264; Einar Hoffstad, *Merkantilt biografisk leksikon. Hvem er hvem i næringslivet?* (Oslo 1935) 714.

116 Grönberg, *Learning and Returning*, 148.

and scientific education; it was especially the high number of well-educated chemists that led to innovativeness. Wengenroth claims: 'Mass-produced research at all levels of the company rather than the mass production of staple goods was the strength and the strategy of the German chemical industry'.[117] This development attracted engineers who wanted to deepen their knowledge of dye-chemistry. Colour works like Witting's workplace in Höchst near Frankfurt-am-Main were admired in the Nordic countries and elsewhere. This 1899 description comes from Britain:

> One great advantage possessed by the German colour works, is in their command of abundant and cheap chemical skill and specially educated operatives. The Universities and Polytechnics of the country constantly turn out a throng of scientific chemists specially trained in the processes of chemical research and willing to work for wages that college graduates and the better class of operatives in England would refuse. The result is that the great aniline and chemical laboratories of Hoechst, Mainhur, Ludwigshafen, Elberfeld, and Berlin employ, in addition to their regular working force, a large staff of young chemists (from fifty to seventy at each establishment), whose sole function is that of research in the tireless quest for something new and valuable, whether it be a new colour, or pharmaceutical product, or a cheaper or more direct method of producing one of the products already known and in use. Mostly young graduates, fresh from the Universities, they work for small salaries in laboratories perfectly equipped by their employers, under contracts which provide that whatever valuable discovery they shall make be patented, and the patents transferred to the company, the inventor receiving a specified percentage of the profit accruing from its manufacture and sale.[118]

Ludwigshafen's Badische Anilin- und Soda-Fabrik (BASF) was a world-leading producer of dyestuff, investing a lot in research which led to the developing of a method for synthetic manufacturing in the 1890s. BASF was the world's largest chemical company and produced new colours that over time began replacing indigo. Several Nordic chemical engineers returned with experiences from Höchst, BASF, and nearby industries. Iivari Karhi[119] later managed a wool-spinning, dry-cleaning, and chemical dyeing company in Oulu. Wilhelm

117 Wengenroth, 'Germany', 143–145, quote from 145.
118 Cf. http://www.colorantshistory.org/GermanDyeIndustry.html, 14 June 2018.
119 SPO, chemical, 1891.

Kahrs[120] returned to start a dye- and clean-works in Bergen, whereas Edward Åstrand[121] was engaged to introduce bleaching and dyeing at a factory in Gothenburg.[122]

Dyestuff and pharmaceuticals were the cornerstones in the success of the German chemical industry.[123] However, there were some Nordic chemical engineers working in other fields that later returned with important knowledge. Birger Carlson[124] had a vital role in the development of a domestic Swedish electrochemical industry as executive engineer and manager of the important establishment near Avesta. He developed a special interest in electrical melting-furnaces and carbide at a gold- and silver-refining company in Frankfurt-am-Main. Upon return, Carlson improved the production of carbide, different kinds of chlorates, and explosives made of ammonium-chlorate.[125] Sugar refining also incorporated many influences from Germany as it was the leading country in processing and refining around the turn of 1900. Several sugar industries in Sweden and Finland saw a need to employ returned engineers from Germany in responsible positions.[126] Danish industries were also leading in sugar refining and had, perhaps, a smaller need to employ engineers from Germany. We can instead find one example from the paper industry. Jørgen Dreyer[127] practised in Brandenburg and Austria and became crucial in the rebuilding of a paper factory in Silkeborg and its re-orientation towards high-quality paper. Eugen Holtan's[128] studies in France and practise in Mannheim contributed to making him one of Norway's most important cellulose engineers.[129]

---

120 BTS, chemical, 1897.

121 KTH, chemical, 1893.

122 Two other Norwegian examples: Gunnar Brænne, owner of a company in Trondheim specialising in wool-spinning, dyeing, and chemical cleaning; A. A. O. Brathole dye-master, Nydalen, Kristiania, see '75-åring', *Åbo Underrättelser*, 10 November 1954; Heiniö, *Matrikel öfver Polytekniska institutets i Finland lärare och elever*, 162–163; Alstad, *Trondhjemsteknikernes matrikel*, 4–5; Bassøe, *Ingeniørmatrikkelen*, 67; Eskedal, *BTS-matrikkelen*, 17–18; Indebetou and Hylander, *Svenska teknologföreningen*, 383.

123 Wengenroth, 'Germany', 144–145.

124 KTH, chemical, 1896.

125 Indebetou and Hylander, *Svenska teknologföreningen*, 425.

126 Sweden: Anthony Grill, Alvar Fritsch, and Ivar Fogelberg also became managers for a sugar refinery near Halmstad; Finland's A. E. Airo (Salo), Toivo Hietanen (Kotka) and Erik Schröder (Töölö, Helsinki) also had expeienced German sugar works, see Gårdlund, *Industrialismens samhälle*, 252–254; Bodman, *Chalmers tekniska institut*, 6; Indebetou and Hylander, *Svenska teknologföreningen*, 261, 368, 430.

127 PL, chemical, 1887.

128 NTH, chemical, 1917.

129 Keld Dalsgaard Larsen, *Dansk papirindustri: mennesker, teknologi og produktion 1829–1999* (Silkeborg 2000) 21–23; Hannover, *Dansk Civilingeniørstat 1942*, 47; Bassøe,

German bridges were, as mentioned, viewed as reliable: their construction was detailed and solid. A nearby city like Hamburg offered interesting objects for study such as the unique iron-concrete pivot bridge over the harbour, built in 1904, and the tunnel under the river Elbe, built in the 1910s. Danish engineer and building contractor Rudolf Christiani[130] described the tunnel as a 'noble and enduring reminiscence of German engineering's high stage and great genius'.[131] The prominent construction company Ph. Holtzmann, recognised in 1902 for the Baghdad Railway, was in charge. Some Nordic engineers were employed with this company. Tor Kempe[132] returned to construct Europe's largest concrete bridge near Stockholm in 1915. Later, he supervised the construction of concrete bridges all over Sweden.[133]

As for architects, we have already revealed that the technical university in Dresden offered interesting studies in architecture, as did the technical universities in Munich and Charlottenburg. Art Nouveau architecture could be observed in the German-speaking countries, but also in other parts of continental Europe. In France and Britain, the style diverged somewhat from the German-Austrian one, which was characterised by dark and powerful colours as well as austere ornamentation. Flowers and other botanical themes were common elements in the buildings. Selim A. Lindqvist[134] became interested in German Art Nouveau when he worked in an architect's office in Berlin. He was one of Finland's foremost spokesmen of Art Nouveau, and his department store in Helsinki's Alexander Street constitutes one example. In Norway, several architects used the experience abroad in the rebuilding of Ålesund after it was largely destroyed in a 1904 fire. Their work resulted in the most distinctive Nordic Art Nouveau town.[135]

Ingeniørmatrikkelen, 220; Brochmann, *Vi fra* NTH, 153; Forsberg and Adlers, *Tekniska föreningen i Örebro*, 472.

130 PL, construction, 1900.

131 Rudolf Christiani, 'Elbtunnelen', *Ingeniøren*, no. 62 (1912) 508. Original in Danish: stolt og varigt Minde om tysk Ingeniørkunsts høje Stade og store Snille.

132 KTH, civil, 1902.

133 C. E. Holmberg, 'Om tyska och amerikanska brosystem', *Tekniska föreningens i Finland förhandlingar* (1894) 77–86; Heinrich Ohrt, 'Den store Jærnbane- og Gade-Drejebro over Oberhafen i Hamborg', *Ingeniøren*, no. 48 (1904) 320–321; Hannover, *Dansk Civilingeniørstat 1942*, 101; Indebetou and Hylander, *Svenska teknologföreningen*, 345, 624; Axel Björkman, 'Skurubron', *Teknisk tidskrift. Afdelningen för väg- och vattenbyggnadskonst*, no. 11 (1915) 120–133.

134 SPO, architect, 1888.

135 Helga Stave Tvinnereim, *Arkitektur i Ålesund 1904–1907: oppatbygginga av byen etter brannen 23. januar 1904* (Ålesund 1981); Indebetou and Hylander, *Svenska teknologföreningen*, 550; Hollberg, 'Storängen'; 'Selim A. Lindqvist, en bortglömd arkitekt', *Hufvudstadsbladet*, 20 May 1967; 'Sjuttioåring', *Hufvudstadsbladet*, 19 May 1937.

Bertel Jung[136] revealed his admiration for German-speaking architects in an article in *Tekniska Föreningen i Finlands förhandlingar* from 1901. The German-speaking countries constituted a culturally awakened part of the world where skilled architects realised new ideas. In Vienna, architect Otto Wagner developed his own style. It contrasted with the classic nineteenth-century Vienna architecture in its use of new materials, new forms, and its aim to proclaim that society was changing. Wagner looked for clearness, simplicity, and clean lines, and his buildings were intended to reflect their actual functions. He made many enemies; they criticised his architectural 'coldness' and what they interpreted as a neglect of traditions. However, he also had many admirers and Jung—describing Wagner as Europe's most exciting architect—was thus definitely one of them. Wagner was interested in city-planning and Jung found a major inspirational source in him. Jung was an obvious choice when Helsinki appointed a city-planning architect in 1908. Later, he had a similar appointment in Turku.

Jung's city-planning was characterised by clear views, broad-mindedness, and sound architectural judgements.[137] Wagner's ideas partly developed in a critique of his colleague and fellow countryman Camillo Sitte, who wanted a break with dullness and mechanic ideals in town-planning and move away from ideas that the street system was the base in a town-plan; streets and market squares were to be surrounded by walls of houses. The street system should be artistically planned, and one idea was the re-introduction of a medieval type narrow and picturesque crooked street. Sitte's ideas were influential long after his death.[138]

2 For Nordic Companies in Austria and Germany

Target mobility was thus overwhelmingly the most common pattern for the German-speaking areas in Europe. Labour-market, or traditional, emigration was next to absent. There were, however, some engineers going to Germany

136 SPO, architect, 1895.

137 Bertel Jung, 'Om nya rörelser på arkitekturens område hemma och i utlandet', *Tekniska föreningens i Finland förhandlingar* (1901) 65–66; Kuusanmäki, *Tietoa, taitoa, asiantuntemusta*, 177–182; 'Bertel Jung 50 år', *Hufvudstadsbladet*, 11 July 1922. Among Jung's most noticed buildings is the Rettig Palace by the Aura River in central Turku from 1928 and the seventy-meter high Hotel Torni in central Helsinki from 1931; the tallest building in Finland until the 1970s.

138 George R. Collins & Christiane Craseman Collins, *Camillo Sitte: the birth of modern city planning* (Mineola, NY 2006).

and Austria for domestic employers. They were almost exclusively from Denmark, which also contributes to weakening further the relative importance of the German-speaking countries as target migration destinations for Danes travelling on individual initiative. These graduates represented the two large companies in this context; F. L. Smidth and Christiani & Nielsen, but building constructors N. C. Monberg and engineering firm Nyboe & Nissen also sent engineers to the German-speaking world.

F. L. Smidth had several activities in Austria and Germany. In the early years of the twentieth century, the company decided to set up a department in the Austrian city Bielitz,[139] as production and other tasks grew at its cement factories in Austria. In our cohort, nine engineers served in Bielitz. Aage Schrøder[140] headed the department in Bielitz in 1902 and remained in that position until the department was closed and moved to Copenhagen in 1910. However, a production unit remained in Bielitz. Peter Carlsen[141] travelled to Bielitz in 1916 following his graduation and stayed until 1921 when he was reassigned to erect an F. L. Smidth factory in China. Individual engineers also worked for F. L. Smidth in Berlin and Lübeck.

Building contractors Christiani & Nielsen, specialising in iron-concrete, were also represented in Germany. An office was established in Hamburg in 1908, and at least eight engineers in our cohort served at the office in the city whose bridges, as mentioned, were described admiringly in Christiani's article in *Ingeniøren*. Herluf Forchhammer, mentioned in chapter one, was one of them. Almost all Christiani & Nielsen technicians stationed in Hamburg were civil and construction engineers. The engineering firm Nyboe & Nissen, which manufactured mechanical products,[142] also had some activity in Germany. Around 1910, this company placed engineers in Hamburg, Berlin, Mannheim, and Mülhausen. Emil Fleron[143] became the manager of Nyboe & Nissen's ironworks in Westphalia.[144] N. C. Monberg was founded by entrepreneur and later Danish minister of trade Niels Christensen Monberg and developed competence in harbour construction. The company built, for example, several ferry berths in Europe and was also involved in this business along the German Baltic Sea coast. Construction engineer Herbert Oscar Lorentzen[145] worked with extensions of the ferry berths in north German port cities such as Rostock.[146]

139 Today Bielsko-Biala in southern Poland.

140 KM, mechanical, 1896.

141 KM, mechanical, 1916.

142 Ole Hyldtoft, 'Perioden 1896–1930', 149.

143 PL construction, 1909.

144 Hannover, *Dansk Civilingeniørstat 1942*, 294.

145 PL construction, 1909.

146 Hannover, *Dansk Civilingeniørstat 1942*, 296–297.

3 Summary

German-speaking Europe, that is, Germany, Switzerland, and Austria, was the most common destination among Nordic technicians. Finnish and Norwegian graduates had a stronger inclination to choose the area compared to Swedish and Danish. The major explanation relates to the long-term lack of higher technical education in Norway and Finland. Technical education—represented to a large extent by German *Technische Hochschulen* and the Swiss Polytechnic in Zurich—was, however, not the only attraction in the German-speaking countries. Germany was viewed as a high-tech country whose rationality and industrial organisation were admirable. Switzerland, besides offering—perhaps—Europe's best quality technical education, was viewed as a country of technology, especially in a field such as power transmission, but there were also some major Swiss workshops. Austria had some of the most prominent architects and high-quality education in mining.

More than 90 per cent of the mobility to this region was bound for Germany, and Switzerland was a somewhat more common destination than Austria. The countries diverged to some extent, but all three were study-trip destinations. Chemical engineers were overrepresented in all three countries, civil and construction engineers in Switzerland, architects in Germany and Austria, and mining engineers and metallurgists in Austria.

Learning mobility to German-speaking Europe can be divided into three types: study trips, migration for enrolment at a technical university or lower technical school, and migration to learn from employment. Study travelling was, as mentioned, considerable to this nearby and interesting area. Employment migration was somewhat more common than enrolment migration on the pan-Nordic level. Finnish migrants and especially Norwegian were, however, primarily on the enrolment side. About every fifth graduate was both enrolled and employed, but only about every tenth from Denmark.

Academic mobility to German-speaking countries had a long historical tradition in the Nordic area, but late nineteenth- and early twentieth-century Nordic enrolments at technical universities were contemporaneous with significant student migration to Germany from many parts of the world. All student migration to the German-speaking countries was not directed to the technical universities; a minor share went to lower technical schools, ordinary universities, and other educational institutes such as weaving and brewing schools. The intermediate school in Mittweida was a model for similar schools in Denmark.

Germany's *Technische Hochschulen* and the Swiss Polytechnic were, however, most important. The most frequently visited German language technical university was in Charlottenburg near Berlin, followed by Dresden, Zurich, Hannover, Darmstadt, Munich, Karlsruhe, Aachen, and the mining university in Freiberg. The pattern corresponds to some previous studies, even if this study's exclusion of graduates who took their entire education outside the Nordic area may distort the picture to some extent. Zurich was the most frequented university for Swedes and Danes, while Finns and Norwegians preferred Charlottenburg.

German-language technical universities were pioneers or at least prominent in some fields: electro-technology, hydraulic engineering, chemical engineering, and Art Nouveau architecture. The Swiss Polytechnic became known for its focus on engineering's practices but also for giving a broader education than many other technical universities. Some returnees from the German-language technical universities became professors and even principals at the newly established technical universities in Helsinki and Trondheim, while others became electrotechnical pioneers or created beautiful buildings in, for example, neo-renaissance architecture. Gustaf Dalèn, the Swedish inventor of lighthouse technology and Nobel Prize winner, studied in Zurich. All these examples suggest that German-language technical universities and other schools were very important in technology transfer to the Nordic countries and particularly to less industrialised Norway and Finland, countries that also had more deficiencies in their technical education systems.

Nevertheless, practise, that is, employment, was also important, especially in the electrical and chemical industries. Berlin hosted several leading electrotechnical companies, which also were known for their rational factories as well as developed practises when it came to social relations between employers and workers. Some of these workshops were influenced by American ideas about efficient and rational workshop organisation. Nordic engineers did not necessarily need to cross the Atlantic to study ideas in the spirit of Taylor and Ford, even if many did. Some engineers returned and rationalised mechanical workshops. Some of ASEA's workshops in Västerås were clearly inspired by rational factories in Berlin and erected after extensive study travelling there. Union specialised in tramways and was a mutual experience for leading tramway engineers in all the Nordic capitals. Berlin was the most common destination for employment from all countries, but especially from Sweden, perhaps because the country hosted the most advanced electrical industry.

Berlin offered more of interest. Different mechanical workshops manufactured a range of unique products such as locomotives and belonged to Europe's leading engineering industries. Some leading engineers in workshops in places such as Stockholm and at the locomotive factory in Falun had practised in Berlin. The machine workshop in the Swiss city Wintherthur was another successful locomotive producer experienced by leading engineers at workshops in Helsinki and Kristiania. Shipyards in Hamburg, Bremerhaven, and Stettin were common experiences of many naval architects in the Nordic countries. The north German shipyards were known for their heavily mechanised arrangements, so were iron and steel plants in the Ruhr district. Besides being enormous, they also implemented far-reaching corporate social welfare practises.

Today's North Rhine-Westphalia and Hesse played important roles in what can be called Germany's chemical success story. Not very Americanised, the German chemical industry developed into a model for many other countries. Nordic engineers were destined for the leading colour works in Höchst near Frankfurt-am-Main. They were also going to the world-leading producer of dyestuff in Ludwigshafen. Returnees from the latter company started and had important positions in industries in Gothenburg, Bergen, and Oulu in northern Finland. There were also other chemical destinations in Germany; sugar plants and paperworks are two examples. Experience from Brandenburg and Austria was crucial to high-quality Danish paper-making. The Swedish electrochemical industry developed through improvements of electric melting-furnaces, carbide production, and chlorates, based on returnee experiences from German gold and silver refining.

Bridges, buildings, and architecture at large also attracted interest such as the unique iron-concrete pivot bridges in Hamburg. Frankfurt-am-Main-based construction company Holzmann was prominent in bridge construction, and an experience there led to the construction near Stockholm of Europe's largest iron-concrete bridge in the mid-1910s and later to bridge construction all over Sweden. Architects not only studied but were also employed in German-speaking Europe, and interest in art nouveau-style, neo-renaissance, and German-Austrian town-planning in the spirit of Wagner and Sitte was expressed in buildings and drawings and town plans for Helsinki and Turku.

A minor share of the mobility to German-speaking Europe suggests technology transfer in the other direction, that is, employment in Germany and Austria for domestic companies. This mobility embraced Danes, many of whom worked at F. L. Smidth's Austrian department and cement factory, but also in iron-concrete in Hamburg, and at an ironworks in Westphalia. N. C. Monberg employed engineers in harbour extensions along the German Baltic Sea coast such as Herbert Oscar Lorentzen in Rostock in the early 1910s.

Where did he work before he came to Rostock? The answer is Sassnitz, the port town with the train ferry connection to Sweden. Lorentzen participated in the preparation to make Sassnitz's port ready to receive the ferries. The same company also participated in the work to make Trelleborg's port ready for the ferries. Danish engineers thus contributed to making possible the connection that the Swedish king described as important for the country in his telegram to the chairman of Trelleborg's town council. In a sense, however, the ferry connection can be viewed as having been established too late to tie Sweden and the Nordic countries even closer to Germany than these nations already were. A few years ahead in time, World War I awaited, and Germany's loss of prestige made technicians look in other directions. France became a major destination for technical university studies in the interwar years. However, one distant country had already long been a focus of technicians' interest, and not the least in Sweden. It was well-known among the masses because people had moved there for more than half a century, but reports from international fairs in the late nineteenth century and other descriptions increased the interest for its technology. We are, of course, referring to the United States.

CHAPTER 5

Journeymen and Traditional Emigrants to North America

> Even as a little boy I used to stand on the heaths of Öland [Ölands Alvar], yearningly looking westward. In that direction lay the large, promising New World, America, where so many people from Öland had gone to seek their fortune and had been successful. There would surely be room for me too.[1]

∴

Naval architect Hugo Hammar[2] recalled in his memoirs his childhood back in Öland. The elliptical island along Sweden's southeast coast belonged to an area identified by Sten Carlsson as a 'coherent nucleus' of transatlantic emigration, that is, one that sent comparably many people to North America.[3] Hammar later went for secondary school in the nearby mainland town Kalmar, one of the ports from which emigrants departed. His wish to become an engineer brought him to Gothenburg.[4] Sweden's second city also hosted the country's main emigration port. Hammar could watch people leave aboard steamers for Hull, for a continued journey to America via the great Atlantic seaport of Liverpool. After an intermission at Palmer's shipyard in Jarrow near Newcastle-upon-Tyne, Hammar realised his childhood dream in 1890, when he reported for his first day of duty at a shipyard in Boston. His memoirs reveal parts of the duality of technicians' migration to America. Hammar emphasises that he was not an ordinary emigrant, but a completed engineer, who was about to reap the fruit of his studies.[5] This duality is also revealed in Kenneth O. Bjork's study of Norwegian engineers in America:

1 Cf. Olsson, 'To See how things were done in a big way', 441–442.
2 CTI, 1888.
3 Carlsson, 'Chronlogy and Composition', 133.
4 Hugo Hammar, *Minnen. 2, Som emigrant i U.S.A.* (Stockholm 1938) 11; Olsson, *Technology carriers*, 50–51.
5 Hammar, *Minnen. 2*, 11.

 | DOI:10.1163/9789004385207_006

> Unlike a large portion of those who left the country districts of Norway to take up land in the Middle West, the engineers burned no bridges behind them; in fact a majority had every intention of returning to the homeland after acquiring experience, perhaps a fortune, and possibly, too, a great reputation. They had no farms to sell and no families to care for. A ticket for the voyage to America, a few dollars to keep them going until they found a job, some articles of clothing—these with exceptions were all that they carried with them. In a short time they would return to visit parents and friends in Europe; a few years more and they would return to take over engineering posts in Norway.[6]

North America shows both differences and similarities with German-speaking Europe. Two differences are rather obvious. The continent lies significantly farther away than German-speaking Europe, which is a main reason why study travels without employment were relatively uncommon. North America was also a destination for technicians aiming to settle for good. The destination's duality thus lay in the combination between the journeyman-like learning mobility patterns characterising German-speaking Europe and patterns resembling traditional transatlantic emigration. However, learning mobility to North America also differed from the streams to Germany, Switzerland, and Austria. It was more one-sided. At least before the 1910s, it was rare to cross the Atlantic to enrol at technical universities. The main similarity between the two major destinations was the trait to practise, that is, to obtain employment and learn from it.

We have revealed how interest in, and admiration of, American technology grew among technicians in the Nordic area as well as other parts of Europe in the decades around 1900. As mentioned, Hughes described the United States as the world's leading industrial country by the turn of the century, in addition to the most innovative one. The United States attracted interest in a variety of fields: the rational, efficient, and cheap organisation of workshops, steel and ironworks, and other industries based on principles that at least loosely can be connected to Taylor and Ford, often combined with corporate social welfare programmes for the employees. America offered the opportunity to see large and modern productions, 'how things were done in a big way', to quote Swedish radio pioneer Ernst F. W. Alexanderson as well as Lars O. Olsson's article on Swedish naval architects in America. There were also interesting developments in road, railway, and bridge building, and, to some extent, architecture, even if

6 Bjork, *Saga in steel and concrete*, 35–36.

architects, as mentioned, had their main inspirational sources more nearby. Canada, too, offered interesting study objects. The southernmost tip of Ontario was part of the so-called 'North American manufacturing belt' as described by Harm J. De Blij and Peter O. Mueller.[7] This area near Toronto hosted Canadian subsidiaries of large American corporations. Canadian power stations also attracted; Niagara Falls offered interesting technological study objects on both sides of the border.

North America was visited by, roughly, somewhat more than every fifth Nordic technical school graduate and was the number two Nordic destination for total mobility. However, if we count only migration, North America was a more common destination than the German-speaking countries. The Swedish share was about the same as the pan-Nordic. Norwegian graduates visited North America to a somewhat greater extent, Finnish and Danish colleagues to a somewhat lesser extent. Nevertheless, if we look only at graduates going abroad, Swedish technicians choose North America to a somewhat greater extent than Norwegian, who, as mentioned, flocked to German-speaking Europe. The United States and Canada were also more frequented than German-speaking Europe among graduates from Denmark, but North America's Danish share was still rather low. In Finland, North America was the fifth destination after German-speaking Europe, the Nordic countries, Europe, and Russia, but the share of graduates going there was still somewhat higher than Denmark's.

Thus, North America was a likelier choice for Swedish graduates. Alongside the fact that Sweden belonged to the European countries that were most struck by transatlantic emigration in general, the pattern may reflect Swedish industrialisation that was more large-scale and mass-production directed. Swedish technicians might have perceived the technological and organisational gap between their country and the United States as narrower than technicians from the other countries. This explanation also relates to why the mechanical, electrical, and naval group, as well as mining engineers and metallurgists, were more likely to choose North America; mechanical workshops and steel and ironworks were often larger-scale and had more potential to be re-organised along American lines. Companies like General Electric in Schenectady in upstate New York, Westinghouse and Carnegie Steel in the Pittsburgh area, and shipyards in Virginia as well as near Philadelphia, Boston, and New York offered a lot to learn in these fields.

Rural-born graduates were also more likely to choose the United States and/or Canada; so were graduates who were natives of the countries they

7 Harm J. De Blij and Peter O. Muller, *Geography: regions and concepts* (New York, NY 1991) 196.

had studied in. The former might reflect Nordic transatlantic emigration's rural character in general, patterns that have been described by Kero, Østrem, and Akenson. The latter are easily explained by the fact that foreign-born graduates had more of a competing alternative in going back home. Technicians from a middle-class background also had more of a preference for North America compared to those from both higher and lower societal layers. Possibly, the middle-class was more inspired by values that might be connected to America, such as resistance to the autocratic privilege society as underlined by Kocka.[8] Lindblad suggested that privileged classes were inspired by continental Europe, primarily Germany, while America and the emigration was the 'people's project.[9] The working class's lack of resources may be one explanation of why graduates from lower social origin also were less prone to choose a distant country like the United States. Also in this context, crossing the Atlantic seems to be a kind of middle-class project, a pattern Wegge noted for mid-nineteenth-century emigration from Hesse-Cassel.[10] Upper-class-born graduates had other inspirational sources and a social capital that abated their need to go to America; working-class-born graduates lacked the family background, the contacts, and the social networks necessary to acquire resources for a longer journey; this pattern also resembles the one Fellman found in Finland. Engineers and architects of working-class origin might have been less able to utilise the financial resources required for a ticket and longer stays in North America. Their parents were probably less able to provide support, and the lack of influential family connections might have made them less successful in applying for grants and the like. Consequently, a period of unemployment in North America was probably more difficult to manage economically for a graduate from the lower ranks of society.

North America was also a likelier choice for those who left school after 1900, and especially in the first decade of the century, when the two countries had recovered and prospered after the recessions in the 1890s; which also was the decade when graduates were least likely to choose to cross the Atlantic. Age of graduation hardly mattered for North America, but it was an unlikelier choice for graduates who spent more than ten years in the domestic labour market before departing. This pattern reveals differences compared to the German-speaking countries, whose period of popularity lasted only three years. On the

8 Kocka, 'The Middle Classes in Europe', 785–786.

9 Lindblad, 'Emigrationen—folkets projekt', 26–38.

10 Wegge, 'Occupational self-selection', 378–389.

one hand, this shows the combination between target migration and traditional emigration in the North American pattern; the earlier revealed divergence in relation to the pattern to Germany and her linguistic neighbours. On the other hand, this also shows that North America, after all, was a 'learning destination'. Areas where the moves of Nordic technicians imply more a permanent migration and technology transfer in the outgoing direction, that is, from the Nordic countries, were generally likelier choices for graduates who departed a comparably long time after graduation. This is valid for overseas destinations outside North America, but to some extent also for intra-Nordic mobility and mobility to mainland Europe, except the German-speaking regions and Russia (see chapter 6). Another very clear difference from German-speaking Europe was that distance made study travelling to North America an unlikely mobility option. The long distance implied a need to invest a lot of time and money. Consequently, there was more often a need to earn the living through employment if a graduate wanted to see how things worked in America: studies had to be performed through migration. Swedes constituted, however, a minor exception as every fourth Swedish visit in the United States and Canada was made by a study traveller.

Table 5 reveals that overwhelming majorities of the graduates going to North America visited the United States. The ones who went to Canada were usually in both countries; only about every fortieth graduate to this region visited Canada without going to the United States. We can, however, note that about every tenth graduate in North America set their feet north of the border. Norwegians comprise, as we can see, the highest share. In general, too, a larger share of North America's Norwegian-born lived in Canada compared to the Swedish-born.

The mechanical, electrical, and naval group was, as mentioned, prone to choose North America, but were much underrepresented in Canada compared to both in the United States and among all transatlantic movers and travellers. It was the other way around for civil and construction engineers, as a group, and chemical engineers. These two groups were less prone to cross the Atlantic in general and were underrepresented in North America in relation to all mobile graduates, but overrepresented in Canada. Civil and construction engineers had interesting study objects such as the waterpower stations at Niagara Falls and in a town like Shawinigan on the border between Quebec and Ontario. Mining engineers and metallurgists were overrepresented in Canada compared to in the United States and among all mobile graduates. Architects were underrepresented in both countries; neither of them probably lived up to what Borgstedt called the 'eternal laws of beauty'.

TABLE 5 Number and percentage of Nordic technical school graduates, 1880–1919, migrating or study travelling to the United States and/or Canada before 1930

	United States		Canada		Total
	Number	%	Number	%	Number
Sweden	1210	98	94	8	1230
Denmark	431	96	51	11	450
Norway	737	97	98	13	763
Finland	183	99	21	11	185
NORDIC	2561	97	264	10	2628

SOURCES: SEE FIGURE 1.

1 Aspects of Nordic Technical Migration to North America

There were some engineers moving to North America as representatives on payrolls of Nordic companies, often for Denmark's F. L. Smidth and Sweden's SKF. The former company established a draughtsman's office in New York City in 1895 and erected a factory in nearby Elizabeth, on the New Jersey side, in the years thereabouts. These establishments help to explain why the region around New York received a larger share of technicians from Denmark than from the other Nordic countries. A total of 37 engineers in the Danish cohort are noted for employment with F. L. Smidth around New York. They had all specialisations and included, for example, Orla Larsen,[11] who stayed for more than twenty years, and Søren Valeur Jensen,[12] head of the draughtsman's office throughout the 1920s.[13]

SKF also established an office in New York, but only in 1909; two years after the company was founded. The number of engineers working for SKF in the United States was much lower than the number working for F. L. Smidth. Uno Forsberg[14] headed the construction and organisational work at the American factories in the 1910s, before he returned to become SKF's deputy manager.[15]

11 PTI, chemical, 1914.
12 KM, mechanical, 1917.
13 Hannover, *Dansk Civilingeniørstat 1942*, 207, 240, 269.
14 KTH, metallurgist, 1907.
15 Indebetou and Hylander, *Svenska teknologföreningen*, 687.

Later, SKF also established a research laboratory in Philadelphia which employed some technicians who later returned to the main factories in Gothenburg.[16]

However, the Nordic technician embarking a ship in North America was likely to be a target migrant, a person on a self-organised placement abroad. Bjork was not alone in his assertion that engineers often possessed different traits than usual emigrants.[17] When Swedish returnee technicians gathered in Gävle in 1901, one speaker claimed that technicians' migration to the United States rarely was to settle there forever, but to return with new experiences and to make use of them at home.[18] One tendency pointing to target migration is the readiness to move within the so-called 'North American manufacturing belt' rather than to traditional Nordic destinations. This belt was demarcated by Milwaukee in the northwest, St. Louis in the southwest, Baltimore and Washington in the southeast, Boston in the northeast, and stretched as mentioned into Canada as it included the southernmost tip of Ontario.[19] Hence, the destinations of the Nordic technicians lay in the same heartland of America's second industrial revolution that attracted the interwar German travellers Nolan has written about.[20] New York, Buffalo, Cleveland, Detroit, Philadelphia, Pittsburgh and Chicago, all within the belt, were also mentioned in a 1927 article in *Ingeniøren* as the 'best' places to go to acquire experience.[21] Schenectady, in upstate New York, with the General Electric headquarters, was another important centre; the power station at Niagara Falls was also interesting; Boston and Philadelphia hosted major shipyards and workshops, Pittsburgh attracted with large steel and ironworks and Westinghouse Electric; Detroit had an emerging automobile industry; and Allis Chalmers in Milwaukee was one of America's largest and most rationalised diversified manufacturers. Chicago's skyscraper era had a lot to offer for those who wanted to learn the construction business. We should point out here that a move within the 'belt' did not necessarily imply minor interaction with migrants of the same nationality. First, there were all kind of societies based on Scandinavian or Finnish origin also in places outside the areas known as 'Nordic'. Secondly, Chicago, Milwaukee, and their surroundings were both major Nordic settlement areas and parts of

---

16 Indebetou & Hylander, *Svenska Teknologföreningen*, 1002.

17 Bjork, *Saga in steel and concrete*, 35–36.

18 'Svensk-amerikanska ingeniörers och arkitekters möte i Gefle', *Teknisk Tidskrift. Allmänna afdelningen* (1901) 209.

19 De Blij and Muller, *Geography*, 196.

20 Nolan, *Visions of modernity*, 17–30.

21 Poul J. Jørgensen, 'Forhold og Muligheder for Teknikere i U. S. A', *Ingeniøren* 8 (1927) 100.

the 'manufacturing belt'. Nevertheless, Minnesota was, relatively speaking, the most 'Scandinavian' state, but located outside the 'belt'.[22] About 70 per cent of the Nordic graduates spent all their time within the 'belt borders'; 20 per cent lived both 'inside' and 'outside', and around 10 per cent only resided 'outside'. Finland and Sweden note somewhat higher shares for 'inside only', Norway and Denmark a little lower.[23]

As indicated above, we can still not entirely separate these technicians' crossings to America from general migration patterns. The connection to traditional transatlantic emigration is, for example, reflected in higher percentages from the 'high-emigration' countries Norway and Sweden. Migrating graduates born within Carlsson's 'coherent nucleus' of Swedish emigration[24] also choose North America to an extent of 52 per cent, whereas technicians born in the rest of the country are noted for 43 per cent.[25] Most of the graduates studied like Hugo Hammar in major port-cities of transatlantic emigration: Gothenburg, Copenhagen, Bergen, Trondheim, and Kristiania. If resources were available, it was only to pack the bag, get a ticket, and walk to the harbour to embark a ship.

Hugo Hammar's first place of residence, Boston, had a Scandinavian sentiment, but graduates could interact with even more countrymen in, for example, New York, Chicago, and Minneapolis. According to Bjork, Norwegian engineers could step right into 'a well-organized Norwegian life in any of the major northern cities'.[26] For the most part, this also held true for their Swedish colleagues. In Chicago, for example, The Swedish Engineers' Society was complemented by Swedish cultural associations and numerous provincial societies

---

22 There were a few exceptions as the significant number of Nordic immigrants around New York, relatively many Danes in California, and people from Sweden and Finland in Massachusetts. However, if we can speak about 'Danish' states in America, they were Iowa, Illinois, and Minnesota in 1920, and if we can talk about 'Finnish' ones we end up in northwest Michigan and Minnesota. It is, however, more relevant to speak of 'Norwegian' and 'Swedish' states. The two nationalities 'shared' Minnesota, and we can characterise Wisconsin and North Dakota as 'Norwegian' and Illinois as 'Swedish', see, for instance, Jouni Korkiasaari, *Suomalaiset maailmalla: Suomen siirtolaisuus ja ulkosuomalaiset entisajoista tähän päivään* (Turku 1989) 29; Torben Grøngaard Jeppesen, *Danske i USA 1850–2000: en demografisk, social og kulturgeografisk undersøgelse af de danske immigranter og deras efterkommere* (Odense 2005) 279.

23 Finland: inside only 75 per cent, both inside and outside 20 per cent, outside 5 per cent; Sweden: inside only 75 per cent, both inside and outside 15 per cent, outside only 10 per cent; Denmark: inside only 65 per cent, both inside and outside 20 per cent, outside only 15 per cent; Norway: inside only 65 per cent, both inside and outside 20 per cent, outside only 15 per cent.

24 Carlsson, 'Chronlogy and Composition', 133.

25 This connection is, however, not equally clear in the other countries.

26 Bjork, *Saga in steel and concrete*, 432.

covering nearly every inch of the old country.[27] Danish- and Finnish-American associations were not as numerous. The presence of 'national infrastructures', whose foundations were the arrivals of 'ordinary' emigrants, contributed to attracting technicians to North America.

1.1 *Experience and Learning in North America*

The following parts will focus on some aspects of learning among Nordic engineers and architects in North America. The survey cannot claim to be complete, but—hopefully—can give a picture of some of the fundamental features in 'American' learning. As we will see later, the United States was not the only place where technicians could learn many of these things. Nordic engineers and architects were not the only foreign technicians who travelled to North America to learn about rational organisation, efficiency, modern and larger constructions, new methods, and progressive building and architecture. Nolan and Braun have, for example, focused on German technicians, and these travels were features for technicians from several countries, but Nordic graduates are in focus for this study.

When a young Swedish engineer crossed the Atlantic with the purpose to study high-voltage engineering in 1907, he wrote to ASEA's manager, J. Sigfrid Edström, for advice. Edström, himself a returnee with many years of experience with General Electric and Westinghouse as well as from studies and employment in Switzerland, responded:

> To organise and run a workshop cheaply is a far more important problem today,[sic!] than to be able to build and run high-tension stations. The one who can run a workshop cheaply also gets money out of it and can claim a corresponding wage. As America is the best place in the world to learn this, I do not want to omit pointing out the big opportunity that is open for you here.[28]

The quote reveals what many technicians viewed as one of the main lessons in North America. Ideals connected to F.W. Taylor's principles of scientific management and Henry Ford's organisations could, as we revealed in the last

---

27 Grönberg, *Learning and Returning*, 77–80; Per Nordahl, 'Lost and found—a place to be: the organization of provincial societies in Chicago from the 1890s to 1933', in: Daniel Lindmark (ed.), *Swedishness reconsidered: three centuries of Swedish-American identities* (Umeå 1999) 65–89.

28 Cf. Grönberg, *Learning and Returning*, 145.

chapter, be studied also in other countries, but America was most interesting.[29] Nevertheless, we should underline that it was not an absolute necessity to go abroad to be inspired by this type of ideas.[30] Some engineers were also critical of Taylor and Ford and called into question the possibility of establishing their methods in the Nordic countries.[31]

Joakim Lehmkuhl[32] was inspired. Francis Sejersted calls him Norway's 'most articulate advocate of Taylorism', but Lehmkuhl also understood that scientific management faced difficulties in Scandinavia.[33] He continued his studies at the Massachusetts Institute of Technology in Cambridge, near Boston, in the late 1910s and tried to introduce scientific management when he started a workshop in Oslo in 1926. Sejersted argues, however, that this was possible only after he had gone back to the United States and started an industry in Connecticut in the 1940s. Lehmkuhl was also one of few in our cohort who crossed the North Atlantic to obtain university studies. The normal practise was, as

29 American rational workshops were admired long before Ford founded his factory in Detroit in 1903 and Taylor wrote his principles of scientific management in 1910. Taylor and Ford did not found the methods for modern factory work, but they were the ones who most clearly expressed them. The field of study was rather rationalisation in a wider sense. The 'American production system' was, as David A. Hounshell has stated, not limited to Taylor and Ford, but included a widespread use of interchangeable parts, modern materials such as steel, as well as the resources like electricity and oil. Production methods were, as for instance Haakon With Andersen and the co-authors of *Fabrikken* have stated, often developed earlier, but combined in new ways by Ford and Taylor; see David A. Hounshell, *From the American system to mass production, 1800–1932: the development of manufacturing technology in the United States* (Baltimore, MD 1984); Håkon With Andersen, et. al. *Fabrikken* (Oslo 2004) 118–119, 127.

30 One of the few persons described as having introduced an almost unchanged Taylor system in the Nordic area, Walter Engel, was a German born engineer with a Danish mother who worked in Germany and Switzerland before he came to the wireworks in Middelfart in 1910. Engel corresponded with Taylor but is not registered for any visits to the United States. Another Danish Taylor pioneer, Carl Erik Carlsen, who was responsible for the introduction of an adjusted system at F. L. Smidth, is not registered for employment and study trips abroad. He had begun at the factory as an apprentice when he was fourteen, see Indebetou and Hylander, *Svenska teknologföreningen*, 472–473; Ole Hyldtoft, 'Perioden 1896–1930', 177–180; Alf O. Johansson, *Arbetarrörelsen och taylorismen: Olofström 1895–1925: en studie av verkstadsindustrin och arbetets organisering* (Lund 1990) 94–99; Georg Brochmann, 'Verkstaden i Kristinehamn', *Teknisk ugeblad*, no. 36 (1923) 296–297.

31 Hans De Geer, *Rationaliseringsrörelsen i Sverige: effektivitetsidéer och socialt ansvar under mellankrigstiden* (Stockholm 1978) 354; Johansson, *Arbetarrörelsen och taylorismen*, 96–99; Michelsen, *Viides sääty*, 204–205; Francis Sejersted, *Demokratisk kapitalisme* (Oslo 1993) 160–161, 183.

32 BTS, mechanical, 1914.

33 Sejersted, *Demokratisk kapitalisme*, 160–161, 183.

mentioned, a self-organised placement abroad at a workplace, and technical studies at the university level became relatively common only towards the end of the period. The institute in Massachusetts was the most popular place to go for studies and Norwegians made up one fifth of the foreign enrolment and outnumbered students from nearby Canada in the early interwar years.[34] The University of Wisconsin was the second educational institute. Truls Wiel Wilson[35] studied in Wisconsin and worked together with Taylor to introduce the system in America. Upon his return, he started western Norway's first car firm in Bergen and imported machines and technical equipment.[36]

Lars O. Olsson argues that one reason behind the popularity of American shipyards was the interest in studying mass production. The template system was one inspirational source and provided a solution to Swedish labour shortage problems.[37] Ford's and Taylor's systems constituted examples of how mass production could be organised. They were subjects of several journal articles and lectures in the Nordic area. Taylor and Ford were described as genial revolutionaries, humanists, and models for a moral working life. Their systems were viewed as caring for the workforce, not as exploitive.[38] Ford's workforce was called 'one happy family' in a 1919 lecture in Finland.[39] In 1920, Jens F. Engberg[40] lectured about American industry in Copenhagen:

---

34 This may partly be due the fact that Quebec, the largely French-speaking part of Canada, is the region closest to Massachusetts.

35 BTS, mechanical, 1908.

36 Eskedal, *BTS-matrikkelen*, 41.

37 Olsson, 'To See how things were done in a big way', 450, 454. Olsson describes the template system as follows: 'templates were made from detailed drawings. The plates were prefabricated from the templates, and when completely finished they were brought to the building berth and assembled on the stocks. At a conventional shipyard, the plates were brought to the ship, marked out, and sent back to the workshop to be further worked'.

38 Some examples are: Edward Martin Eliassen, 'Modern Driftsledelse ved amerikanske Maskinverksteder', *Teknisk ugeblad*, no. 15, 16 (1913) 178–180, 182–183; Alex. Engblom, 'Rationell arbetsledning och dess praktiska tillämpning, Föredrag vid Svenska Ingenjörskongressen i Chicago 1915 af civilingenjör Alex. Engblom', *Teknisk tidskrift. Veckoupplagan*, no. 39, 40 (1916) 352–357, 362–366; Thorvald Eilertsen, 'Moderne amerikansk Værkstedsledelse (Scientific Management)', *Ingeniøren*, no. 50 (1913) 329–336; Walo von Greyerz, 'F. W. Taylor—revolutionären', *Teknisk tidskrift. Veckoupplagan*, no. 50 (1915) 510–513; Robert Lavonius, 'Pohjois-Amerikan konepajojen sisäisestä työnjärjestämisestä', *Teknillinen aikakauslehti*, no. 6–7 (1913) 131–135; C. Selim Westman, 'Om verkstadsmetoder hos Ford Motor C:o', *Tekniska föreningens i Finland förhandlingar*, no. 1 (1920) 37–42.

39 C. Selim Westman, 'Om välfärdsordningar inom den amerikanska storindustrin', *Tekniska föreningens i Finland förhandlingar*, no. 9 (1919) 165–175. Original in Swedish: en enda lycklig familj.

40 PTI, construction, 1915.

> It is not only Dr. Taylor's principles of scientific labour management one can learn over there, but also the Americans' inborn talent to do the things in an easy and quick way. 'Efficiency' is the slogan in an American factory, and the one engineer exceeds the other in thinking of things that make the production more economic and easy. The Americans possess the Germans' talent for system, but have received practical talent instead of the Germans' heavy thoroughness as a special gift from above.[41]

Engberg described the extensive use of modern tools, machines, and practical arrangements of the tool machines as well as the brilliantly conceived locations of the different workshop departments as a 'great science' and recommended that young engineers work in the production, 'where things are done'. The admiration for efficient tools and machines and how they were used to organise the production in a labour-saving way was a recurrent theme in Nordic reports about American technology since the mid-nineteenth century. Interchangeable parts impressed Nordic, and of course also other, visitors. The ways which companies like the tool maker Pratt & Whitney in Hartford, the J. P. Morris and William Sellers mechanical workshops in Philadelphia, the Baldwin locomotive works, and shipyards around the same city as well as in Boston, Allis-Chalmers in Milwaukee, the two electrical giants, and Carnegie Steel solved organisational problems through automatic machines and other devices were seen as excellent.[42] Engberg called American factories exemplarily when it came to organisation, and the different 'wheels' of the factory

41 J. F. Engberg, 'Danske Ingeniører i Amerika', *Ingeniøren*, no. 23 (1920) 189. Original in Danish: Det er ikke alene Dr. *Taylor's* Principper for videnskabelig Arbejdsledelse, man kan lære derovre, men ogsaa Amerikanernes medfødte Sans til at gøre Tingene paa en let og hurtig Maade. 'Efficiency' er Slagordet i en amerikansk fabrik, og den ene Ingeniør overgaar den anden i at hitte paa Ting, der gør Produktionen mer økonomisk og lettere. Amerikanerne besidder Tyskernes Sans for System, men har som særlig Naadegave faaet praktisk Sans i Stedet for Tyskernes tunge Grundighet.

42 Fagerberg, 'Det amerikanska systemets införande', 161–188; W. H., 'Meddelanden från verldsutställningen i Filaldelfia. 33. William Sellers & Co's verkstäder och utställning af verktygsmaskiner', *Teknisk tidskrift.* (1876) 12–15, 36–37, 87–88, 159–160; G. Sellergren, 'Nyare verktygsmaskiner på Chicago-utställningen', *Teknisk tidskrift. Afdelningen för mekanik och elektroteknik* 9 (1893) 16–18; G. Sellergren, 'Nyare verktygsmaskiner på Chicago-utställningen', *Teknisk tidskrift. Afdelningen för mekanik och elektroteknik* (1894) 29–32, 53–57; G. Sn., 'Moderna arbetsmetoder och verktygsmaskiner i Nordamerikas Förenta Stater. Ur en reseberättelse af ingenjör W. Fagerström', *Teknisk tidskrift. Afdelningen för mekanik* 1 (1915) 3–4; Grönberg, *Learning and Returning*; Lars O. Olsson, *Engineers as system builders: the rise of engineers to executive positions in Swedish shipbuilding and the industry's emergence as a large technological system, 1890–1940* (Göteborg 1995) 50–51.

worked in a 'marvellously safe and elastic' interconnection with each other. More skilful workers and mechanics did not explain industrial production, rather it was the use of advantageous tools and the workshop engineers' ability to arrange the work process practically.

Engberg introduced several ideas on rationality and organisation as works engineer at the Thomas B. Thrige automobile factory in Odense. There are several other examples of returned engineers who were inspired by American rationalism, new machinery, and organisation, for example, Knud Zimmer,[43] who managed the Rosenberg Mechanical Workshop in Stavanger upon his return from America.[44] Edström used 'American principles' to reorganise ASEA and employed engineers and foremen with experience in American mass production. Edström's roommate in Pittsburgh, Emil Lundqvist,[45] introduced it in ASEA's workshops.[46]

The Northeast United States and adjacent regions in Canada were important both for pulp and paper and textile technicians. The latter came almost exclusively from Finland and specialised in cotton-milling, wool-weaving, and yarn-milling. With experience from industrial centres like Lowell, Fall River, and New Bedford in Massachusetts, some of them returned to responsible positions in Finland.[47] Pulp and paper technicians arrived from Sweden,[48] but also within

43 KTS, mechanical, 1888.

44 For example: Nørregaard, 'Ingeniørernes indsats' 194–195; Hyldtoft, 'Perioden 1896–1930', 180; Eskedal, *BTS-matrikkelen*, 24, 41; Georg Brochmann, 'A/S Myrens Verksted', *Teknisk ukeblad*, no. 13 (1924) 114; K., 'Nordmand hædret i Amerika', *Teknisk ugeblad*, no. 36 (1920) 480; R. F. R, 'Planmæssig Arbeidsordning. Frederik Winslow Taylor. Kortfattet referat av advokat Ludvig Meyers foredrag i Polyteknisk forening den 14de november 1916', *Teknisk ugeblad*, no. 47 (1916) 527; Grönberg, *Learning and Returning*, 226–227; Walter Pollock, *The Bolinder book* (Stockholm 1930) 5. http://www.byhistoriskforening.org/AArringer-i-byhistorien/1900/1921, (19 November 2017).

45 CTI, 1893.

46 Another returned engineer was Gustaf Blume, who made ASEA's foundry 'devoted to mass production', see Grönberg, *Learning and Returning*, 116–122, 125–128, 150–151; Jos. Svartz and A. Norstrand, 'Westinghouse-kompaniets elektriska verkstäder i Pittsburgh', *Teknisk tidskrift. Afdelningen för mekanik och elektroteknik* 9 (1904) 151–156.

47 For example: Emil Simola, who was appointed professor in textile technology in 1919; Nikolai Uschanoff and Pentti Ignatius got prominent engineering positions in Tampere's clothing industry, see: 'Förteckning öfver i Amerika vistande finska tekniker inom textilbranschen', *Tekniska föreningens i Finland förhandlingar* (1914) 212.

48 Swedish examples of technicians who had been there are: B. M. C. Laurell was in Maine and became executive engineer for paper works in Örebro, Vargön, and Bengtsfors in central and southern Sweden from 1907; Arvid Hålldèn, was in the United States and at a sulphite factory in Quebec and became a leading engineer at a sulphate factory in Kramfors, northern Sweden. Thorild Folin worked in France and Quebec and became works engineer and executive engineer at the Bergvik sulphite factory in northern central Sweden

this field, this region appears as more important in Finland. Georg Holm[49] became acquainted with advanced apparatus in the so-called paper city, Holyoke, Massachusetts, and introduced pulp technology, machines, and modernisations at several Finnish paper works upon his return. Johannes Andersin[50] was another returned chemical engineer who introduced many improvements and became known as Finland's most skilful fine-paper technician.[51]

The American steel and iron industry as well as mining also attracted interest because of its rationalism, automation, dimensions, and mass production. Kenneth Warren and Joel Sabadasz have stated that Carnegie always was ready to equip their plants, primarily located in the Monongahela Valley near Pittsburgh, with the most modern and automatic equipment. Steel and iron plants in Pennsylvania were generally more efficient than the American steel industry as a whole.[52] Many engineers in responsible positions in the early twentieth-century Swedish steel and iron industry had experience with Carnegie and other American steelworks.[53] Eric Esselius[54] rationalised the Sandviken ironworks' rolling mills based on experiences from twelve years of employment at Carnegie Steel.[55] He contributed to the impression revealed in a 1925 letter: 'Sandviken is probably, when it comes to modern works and facilities, Sweden's America'. Albert von Julin[56] returned to modernise the Koskis ironworks in southern Finland.[57] Within mining companies in the Adirondack

from 1917, see Forsberg and Adlers, *Tekniska föreningen i Örebro*, 364, 385; Indebetou and Hylander, *Svenska teknologföreningen*, 655, 832; Jan Ringström, *Wäija sulfatfabrik och pappersbruk: en tillbakablick på 80 års papperstillverkning i Dynäs: AssiDomän* (Bjästa 1998) 31.

49 SPO, chemical, 1886.

50 SPO, chemical, 1897.

51 'Dödsfall', *Tammerfors aftonblad*, 1 February 1928; Michelsen, *Viides sääty*, 162, 175. John Johansson is another example, but he worked in New York. As technical manager, he performed many extensions and when he was engaged to erect, and later as technical manager, for a sulphite factory close to Tampere, he developed it into Finland's largest and most modern.

52 Kenneth Warren, *Big steel: the first century of the United States Steel Corporation, 1901–2001* (Pittsburgh, PA 2001) 83; Joel Sabadasz, 'The Mon Valley—Discovering the Genesis of the Modern American Steel Industry', *Cultural Resource Management* (*CRM*) 16:3 (1992) 28; William Sisson, 'A Revolution in Steel: Mass Production in Pennsylvania, 1867–1901', *The Journal of of the Society for Industrial Archeology* 18, no. 1 & 2 (1992) 92.

53 Grönberg, *Learning and Returning*, 172–175, 243–245.

54 CTI, mechanical, 1893.

55 Grönberg, *Learning and Returning*, 178–190, 226–227; Pollock, *The Bolinder book*, 5; Bengt Eiserman, *Minnen och episoder* (Stockholm 1937) 137.

56 SPO, chemical, 1900.

57 Wilhelm Haglund, *Levebröd: strövtåg i minnet och brukshistorien* (Stockholm 1978) 200; *Suomen Insinöörejä ja arkkitehtejä*, 637; 'Albert von Julin', *Åbo Underrättelser*, 14 April 1918.

Mountains in upstate New York as well as in Minnesota, mining underwent rapid technological developments in the early twentieth century and offered valuable experiences for men like Bertel Skjerdal,[58] who became manager at Dunderland in northern Norway, known for 'American methods'. Georg Fagerberg,[59] executive engineer and manager at LKAB in northernmost Sweden, utilised experiences from America when the mines became increasingly mechanised and modernised. Bror Andersson[60] had similar experience and was leading engineer at the Grängesberg mine in central Sweden, described as 'a piece of America' referring to its arrangements for transport and hauling.[61]

Another interesting trait to study was the organisation of industrial research laboratories at larger American corporations in the early twentieth century. Axel Appelberg[62] described this development in impressive terms, after having been employed in Schenectady between 1904 and 1908.

> The circumstance that laboratories of this kind have been erected not only at General Electric but also at several other large electrical and chemical companies shows that also in America one has learned to realise the importance to use scientifically as well as technically educated engineers for the elaboration of technical problems.[63]

Around 25 engineers and several helpers were employed exclusively to improve and work out new methods for electric power and electric furnaces. General Electric's employment of Willis R. Whitney in 1900 implied the creation of

---

58 UiK, mining, 1907.

59 KTH, mining, 1903.

60 KTH, mining, 1904.

61 Indebetou and Hylander, *Svenska teknologföreningen*, 31, 576, 616–617, 953; Skard, *USA i norsk historie*, 180; Ulf Eriksson, *Gruva och arbete: Kiirunavaara 1890–1990. Avsnitt 1, 1890–1920* (Uppsala 1991) 103, 126–138; Sven A. Anderson and Augustus Jones, 'Iron in the Adirondacks', *Economic geography* 21, no. 4 (1945) 278–279. Another example from a later period concerns Sven Schwartz, who had been in Kansas and came to Grängesberg as a mining engineer in the mid-1920s. Schwartz remained two years before leaving to work for a French company in Madagascar. Upon his second return in the 1930s, he became vice president for Boliden

62 CTI, chemical, 1900.

63 Axel Appelberg, 'Elektriska ugnar vid General Electric Co:s försökslaboratorium, Schenectady. Ur reseberättelse af fil. dr A. Appelberg', *Teknisk tidskrift. Afdelningen för kemi och bergsvetenskap* 1 (1911) 7. Original in Swedish: Den omständigheten, att laboratorier af detta slag upprättats ej endast vid General Electric, utan äfven vid flera andra andra stora elektriska och kemiska bolag, visar, att man äfven i Amerika lärt sig inse vikten af att vid utarbetande af tekniska problem använda sig af vetenskapligt såväl som tekniskt utbildade ingenjörer.

America's first industrial laboratory devoted to scientific research. Whitney carried out pioneer experiments in the industrial use of scientists; he used 'scientists as scientists—not as inventors, consultants, engineers, or testers' and made important inventions. Whitney's team developed 'vacuum tubes, tungsten filaments for better lighting, x-ray equipment, improved plastics for insulation, and metal alloys for better-performing equipment' within roughly ten years.[64] William D. Coolidge joined Whitney in the laboratory in 1905 and became the first to use tungsten in light bulbs and produced X-ray tubes that revolutionised medical science.[65] Karl-Erik Eriksson[66] followed the development for two years and became responsible for the establishment of a similar laboratory at ASEA upon his return in 1914.[67]

Taylorism, Fordism, and other rational systems in manufacturing organisations were often combined with ideas on social relations between employers and labourers. American 'welfarism' or 'welfare capitalism' impressed Engberg and many other engineers. Employers in the United States had started to acknowledge demands for a decent living standard and to emphasise increased productivity and profit as mutual interests.[68] Workers received material advantages, were integrated into the companies, and given limited influence. Programmes consisted of employment guarantees for strategically important groups, pension systems, dwellings, libraries, schools, kindergartens, healthcare institutions, cinemas, theatres, and sports activities, everything collected under the company roof. There were often also facilities inside the workplaces. All buildings at the shipyard in Newport News, Virginia, had, for instance, baths, toilets, and could be warmed up in the winter and chilled during the summer.[69] Rockefeller, General Electric's Gerald Swope, and Charles Schwab at Bethlehem Steel advocated 'welfare capitalism', and the ideas gained popularity. A course started at the Chicago Institute of Social Science in 1906, and 80 per cent of the American companies had welfare programmes in 1926.[70]

---

64 Alfred D. Chandler, 'The United States: engines of economic growth in the capital-intensive and knowlegde intensive industries', in: Franco Amatori et al. (eds.), *Big business and the wealth of nations* (Cambridge 1997) 75.

65 George Wise, *Willis R. Whitney, General Electric and the origins of U.S. industrial research* (New York, NY 1985); Leonard S. Reich, *The making of American industrial research: science and business at GE and Bell, 1876–1926* (Cambridge, MA 1985).

66 KTH, electrical, 1908.

67 Grönberg, *Learning and Returning*, 132.

68 Engberg, 'Danske Ingeniører i Amerika', 190; 'Svensk-amerikanska ingeniörers och arkitekters möte i Gefle.'

69 Torbj. Hermanrud, 'Newport News Shipbuilding and Dry Dock Co', *Teknisk ukeblad*, no. 47 (1915) 584.

70 Stuart D. Brandes, *American welfare capitalism, 1880–1940* (Chicago, IL 1976) 38–83.

Nordic industry had a lot to learn when it came to the evaluation of people in what Engberg called 'a democratic spirit'. Cooperation between management and workers was essential to meet post World War I industrial challenges, and many American employers had initiated profit-sharing systems. Jens F. Engberg had also witnessed how the workers elected a 'parliament', whose members accessed power at the shareholders' meetings. He called these ideas 'a path to a promised land', where workers and engineers worked together towards the perfection of technology.[71] Hugo Hammar was also influenced by welfare capitalism and established a spirit of cooperation at the Götaverken shipyard in Gothenburg where he became manager upon his return.[72] Oscar Falkman[73] made another experience. He witnessed how hot, hard, twelve-hour workdays wore workers down at Carnegie Steel's plants and stated, 'I got a solid opinion about the necessity of limiting the working-hours to eight hours per day'.[74] This experience led Falkman to the same conclusion as Engberg and Hammar. As manager for the north Swedish mining company Boliden, he became known for his cooperativeness.

All learning was, of course, not connected to rational organisation and labour relations. Ludvig Thorsen[75] returned to head the *Stavanger Steamship Company* and 'revolutionised' Norwegian coastal traffic with his introduction of 'American' car-decks.[76] Hammar also learned warship construction and was awarded the main responsibility in the shared Swedish shipyard project to construct the warship 'Sverige' in 1912.[77] General Electric, the United States' largest electrical manufacturer, offered, according to Bjork, 'excellent opportunities

71 J. F. Engberg, 'Social Udviking(sic!) i amerikansk Industri', *Ingeniøren*, no. 6 (1920) 32–34. We should note that many contemporary observers and researchers claim that American companies, like the National Cash Register Co., introduced these welfare institutions because they were forced to, not because they wanted to.

72 Jonny Hjelm, *Begåvningsreserven inom industrin: förslagsverksamhet i Sverige under 1900-talet* (Lund 1999) 239–240; Thommy Svensson, 'Japansk företagsledning och svenska bruk: en felande länk?', *Arkiv för studier i arbetarrörelsens historia*, no. 33 (1986) 31; Gösta Carlestam, *Samhällsbyggarna vid Storsjön: en plats i utkanten blir en stad i världen: en diskussion om stålstaden Sandviken—och—om mål och medel i samhällsplaneringens Sverige 1862–1984* (Gävle 1986) 378–379.

73 KTH, mining, 1900.

74 Cf. Grönberg, *Learning and Returning*, 203.

75 TTL, mechanical, 1904.

76 Per Surnevik, 'Ingeniør Ludvig Thorsen i DSD ble mannen bak en epoke i lokalfarten på norskekysten', *Stavanger Aftenblad*, 31 July 1971; 'Sjøbussens skaper er død', *Stavanger Aftenblad*, 22 January 1980.

77 Olsson, 'To See how things were done in a big way Swedish naval archtects in the United States, 1890–1915.', 436.

for employment and practical experience'.[78] The same held true for the other large American electrical manufacturer, Westinghouse. Allis-Chalmers was another renowned company.[79] Elov Englesson[80] was employed at the turbine department in Milwaukee in the 1910s. Upon return, he became executive engineer at a mechanical workshop in Kristinehamn in western Sweden and headed the manufacturing of the largest water turbines in the world.[81]

General Electric attracted engineers from Europe. Executive engineer and German-born Charles P. Steinmetz is described as a magnetic force who 'embodied the entire spectrum of electrical knowledge', even if he was specialised in alternating current and electric motors.[82] Edwin W. Rice Jr., who furthered the growth of long-distance transmissions and Elihu Thomson, inventor within arc lighting and alternating current, are other examples of attractive potential partners. Nordic General Electric engineers often came from Sweden, perhaps since the country had a relatively large-scale and developed electrical industry and maybe also because radio-pioneer Ernst F. W. Alexanderson[83] was a countryman.

The connections between General Electric and ASEA are clear and we have already mentioned Edström and Lundqvist.[84] In 1903, Edström and Carl Silvander agreed that Silvander was going to start work for ASEA within one year. Meanwhile, ASEA paid for a journey to Schenectady to search for a job at General Electric, and he was instructed to pay attention to all sorts of constructions. Silvander's stay in America was surrounded by a lot of secrecy. Edström carefully declared that the news of Silvander's engagement with ASEA was not to be spread before his return. He also instructed Silvander to say that he was going to go westwards to look for a new position when he was about to leave Schenectady as he did not want General Electric to know.[85] Silvander participated in the construction of the largest capacity generators ever and was in charge when ASEA later delivered the largest generators in Europe.[86] Gunnar

78 Bjork, *Saga in steel and concrete*, 442.
79 Chandler and Hikino, *Scale and scope*, 203.
80 CTI, mechanical, 1904.
81 Indebetou and Hylander, *Svenska teknologföreningen*, 738; Per-Olof Grönberg, 'Tillbaka till framtidslandet: ingenjörsmigrationen mellan Nordamerika och Norrland före 1940', *Oknytt* 21, no. 3–4 (2000) 46.
82 Kline, *Steinmetz*.
83 KTH, electrical, 1900.
84 Glete, ASEA *under hundra år*; Fridlund, *Den gemensamma utvecklingen*; Grönberg, *Learning and Returning*, 113–153.
85 Fridlund, *Den gemensamma utvecklingen*, 50–51; Grönberg, *Learning and Returning*, 128.
86 Edström and Silvander were far from the only returnees at ASEA: Harald Håkansson, responsible for patents; Ernst Danielson, developer of three-phase machines that made

Dahlbäck[87] became the first works engineer at the Porjus power station in Swedish Lapland and had also been in Schenectady. Hugo Mäklin[88] became the manager of the Helsinki electrical lightning company and a Finnish pioneer in high frequency and wireless telegraphy upon return from Schenectady.[89]

At Westinghouse in East Pittsburgh, engineers could closely follow the work of inventor George Westinghouse. He introduced alternating current in the United States. Its commercial application was also to a large extent due to Westinghouse's introduction of high-frequency systems that made alternating current useful for purposes other than incandescent lighting. A Swedish Westinghouse engineer was instructed by Edström to study commuting motors for one-phase alternating current, utilised at several early twentieth-century American electric railways. The engineer stated that Westinghouse's construction was superior to any other existing alternating current motor. To work close to Serbian-born engineer Nikola Tesla was also attractive; it was an important electrotechnical mark when Tesla's induction motor was brought out around 1890. Emil Lundqvist obtained employment through Tesla's concern and participated in drafts of Tesla's three-phase machines. Westinghouse also achieved world records when it came to the size of turbogenerators in the 1920s. Carl Richard Söderberg[90] brought this experience to ASEA's construction of turbogenerators.[91] Examples of Norwegian returnees from Westinghouse include

---

possible the first long distance three-phase transmission of electricity in Sweden; Ivan Öfwerholm, who modernised ASEA's oil-breakers and introduced uniformity in the constructions are more examples; see Fridlund, *Den gemensamma utvecklingen*; Grönberg, *Learning and Returning*, 113–153.

87 KTH, electrical, 1908.

88 SPO, mechanical, 1888.

89 There were several example of returnees from General Electric, in Denmark: Ove Holm, inspector of Ford's Danish factory, Copenhagen; in Finland: Kaarlo Ilmari Levanto, head of electricty dept., later manager, Outukumpu mine, eastern Finland; Hjalmar Svanström, town engineer Jyväskylä, manager, electrical companies in Loviisa and Vaasa; in Norway: Einar Schjølberg Henriksen, employed with *Norsk Hydro* at Rjukan power station; Johannes Rørvik, who came to work with railroads in the Norwegian capital; in Sweden (not ASEA) Göran Kärne, founder of electrotechnical bureau, Uppsala; Fritz Forsman, Robert Fredriksson, Thor Björnbom and Rudolf Bergman, heads of the municipal electricity works in Gävle, Umeå, Skellefteå and Luleå: see '60-åringar', *Åbo underrättelser*, 22 February 1937; 'Sextioåringar', *Hufvudstadsbladet*, 24 November 1955; *Matrikel öfver Tekniska realskolans* (Helsingfors, 1899) 271; Georg Christiernin, 'Elektroteknikens pionjärer i Finland', *Tekniska föreningens i Finland förhandlingar* 4 (1930) 83.

90 CTI, naval architect, 1919.

91 Grönberg, *Learning and Returning*, 130–131, 136. Söderberg later became professor at the Massachusetts Institute of Technology.

the head of Kværner Brug's turbine department in Kristiania and the works engineer at the Rjukan power station.[92]

The earlier-mentioned Axel Appelberg—later a founder of a wire lamp factory in Stockholm—was in America to study electric furnaces. General Electric developed several furnaces for experiments with high temperatures. At the plant in Harrison, New York, the furnaces were used to metallise filaments for electric lamps, a process that raised the filaments' efficiency.[93] Electric furnaces were also used in steelmaking and America offered solutions to the questions of how electric furnaces could be used in high-quality steel production. In 1912, *Teknisk Tidskrift* stated that electric steel was cheaper than crucible and was dispersed in America.[94] Eleven years later, a manager of a Swedish ironworks was surprised by electric steelmaking's low priority. It was however only a matter of time:

> We ought to scope from the large source of experience that the Americans have collected in the field. I am hereby aiming at their experiences of constructions and constructive details In this field, the Americans have put down enormous costs and have also been succeeding in working out types of furnaces, that hardly leave anything else to wish when it comes to running them. We cannot afford to experiment with new furnaces, something that is still going on to quite a large extent.[95]

Electric furnaces were, according to Thomas J. Misa, the 'last major production technology that American steelmakers adopted in advance of external competitive pressures'.[96] The fifteen-ton electric furnace at Illinois Steel's plant in South Chicago impressed an engineer who later became a lecturer at the Norwegian Institute of Technology. The most striking feature in the American steel boom after World War I was the development of electric steel. The Naval Ordnance Plant in South Charleston, West Virginia, was an interesting place to visit. The arrangement could be uplifted when the tapping began. This was a

---

92 Arne Ulstrup, factory manager, Høvik glassworks near Kristiania; Thomas Hjort, constructor, North Western Cyanamide Company, Odda, southwest Norway, had also been at Westinghouse, see Alstad, *Trondhjemsteknikernes matrikel*, 165, 199; Bassøe, *Ingeniørmatrikkelen*, 209; Brochmann, *Vi fra NTH*, 112; Eskedal, *BTS-matrikkelen*, 12.

93 Indebetou and Hylander, *Svenska teknologföreningen*, 545; Appelberg, 'Elektriska ugnar vid General Electric', 7–9.

94 'Elektroståltillverkning i Amerika', *Teknisk tidskrift. Afdelningen för kemi och bergsvetenskap*, no. 10 (1912) 136–137.

95 Cf. Grönberg, *Learning and Returning*, 191.

96 Misa, *A nation of steel*, 251.

feature in a construction made by a returnee engineer at the Sandviken iron works in the 1930s.[97]

Anton Martin Grønningsæter[98] experienced American and Canadian mining including an electrolytic refining process invented by the Swedish-American engineer Victor Hybinette.[99] He returned to become manager of the nickel refining plant in Kristiansand. Based on Hybinette's process, this plant on the southernmost tip of Norway developed into the largest for nickel refining in the Western world.[100] Anaconda in Montana, described as 'the copper El Dorado of the West' by Bjork,[101] attracted some Norwegian and Swedish technicians in our cohort. The impressive flame furnace was described as a prototype most copper works in western America used that was far from the Welsh original. The writer Paul Palén[102] also experienced the heated conflict on pollution in 1906 and 1907; farmers in the area filed complaints over destroyed land, water, crops, and livestock. The company's response was to try to disperse pollution in the atmosphere by constructing taller smokestacks: a less than successful strategy. Palén later continued to Globe, Arizona, and the world's largest smelting plant at the time. When Palén became head of the Rönnskär smelting plant in northern Sweden, he copied several of the ideas he had seen in America. Rönnskär became the largest Swedish smelting plant ever, with modern flame furnaces and the tallest smokestack in Europe. The company wanted, as it jestingly was written, to send the smoke to the old arch-enemy Russia, but poisoned land and water were soon also found around Skellefteå.[103]

Civil and construction engineers were, as mentioned, less attracted by North America, although a Finnish visitor to the Chicago fair claimed that American bridge building was ahead of European. Several remarkable railway bridges over the great rivers had been constructed.[104] Axel Björkman[105] returned to

97 Grönberg, *Learning and Returning*, 191–192.
98 TTL, chemical, 1900.
99 CTI, 1887.
100 Sandvik, *Falconbridge nikkelverk.*
101 Bjork, *Saga in steel and concrete*, 246.
102 KTH, mining, 1903.
103 Per-Olof Grönberg, "Skorsten kallar vi det rör som mot Ryssland röken för': om konstruktionen av Europas högsta skorsten vid Rönnskärsverken', *Polhem* 2 (2005) 25–42; Paul Palèn, 'Skärstensmältning i flamugn', *Teknisk tidskrift. Afdelningen för kemi och bergsvetenskap*, no. 4 (1908) 70–72.
104 C. E. Holmberg, 'Om tyska och amerikanska brosystem', *Tekniska föreningens i Finland förhandlingar* (1894) 77–98.
105 KTH, civil, 1893.

construct all bridges along the railway between Stockholm and Gothenburg, and his 1903 iron arch bridge was Sweden's first of its kind.

Companies based on foreign experience with reinforced concrete were a relatively common pattern in the Nordic countries.[106] Kreuger & Toll was founded on mechanical engineer Ivar Kreuger's[107] experience of Detroit entrepreneur Julius Kahn's technology in New York and Syracuse. Kreuger was enthusiastic; he claimed that reinforced concrete had proven fireproof and reduced costs, and he described the somewhat cautious Swedish attitude as 'hardly worthy a country that often views itself as the owner of the best engineers in the world'.[108] Kreuger's and Paul Toll's company quickly became profitable; work methods that enabled faster completions were impressive. Several of their buildings set Swedish time records, and the company was awarded the prestigious task of constructing the stadium for the 1912 Olympic Games.

The 1915 Stockholm warehouse for Nordiska Kompaniet was another of Kreuger & Toll's success stories. Nordiska Kompaniet's founder Josef Sachs was inspired by large American warehouses after a study trip he had made together with the famous architect Ferdinand Boberg.[109] Statistically speaking, architects were least prone to go to the continent that did not live up to the 'eternal laws of beauty', but some, nevertheless, went both back and forth. Thor Lagerroos[110] returned to design several buildings in Vaasa, Turku, and Helsinki. Eystein Michalsen[111] returned to participate in the exhibition *Nye Hjem* (New Homes). This 1920 Kristiania exhibition was an attempt to start serial furniture production by prominent designers and a breakthrough for progressive furniture with simplicity in the construction, modest décor, and better concordance between form and function. The furniture has been viewed as some of the country's best products in the 1920s. All Swedish returnee architects became more or less famous, and Boberg is the best example. He spent some

106 Examples are: Per Hallström was co-founder of a Sundsvall company erecting houses, etc. in primarily northern Sweden; Sigurd Bruun founded a Copenhagen company that took several patents; see *Vi bygger vidare: byggnadsaktiebolaget Hallström & Nisses verksamhet under ett kvartssekel: en kort berättelse i ord och bild* (Sundsvall 1955) 8–10; Indebetou and Hylander, *Svenska teknologföreningen*, 969–970; Hannover, *Dansk Civilingeniørstat 1942*, 115.

107 KTH, mechanical, 1899.

108 Cf. Lars-Erik Thunholm, *Ivar Kreuger* (Stockholm 1995) 29. Original in Swedish: knappast värdigt ett land, som gärna vill anse sig äga de bästa ingeniörer i världen.

109 KTH, architect, 1882.

110 SPO, architect, 1904.

111 NTH, architect, 1917.

months with the Chicago office of Adler & Sullivan, called America's most creative architects, after the 1893 fair.[112] Boberg's main inspiration was Henry Hobson Richardson's round-arch architecture. Stockholm's main post office from 1903 is one example.[113] Similar influences were present among other returned Swedish architects.[114]

2 Nordic Technicians as Traditional Transatlantic Emigrants

Three flags were hanging over the entrance of the Swedish Engineers' Society mansion in Wrightwood Avenue on Chicago's near north-side in the 1910s. If a society member looked up to the left, he saw the flag of his native country. This flag was to remind him of his background, his youth and his education in the country that had made the 'preparations' for his professional life. When he looked up in the middle, he saw the Star-Spangled Banner. It symbolised his present, his life as an active, industrious, and successful engineer in a country that made him what he was at that point. Looking up to the right, finally, he once again saw a Swedish flag. This flag symbolised a distant future when he was going to enjoy the autumn of his life back in his country of birth.[115] Some graduates aimed to settle for very long, sometimes forever. The earlier-mentioned

112 Staffan Nilsson, 'Brevborgen vid Vasagatan', in *SFV kulturvärden* 3 (2005) 31; Rygert, *Svenska arkitekter i USA*, 28–30.

113 This house was not only 'American' on the outside, but also on the inside: equipped with all embracing electrical lighting, sanitary arrangements, and elevators from the Stockholm company *Graham Brothers,* founded on experience from Otis Elevator Company. Another new-fangled thing was 'American post office boxes', the first of its kind in Sweden. In 1915, an extension making use of the whole quarter was performed with front drawings by Boberg. Victor Bodin headed the work. He had been eighteen years in New York and designed several buildings around Stockholm. Rygert claims that Bodin left few, if any, American traces in his post-return architecture. The only indication he discusses is a parish house in Stockholm, where a church and assembly hall were placed in the yard below street level. His work on the Stockholm Post Office meant that he at least worked *with* another American-inspired building, see Nilsson, 'Brevborgen vid Vasagatan', 31; Rygert, *Svenska arkitekter i USA*, 26; Indebetou and Hylander, *Svenska teknologföreningen*, 239.

114 Ture Stenberg, designing several residential houses in Stockholm and the Bank of Sweden building in Härnösand with corner towers, hipped and steep sloping roofs. One of Stenberg's buildings at Uppsala University resembles Richardson's Trinity Church in Boston. Stenberg's lighthouse on an island near Umeå has tarred shaving coverage, inspired by American East Coast houses and was unique for Sweden, see Indebetou and Hylander, *Svenska teknologföreningen*, 164; Rygert, *Svenska arkitekter i USA*, 29.

115 Grönberg, 'My kind of town?', 121–142.

Steinmetz had a clear opinion on where the future for European engineers lay. This is from an 1895 letter he wrote to Swedish engineering student Eskil Berg:

> Now the matter is this: come over at once, without waiting any more time in Europe. There is nothing to be got there, everything here. You better start with the idea, that you will never go back except for a visit I never saw a sensible man, who had lived a few years in the United States, willing to go back to Europe to stay.[116]

Eskil Berg followed Steinmetz's advice; he did not even finish his education at Chalmers, but departed immediately, reunited with his brother Ernst Berg[117] in Schenectady and stayed in America for the rest of his life. So far, we have primarily discussed learning and experiences of future returnees. Here, we will mention some of the Nordic technicians who made their entire or at least a large part of their careers in North America. Some became well-known personalities in their local communities and sometimes across America and internationally.

The Berg brothers, from the northern town Östersund, constitute one example. This is especially true about Ernst, who became a professor at Schenectady's Union University and won worldwide recognition as the author of a book projecting practical mathematical solutions of complex electrical matters and made a lot of research in voice transmission for radio and telephone. His namesake and countryman Alexanderson became, as indicated, even more famous within radio technology. He invented an alternator for high-frequency long-wave transmissions that made the 1906 broadcast of the world's first radio programme possible and made modulated radio broadcasting more practical; this was a key development in electrical engineering. General Electric exerted, as mentioned, the strongest attraction on Swedish technicians, but Norwegian Ludvig Severin Walle[118] became an important innovator of automatic power stations.[119]

Breslau[120] native Steinmetz was, of course, an even more momentous General Electric technician. The earlier-mentioned Serbian-born Westinghouse engineer Nikola Tesla had a huge impact on American electro-technology

116 Charles Proteus Steinmetz, Excerpt from a letter to Eskil Berg, 1895, General Electric Historical Achives, Schenectady, NY, USA.

117 KTH, mechanical, 1892.

118 BTS, mechanical, 1898.

119 Bjork, *Saga in steel and concrete*, 319; Eskedal, *BTS-matrikkelen*, 25.

120 Today the Polish city of Wroclaw.

with his induction motor. A small region like the Nordic area was not, of course, the most important contributor to technological development in North America, but there were still influential Nordic technicians in places other than Schenectady. The USS Monitor's builder, John Ericson, is, of course, one of the most well-known, but he also had a namesake[121] who was Chicago's city engineer from 1897 to 1928. Ericson modernised the water-supply system, constructed more than fifty bridges, and employed several Scandinavian engineers in the engineering office.[122] Illinois Steel's executive engineer Albin G. Witting[123] was also known as a skilled technician in Chicago's Swedish community.[124]

In 1913, the sign *SVEA, moder åt er alla* (SVEA, mother to all of you) wrote the following greeting to Swedish engineers in Chicago:

> In the dream, my thoughts hasten to the main stronghold of Swedishness in America, to Chicago, where the Swedish engineers possess all official positions of importance, where a Swede is minister of justice and a manifold of Swedes are presidents.[125]

The writer claimed that Swedish had crowded out other languages from Chicago's schools; that Swedish monuments were raised, and that Chicago's Swedes resided in magnificent palaces of gold and marble.[126] This was extremely exaggerated, but Nordic engineers exercised more influence around Chicago and other cities in the Midwest than elsewhere. The Swedish contribution to Chicago's development was cherished in a 1943 letter from the city major to the chairman of the Swedish Engineers Society:

> I have known your Society long and favourably and regards its membership as being representative of all our city's finest engineering talent. Among your membership during the past 35 years have been men who have brought international fame and attraction to the City of Chicago.[127]

121 KTH, civil, 1880.

122 Carlsson, 'Swedish engineers in Chicago', 185.

123 KTH, mining, 1896.

124 Allan Kastrup, *The Swedish heritage in America: the Swedish element in America and American-Swedish relations in their historical perspective* (Minneapolis, MN 1975) 384; Carlsson, 'Swedish engineers in Chicago', 187.

125 Cf. Grönberg, *Learning and Returning*, 79–80.

126 Cf. Grönberg, *Learning and Returning*, 80.

127 Cf. Grönberg, *Learning and Returning*,

Swedish technicians were not the only successful ones in Chicago. The Norwegian technical society has been described by Odd S. Lovoll:

> The members of the Norwegian Technical Society possessed skills much in demand as architects and engineers. The latter group comprised individual specializations such as civil, electrical, structural and mechanical. They found positions of responsibility as consultants and directors in the building of bridges, streets, skyscrapers, and apartment buildings.[128]

Isak Rasmussen Faleide[129] developed a special jack that still was in use in the 1970s. In 1916, he started the Folwell-Ahlskog Company for engineering and structural work and served as its president and chief engineer. Faleide had charge of design and supervision of grain-elevator construction as well as the construction of flour and feed mills, industrial plants, and bridges in concrete. Alfred Alsaker[130] was chief engineer with the Delta Star Electric Company and published *The Capitalist System and the Nature of Unemployment,* which, according to Bjork, was a unique study of unemployment based on mathematical equations.[131]

There are also examples of well-known technical school graduates from Finland in the Midwest. Architects were, as mentioned, less attracted by the United States, but one of the most famous Finnish-Americans still had that profession. Eliel Saarinen[132] was noticed for his second-placed competition entry for the Chicago Tribune Tower when he arrived to take up a professorship at the University of Michigan in the 1920s. He became, writes Reino Kero, an institution in American city building and planning.[133] Bruno Nordberg designed steam engines for power plants and waterworks as well as mining equipment. In the 1880s, he established a manufacturing company in Milwaukee, which became one the city's largest employers. Johannes Åström[134] advanced to become Nordberg's head builder and made inventions within air compression and water purification. In 1907, Åström founded a manufacturing company in Fort Wayne, Indiana, that quickly rose in importance.[135]

---

128 Odd S. Lovoll, *A century of urban life: the Norwegians in Chicago before 1930* (Northfield, MN 1988) 281, 284.

129 BTS, mechanical, 1902.

130 BTS, mechanical, 1904.

131 Bjork, *Saga in steel and concrete,* 441–442, 478–481; Lovoll, *A century of urban life,* 284.

132 SPO, architect, 1897.

133 Kero, *Suureen länteen,* 250–251; Laine, 'Americanism and Architecture in 1920s Helsinki', 28.

134 SPO, mechanical, 1896.

135 'Johannes I. Åström', *Hufvudstadsbladet,* 20 August 1933.

Minnesota was both the 'Norwegian' and the 'Swedish' state. Bjork stated that Minneapolis and St. Paul were among the major destinations for Norwegian engineers, while Carlsson concluded that the Twin Cities were less important for their Swedish colleagues.[136] Minnesota had a demand for civil and construction engineers, and this group constituted a larger share in Norway. Bjork writes the following:

> The Twin City area has a heavy Scandinavian population. It is therefore not surprising that many Norwegian engineers should have been attracted to this metropolis center, which after the dull 1870s experienced a rapid growth. They were employed both by the many railroads that serve the Northwest and by engineering departments in the two cities.[137]

Carl Illstrup spent fifty years with Minneapolis's city engineering office and modernised the city's sewer system as well as pumping stations and tunnels.[138] Kristoffer Olsen Oustad[139] had a major role as a bridge builder in the area, and Bjork describes him as an able engineer. Ole Ingenius Tolaas[140] became the long-time head architect for the Northern Pacific Railway Company. Stork Johan Bratager[141] was another Norwegian pioneer with this company, developing the railway for an increasingly heavy traffic. Bratager established distance schedules, performed the first valuation, and designed steel rollers and screws.[142]

New York, too, hosted Nordic technicians. Danish mechanical engineer Louis Birk[143] was one of the founders and technical manager for a foundry that made bronze ornaments for larger prestigious buildings. Finnish mechanical engineer Bruno Törnroth[144] won several prises for his sailing ship constructions, and the Swedish engineer David L. Lindqvist 'revolutionized the whole system of vertical transportation'. Lindqvist, executive engineer with the Otis Elevator Company, developed elevators that made structures like the Empire State Building practical.[145]

---

136 Bjork, *Saga in steel and concrete*; Carlsson, 'Swedish engineers in Chicago', 182.
137 Bjork, *Saga in steel and concrete*, 140.
138 Bjork, *Saga in steel and concrete*, 52–53.
139 TTL, contructional, 1880.
140 TTL, contructional, 1881.
141 BTS, 1880.
142 Bjork, *Saga in steel and concrete*, 55–56, 140–141.
143 KM, mechanical, 1906.
144 SPO, mechanical, 1900.
145 Nørregaard, 'Ingeniørernes indsats', 53; Jason Goodwin, *Otis: giving rise to the modern city* (Chicago, IL 2001) 127–132; Kastrup, *The Swedish heritage in America*, 552–553; T. R. H., 'Im memoriam. Ingenjören George A. Palmgren', *Hufvudstadsbladet*, 7 September 1927.

Bernt Berger[146] is described as 'one of the ablest Norwegian engineers in America', known for his ability to solve difficult structural problems connected to bridge building. He was recognised as a leading authority on steel construction and contributed, for instance, to Fifth Avenue's public library and the city's elevated railroads. Berger's countrymen Kort Berle[147] and Gunvald Aus[148] designed, for example, the Woolworth Building at 233 Broadway; the world's tallest building from its inauguration in 1913 until the early 1930s, and several other well-known buildings.[149]

We can also mention some examples from other areas in North America. Adolf Frederik Krabbe[150] from Denmark managed a cement factory in Bellingham, Washington; the only one of its kind in America, using a modern method for cement making. His countrymen Edwin Struckmann[151] and Holger Struckmann[152] were two of the founders of International Cement Corporation, and Holger became its president.[153] Finally, Norwegian Robert Kock Lepsøe,[154] represents an example from Canada. As the head of research at a mining and smelting company in Trail, British Columbia, Lepsøe developed new methods to produce zinc. He was awarded a medal from Canada's mineral industry.[155]

3 Summary

The United States and Canada constituted the second most common destination among all transnationally mobile technical school graduates in the Nordic countries, but the most frequented one for migrants. We have suggested the distance as one explanation of the low likelihood for study travel across the Atlantic, and the prosperity in the earliest years of the twentieth century for the high numbers crossing the Atlantic among graduates after 1900.

146 TTL, mechanical, 1885.

147 KTS, mechanical, 1886.

148 BTS, 1879.

149 Bjork, *Saga in steel and concrete*, 70; Mølmen, 'Kvinnelige pionerer', 70; Arne Odlaug, 'Ingeniør Bernt Berger. In Memoriam', *Teknisk ugeblad*, no. 5 (1919) 62; Arne Kildal, 'KORT BERLE. En av de mest fremtredende nordmenn i Amerika er død', *Nationen*, 7 May 1934.

150 KM, mechanical, 1899.

151 KM, mechanical, 1896.

152 KM, mechanical, 1902.

153 Nørregaard, 'Ingeniørernes indsats', 101–102.

154 NTH, chemical, 1916.

155 Brochmann, *Vi fra NTH*, 109–110; Eskedal, *BTS-matrikkelen*, 45. Lepsøe also had a chemical engineering degree from Bergen (1910) and returned to become professor of metallurgy at the Norwegian Institute of Technology in 1946.

North America was a ‘dual’ destination; ‘learning mobility’ was the major pattern, but the transatlantic streams also consisted of many engineers and architects who can be considered ordinary emigrants. Hammar‘s childhood memories indicate that technicians also were influenced by ‘America fever’ in general. Norway and Sweden were ‘high emigration’ countries, and their graduates also went to North America to a greater extent. The share of the total graduation going to North America was highest in Norway, but it still constituted a higher percentage of the transnationally mobile graduates from Sweden. In Finland, North America was only the fifth destination and the Nordic countries, other European destinations, and Russia also attracted higher numbers. The rural character of transatlantic migration in general could have had an impact also on rural-born technical school graduates. The dual destination character is also reflected in the longer popularity compared to the German-speaking countries; it remained an alternative also for graduates who departed relatively many years after leaving school.

Target migration or placements abroad was still the dominant pattern; the tendency to settle within the so-called manufacturing belt rather than in traditional Nordic-American areas like Minnesota is one sign. One major difference compared to German-speaking Europe was the fact that North America was not an enrolment destination in the same way; studying at American technical universities became common only in the interwar years, and the Norwegian enrolment at Massachusetts Institute of Technology constitutes one example. The attraction lay in the rapid technological development that made contemporary lecturers describe America in admirable terms upon returns from events such as the 1893 Chicago fair. Rational organisations like Taylor‘s and Ford‘s were of immense interest for some Nordic technicians and may explain why the mechanical, electrical, and naval group and mining engineers and metallurgists were more prone to choose North America. These two specialisations were supposed to serve at electric or other mechanical workshops, shipyards, mining industries, steel and ironworks and the like. We may assume that these kinds of industries were suitable for rational organisation in other ways than architect offices and so forth. General Electric, Westinghouse, Allis-Chalmers in Milwaukee, shipyards in Virginia and around Philadelphia and Boston as well as Carnegie Steel’s plants near Pittsburgh offered interesting study objects in these fields.

Sweden’s industrialisation was more large-scale than the neighbouring countries’. Therefore, ideals partly aiming at mass production might have been more relevant in a Swedish context. J. Sigfrid Edström and his reorganisations of the country’s largest electrotechnical company ASEA along American principles is one of the most obvious examples. Edström had nine years of experience with America's two electrotechnical giants. He employed a friend from Chalmers and Pittsburgh to reorganise the workshop and made it profitable after a short time. There were many technicians from the other countries that had Taylor, Ford, and American rationality as inspirational sources; Lehmkuhl, in Norway, Julin in Finland, and Engberg in Denmark are examples. They all tried to introduce American organisation in several places in the Nordic area. Most of the names above belonged to the mechanical, electrical, and naval group or were mining engineers and metallurgists. Some industrial arrangements in the Nordic area underwent reorganisations after they had employed engineers who had experienced rational organisation, modern machinery, and the like in the United States. Edström's ASEA is perhaps the most obvious example. However, Swedish and Norwegian shipyards, Danish automobile factories as well as steel and ironworks and mining industries were also influenced by this American style rationalisation. Chemical engineers were also interested in modern industrial settings like the Finnish technicians who went to pulp and paper centres in the northeast United States and returned to modernise and improved similar industries in Finland. To some extent, these rational organisations were also combined with ideas on corporate social welfare, which began to diffuse in larger American corporations in the early years of the twentieth century. Denmark's Engberg cherished the relations between management and workers in the United States including profit-sharing systems and a workers' 'parliament' that accessed some power at the shareholders' meetings. In Sweden, parts of these ideas were applied at for example shipyards, steel and ironworks, and mining companies.

North America was an area where Nordic technicians picked up other applicable ideas, which partly were behind actions like the introduction of car-ferries in western Norwegian coastal regions and manufacturing of large water turbines in Sweden. America's electrotechnical industry with its famous engineers such as Westinghouse, Tesla, Steinmetz and Swedish-born radio-pioneer Alexanderson attracted attention. Many engineers in different positions in the Nordic area were experienced in the American electrotechnical industry including the manager of Helsinki's electric lighting company and the founders and leaders of different electric businesses all around the Nordic area. America also offered solutions for the usage of electric furnaces in the production of high-quality steel. Types used in a West Virginia facility were 'imitated' by a returnee engineer at the Swedish ironworks Sandviken around 1930. An electrolytic refining process by a Swedish-American inventor was behind a returnee engineer's development of the nickel works in Kristiansand as one of the largest in the world. One returnee from Montana introduced flame furnaces and constructed Europe's tallest smokestack at a smelting-plant along the northern

Swedish coast. The purpose of the tall stack was to disperse pollution away from the immediate neighbourhood. In construction, we can observe some influence in reinforced concrete such as companies founded on experiences with Detroit entrepreneur Julius Kahn‘s technology. American bridges were sometimes described as remarkable and were models for bridges constructed by returnees. Architects were, as Borgstedt‘s account of the lack of respect for the *eternal laws of beauty* reveals, divided in their view of North American architecture and more directed towards Europe. Some architects returned however from the United States and designed buildings inspired by Henry Hobson Richardson‘s round-arch architecture, and there are evidences of interest and influence from American furniture architecture.

We cannot forget that some of these technicians did not view America as a short-term temporary place of residence but a permanent residence. The two Swedish flags that flanked an American flag above the entrance to Chicago‘s Swedish Engineers Society building symbolised left to right the past, the present and the future. However, the engineer's society mainly referred to a distant future, when the member was going to return to retire. Some Nordic immigrant technicians became well-known locally, nationally, and sometimes even globally. The Nordic area cannot claim to be more important than other countries and areas in this sense. However, Sweden, Denmark, Norway, and Finland not only *acquired* competence from North America; sometimes, these countries also were the givers. Development in radio technology, the revolution in elevator transport, innovations in bronze ornaments and cement-making, automatic power stations, water purification, air compression, many important railway and bridge builders, zinc producers, steel construction specialists, builders of sailing ships, and one of America's most influential architects in city building and city planning were immigrants from the Nordic countries.

Like the German-speaking countries, Nordic technicians also went to America, learned about technology and technical development, and returned to contribute to development back home. Some stayed and contributed to development in North America. However, this exchange was not limited to the two leading industrial regions of the world; it embraced many countries in all continents. We will continue our journey together with our technicians, now to destinations outside the two major ones.

CHAPTER 6

A Worldwide Labour Market

> They travel [...] because the market for their competence has since the mid nineteenth [sic!] century been world wide [sic!].[1]

We have already stated that the German-speaking countries and North America were far from the only destinations for Nordic technicians. The graduates in this cohort went to at least 93 countries and dependencies; including 21 in the Americas south of the United States and 39 in Africa, Asia, and Oceania. The graduates from the Norwegian schools went to 71 countries and dependencies, their Danish colleagues to 68, those who had studied in Sweden to 67, whereas Finnish technicians went to 46 countries.

In all, 4,165 graduates, about one-third of the total graduation and nearly two-thirds of the transnationally mobile technicians, went to a country outside the two main regions at least once. It was most common among the Finns of whom almost all mobile technicians visited at least one country outside Germany, Austria, Switzerland, the United States, and Canada. However, this mobility also embraced three-fourths of the mobile Danes, almost 60 per cent of the mobile Swedes, and somewhat below every second mobile Norwegian technician. This section will focus on mobility to these other destinations: the Nordic neighbour countries, the British Isles, Russia, other European destinations, Latin America and the Caribbean as well as Africa, Asia, and Oceania.

1 Intra-Nordic Studies and 'Expertise' Migration

According to the division in table 3.1, intra-Nordic mobility was the third pattern, embracing every fifth mobile technician, which is the same as every ninth graduate. Transnationally mobile Norwegian, Danish, and Swedish graduates lay between 12 and 18 per cent, which represented between 7 and 9 per cent of

1 Stang, 'A measure of relative development?', 93.

 | DOI:10.1163/9789004385207_007

the total graduation. Finland diverged significantly; the Nordic countries constituted the second destination after German-speaking Europe. Almost 60 per cent of the mobile Helsinki graduates visited a neighbouring Nordic country at least once, and this represents more than 40 per cent of the total graduation. Three-fourths of the Finnish mobility to neighbouring Nordic countries was study travelling; the Finnish travellers almost always had Sweden, and often also Denmark, on their itinerary, and about 70 per cent of them visited another Nordic country. Many Finnish graduates were also attracted by these two countries, not least the architects.

Norway notes the lowest Nordic shares: 12 per cent of the mobility and 7 per cent of the total graduation. The pattern can partly be explained by a very low number of students from the neighbouring countries at the Norwegian technical schools. Even if the schools in Trondheim, Bergen, and Kristiania educated engineers who performed work on a par with most colleagues from other countries, Norway was rarely the first choice of a future engineering graduate who wanted to go abroad for studies. Being born outside the country of education was another characteristic that was positively correlated with intra-Nordic mobility. This was a pattern primarily embracing Norwegian and Finnish students in Sweden and Icelandic students in Denmark.

1.1 *Architecture, Food, Industrialisation and Exchange of Expertise in Sweden and Denmark*

Finnish graduates contribute significantly to the pattern of Sweden and Denmark as study-trip destinations, even if Norwegian study travelling to Sweden also was significantly stronger than Norwegian study travelling in general. In Denmark, Swedish graduates were somewhat more represented than Norwegian. Both countries became more common destinations after the turn of the century, especially among the graduates of the 1910s.

Architects choose Sweden and Denmark to the greatest extent. In many cases, visits in the countries were, as mentioned, a part of the Finnish road study trip. Exciting and interesting architects like the Nordic classicists Tengbom, Asplund, and Lewerentz attracted in Sweden, and the Finland-Swedish Ekelund-Kuhlefelt couple worked, as mentioned, together with some of them in Stockholm in the 1920s. Danish national romanticism and brick architecture were also interesting study objects. Martin Nyrop, whose main work was the internationally recognised Copenhagen Town Hall, was one source of inspiration, and other Nordic architects practised with him and returned to major positions. Swedish architect Magnus Dahlander[2] became city architect in

2 KTH, architecture, 1884.

Örebro and later county architect in Dalarna and is described as an opponent of functionalism. Martin Borch was another representative of Danish national romanticism. Fritz Morten Anker Bachke[3] returned to Trondheim and used influences from Borch to design brick buildings.[4] Some influences within food processing in Sweden, Norway, and Finland were also mediated via Denmark. Alfred Jørgensen's laboratory in Copenhagen was a source of knowledge within the Nordic fermentation industry, and many engineers in Norway and Sweden worked in brewing upon return from the laboratory.[5]

Sweden was, as appendix 1 shows, the most frequented Nordic destination, and its status as a destination becomes even clearer when we consider the significantly lower numbers of graduates in Denmark, Finland, and Norway. We may point to Sweden's central geographical location in the Nordic area. It was the most easily accessible Nordic country from Denmark, Norway, and Finland. The Swedish-speaking domination among the Helsinki graduates also suggests a strong identification with the western neighbour. Sweden had more large-scale industrialisation than the other Nordic countries. A Norwegian engineer expressed his admiration for a Swedish mechanical workshop in the 1920s and claimed that he had witnessed the ideal adjustment of Taylor's system to Scandinavian conditions.[6] Another Norwegian article described a visit to ASEA as a must for everyone with interest in the electrical industry. Some Nordic engineers practised and took training courses in Västerås. They later were employed in the testing and assembly of electrical equipment at ASEA's Norwegian and Danish branches and also as professors at technical universities such as the Norwegian Institute of Technology, managers for municipal electricity works in cities such as Helsinki, and as founders of electrotechnical companies.

Danish engineers, as mentioned, often worked abroad for Danish companies, but other Nordic engineers rarely worked in Denmark for companies from their respective countries. The Norwegian classification society *Det Norske Veritas* (DNV) employed some Norwegian engineers at their Copenhagen office in the 1910s and the 1920s. ASEA also had some Swedish engineers representing them in Denmark.[7] There were, however, many more Danish engineering activities in Sweden. Many took place in Scania, just across the narrow strait from the Copenhagen area. N. C. Monberg's involvement in preparing Trelleborg's

---

3 TTL, architecture, 1906.

4 Torgeir Suul, *Arkitektur i 1000 år: en arkitekturguide for Trondheim* (Trondheim 1999) 124.

5 Indebetou and Hylander, *Svenska teknologföreningen*, 344.

6 Georg Brochmann, 'Verkstaden i Kristinehamn', *Teknisk ugeblad* 36 (1923) 296–297.

7 Indebetou and Hylander, *Svenska teknologföreningen*, 555, 708–709, 799.

harbour for the train ferry connection to Sassnitz has been mentioned, but Danish engineers were also involved, for example, in the construction of Scandinavia's largest dry dock in Landskrona. Danish engineering in Sweden was, however, not limited to the southernmost province, but also, for example, in harbour and railway building on the island of Gotland. Danish engineers also did the pillars to Stockholm's Saint Erik Bridge in the early years of the twentieth century.[8] However, most journeys were made on the payroll of F. L. Smidth, and engineers travelled to manage a brick works near Gothenburg, which applied Danish technology.[9]

1.2 *Students Coming Back from Sweden and Swedish Expertise in Norway and Finland*

The mobility to Norway and Finland share a few characteristics, but Norway attracted almost twice as many graduates. Mining engineers and metallurgists often went there, both countries received native-born students coming back from studies in Sweden, and Swedes worked for Swedish companies more often than Danes worked for Danish. There are some differences too; architects more often went to Finland. Norway's Harald Wildhagen[10] worked with the famous Alvar Aalto in Turku in the late 1920s, and Swedish architects sometimes studied with leading Finnish colleagues such as the national romanticist Lars Sonck. Swedes constituted the significantly largest group in Finland, whereas Finnish study travelling was so extensive that Finns, relatively speaking, also went more often to Norway; hydroelectric power stations were one of their study objects. However, study travellers from the other Nordic countries were less likely to choose Finland; five-sixths of the mobility was migration. Another difference was that timing hardly mattered for Finland, whereas Norway became a more interesting destination from the 1890s and onwards.

Chalmers was an important educational institution for both Norwegians and Finns. There were 67 Norwegian-born and 45 Finnish-born graduates in the Swedish cohort of whom 57 and 35, respectively, studied in Gothenburg. In Finland, Sweden was second to Germany when it came to foreign studies among early electrical engineers: 14 per cent had studied there, but some of them were Swedish immigrants.[11] Geographically, Chalmers was relatively close to Norway, but it is perhaps more unexpected that Finns tended to

---

8 Hannover, *Dansk Civilingeniørstat 1942*, 98, 170.

9 Hannover, *Dansk Civilingeniørstat 1942*, 78, 176, 240. Also see 80, for Holger Stenbjørn who served 1896–97.

10 NTH, architecture, 1919.

11 Myllyntaus, 'The Best Way', 143.

choose an institute in Gothenburg rather than in 'nearby' Stockholm. One reason may be that the Royal Institute of Technology was less inclined to accept students from abroad. A majority of the Norwegian and Finnish students went back and were involved in fields such as shipping, shipbuilding, mechanical workshops, and the electrical industry in both countries as well as in the textile industry and architecture in Finland, and mine and stone industries and pulp and paper in Norway.

Many made successful careers. In Finland, Chalmers graduates became managers of a spinning-mill in Tampere, a chemical-technical factory in Turku, a mechanical workshop in the southwest town of Uusikaupunki, a wool factory in Helsinki, and one of the capital's largest shipyards. Victor Buur[12] founded a shipyard in Vaasa. One mechanical engineer stayed eight years with a Swedish granite company before he became director of a granite-exporting business in eastern Norway, and later also for a limestone quarry. One electrical engineer worked some years in the Swedish electrotechnical industry before he returned to plan and erect or build Norwegian power transmission lines, generator and transformer plants, and electric smelting furnaces. Only a few Norwegian Chalmers graduates went straight home, among them Erik Jacobsen Rotheim[13] who started a machine shop and motor factory and became Norway's first authorised car inspector and a pioneer for cars as a means of communication.

Employment for domestic companies was an important trait in the mobility also to Norway and Finland. Contrary to most other destinations, Swedish interests were more strongly represented than Danish in both countries. Danish journeys to Norway often set out for Christiani & Nielsen's branch in Kristiania, and only a few F. L. Smidth engineers passed through Norway as one of several countries. Most Danes travelling to Finland went to F. L. Smidth's cement factory south of Turku and the office in Helsinki. Many Swedes crossed the Baltic Sea to work for building contractors Kreuger & Toll, who expanded in Finland and Russia in the early 1910s,[14] at the branch office in Helsinki.[15] Cement producer Skånska Cementgjuteriet and the ASEA branches in cities like Helsinki

12 CTI, mechanical, 1915.

13 CTI, construction, 1892.

14 Thunholm, *Ivar Kreuger*, 34.

15 Bodman, *Chalmers tekniska institut*, 200. For other Kreuger & Toll engineers in Finland, see: Bodman, *Chalmers tekniska institut*, 184 (Malmcrona); Indebetou and Hylander, *Svenska teknologföreningen*, 690 (Hesser), 910 (Eneroth), 932 (Tiderström), 963 (Asperen).

and Viipuri were other places where Swedish engineers worked during longer and shorter periods.[16]

ASEA also had many activities in Norway; some Swedish engineers worked for example at the subsidiary in Kristiania. Wilhelm Schmidt[17] was the director of the subsidiary between 1908 and 1917. He participated in the negotiations to take over and merge with the firm owned by electrical engineer Per Kure,[18] which was realised after World War I.[19] ASEA often participated in extensions and the building of power stations like the one at Rjukan in the 1910s and delivered, as mentioned, the giant generators to the power station at Svælgfoss.[20] Another ASEA project involving Swedish engineers was the electrification of the railway between Kristiania and Drammen in the 1920s.

Some Swedish engineers were also involved in mining, often at Swedish-owned establishments like industrialist Nils Persson's ventures in northern Norway at the turn of the century and the Svea pit coal mine in Svalbard around 1920. Gustaf Örn[21] owned several ironworks and served as long-time Swedish consul in Trondheim. In Finland, Swedish mining engineers served as specialists at establishments such as the steelworks in Åminnefors and the zinc and lead mine Orijärvi in the southwest, and the mine in Pitkäranta by Lake Ladoga. Some foreign engineers also arrived in Finland as specialists in other fields of technology. Most came from Germany and Sweden. Finland's government did not actively recruit foreigners with special skills. Foreign specialists were either drafted by managers in desperate need of technical expertise, or they migrated to Finland on their own initiative. The Swedish civil engineer Bror Sjögren[22] owned a large engineering office and was the designer of many interwar hydroelectric power plants. Sjögren was an advocate for hydropower who often participated in public debates about the country's energy policy. Another Swede, Oskar Faith-Ell, who had studied in Borås, operated an electrotechnical consulting office.[23]

16 Indebetou and Hylander, *Svenska teknologföreningen*, 708–709, 799.

17 CTI, mechanical, 1895.

18 Kure was a graduate of the intermediate technical school in Horten, but had also studied at Mittweida, Saxony, Germany, between 1894 and 1897, see: Trond Smith-Meyer, 'Per Kure—utdypning (Norsk biografisk leksikon)', https://nbl.snl.no/Per_Kure, (15 November 2017).

19 Sverre A Christensen and Harald Rinde, *Nasjonale utlendinger: ABB i Norge 1880–2010* (Oslo 2009) 68; Indebetou and Hylander, *Svenska teknologföreningen*, 690.

20 On ASEA in Norway, see: Christensen and Rinde, *Nasjonale utlendinger*.

21 TESÖ, mechanical, 1881.

22 KTH, civil, 1916.

23 Myllyntaus, *The gatecrashing apprentice*, 76–77; for Faith-Ell, also see: *Suomen Insinöörejä ja arkkitehtejä*, 624.

1.3 *Infrastructure Building in Iceland and Greenland*

Icelandic youngsters interested in technology had to go abroad, and this stream went almost exclusively to Copenhagen. Denmark hosted twelve Icelandic-born students. There were strong ties between Danish and Icelandic technology, and all graduates went back. Jon Hallson Isleifson[24] studied in Trondheim and was the only Icelandic-born in the cohort who graduated elsewhere.[25] Isleifson, however, followed a typical pattern as a construction engineer working for the National Road Administration. Most engineers had similar tasks, and this is connected to the build-up of Icelandic infrastructure from a minimum level before World War II.[26] Guðmundur J. Hlìðdal took his education in Mittweida and Charlottenburg, practised with Siemens, and followed a similar pattern after returning in the mid-1910s. Hlìðdal participated in harbour construction, the erection of a water power station, and became the director of the Icelandic telegraph services. He is the only engineer educated outside the Nordic countries in an Icelandic register.[27]

Sigurdur Jónsson Thoroddsen[28] became the first fully educated Icelandic engineer and was later appointed 'country engineer'. He initiated the process to make roads more rational and adjusted to ascents, bends, and dewatering. Most important was his initiative to set up smaller and larger bridges to increase usefulness.[29] His successor, Jon Thorlaksson,[30] developed coach traffic with the help of horses, increased the number of iron and iron-concrete bridges, and modernised the road service's repair workshop. Later, he was one of the initiators of Reykjavik's electricity works.[31] Geir Zoëga[32] continued to develop roads, was responsible for creating a nationwide network, and furthered the use of iron-concrete in bridge building. He was also a governmental advisor in several technical matters and served as chairman of the Engineers' Association from the 1920s.[33]

---

24 TTL, construction, 1911.

25 Alstad, *Trondhjemsteknikernes matrikel*, 283; Bassøe, *Ingeniørmatrikkelen*, 240.

26 Magnússon, *Iceland in transition*, 59.

27 Jón E. Vestdal et al., *Verkfræðingatal: Æviágrip íslenzkra verkfræðinga og annarra félagsmanna verkfræðingafélags Íslands* (Reykjavík 1956) 146–148.

28 PTI, civil, 1891.

29 Hannover, *Dansk Civilingeniørstat 1942*, 57; Krabbe, *Island og dets tekniske udvikling*, 18, 347.

30 PTI, construction, 1893.

31 Krabbe, *Island og dets tekniske udvikling gennem tiderne*, 18–20, 72, 229, 233.

32 PTI, construction, 1911.

33 Bjarni Reynarsson, 'The Planning of Reykjavik, Iceland: three ideological waves—a historical overview', *Planning Perspectives* 14 (1999) 49–67; Krabbe, *Island og dets tekniske*

Not all engineers worked with roads. Danish-born Thorvald Krabbe[34] was appointed ‘country engineer’ in 1906 and director of the Icelandic lighthouse service from 1918 to 1937. Other engineers built and managed herring factories, water purification plants, and served as city engineers in Reykjavik. Knud Zimsen[35] founded Reykjavik's first local telephone company. He was later appointed city engineer, became the mayor, and marked the beginning of Reykjavik's infrastructural modernisation; introduced drainage, made better use of the hot springs, introduced water and gas supply and extended the use of concrete in the construction of residential houses.[36] A colleague became leading engineer when the municipality was setting up an electricity works based on water power in the early 1920s. Thòrarinn Kristjansson[37] managed the newly erected harbour in Reykjavik in the 1910s and was later employed with N. C. Monberg—a well-known Danish employer in Germany, Sweden, and Iceland—in the Westman Islands together with Danish-born Nils Pedersen Kirk.[38] Kristjansson and Kirk were two of the few engineers working outside Iceland's capital area. Swedish Otto Frick[39] was another. He measured the power of the waterfall Dettifoss in northeast Iceland in the early 1910s. Frick was unusual in two ways: He served in the east and was not educated in Denmark. He was one of two Swedish engineers in Iceland.

Greenland, too, received two Swedish engineers, but one of them was there in the capacity of an ethnologist. Einar Krantz[40] was employed by a Danish consortium to investigate Greenlandic ore in the summers of 1903 and 1905. The remaining graduates were educated in Denmark, and civil and construction engineers and chemical engineers dominated. Greenland attracted the most technicians after the turn of the century and especially among graduates from the 1910s.

---

udvikling gennem tiderne, 22–23, 49–50; Sveinbjörn Björnsson, ‘Menntun íslenskra verkfræðinga. Sveinbjörn Björnsson tók saman úr frumdrögum Jóns E. Vestdal og ýmsum öðrum heimildrum’, in: Jón E. Vestdal et al.(eds.), *Verkfræðingatal: Æviágrip íslenzkra verkfræðinga og annarra félagsmanna verkfræðingafélags Íslands* (Reykjavík, 1956) X; Hannover, *Dansk Civilingeniørstat 1942*, 233.

34 PTI, construction, 1900.

35 PTI, construction, 1900.

36 Hannover, *Dansk Civilingeniørstat 1942*, 113–114; Krabbe, *Island og dets tekniske udvikling*, 210, 212, 249; Guðbjörg Lilja Erlendsdottir, *Trafiksäkerhet och tätortsplanering: en analys av Reykjavíks lokalgatunät med GIS* (Lund 2003) 19.

37 PTI, construction, 1912.

38 PTI, construction, 1906. Krabbe, *Island og dets tekniske udvikling*, 164–165, 170.

39 KTH, mechanical, 1888.

40 KTH, mining, 1906.

Telecommunications was another field of development. Engineers worked with wireless telegraphy, and Albrecht Fischer[41] arrived in the mid-1920s to be superintendent of the radio station in Julianehåb. Later, he was also the controller for the quarry in Ivigtut, the world's only significant source of cryolite, a mineral used in aluminium production as well as in the glass and porcelain industry. This quarry was the leading employer of Danish engineers. Train oil production was Knud Fæster's[42] field. He wanted to improve methods and went to Greenland where he stayed for two years. He is, however, best known as the manager of a factory erected in Copenhagen to develop seal, whale and cod liver oil. The factory opened new possibilities for Greenland's fishermen and whalers. Fæster developed methods to extract oil from marine animals and patented some of them. He planned factories and methods for refining fish products and is believed to be behind the high-class Greenland salt fish, as he developed production manuals with higher qualitative demands than ever before.[43]

2 Study Travelling Dominated the Mobility to Continental Europe

In this group, we have counted European countries outside the Nordic area, the British Isles and Russia. The most targeted destination was France, followed by Italy, Belgium, the Netherlands, and Spain. The region was the fourth largest according to our division. Every tenth Nordic graduate visited one of these countries, but more than every fourth from Finland. The Scandinavian shares lay between seven percent in Norway and nine percent in Denmark.

This was the only region according to our division in which the number of study travellers was higher than the number of migrants. This implies that the region was most attractive for Finnish graduates. Apart from Spain—where many Swedes worked for Swedish companies—and a few smaller destinations, Finns comprised the highest shares to all countries in this 'region'. Almost 60 per cent of the mobile architects went to these destinations.

These countries were also architecture destinations. Poland was the only relatively large destination that had a stronger attraction for another specialisation. Architectural education had an early focus on study tours, and one example is the *Grand Prix de Rome* from the French architectural academies; the tradition of sending the best students to Rome for many years began in

41 PTI, construction, 1913.

42 PTI, mechanical, 1913.

43 Willy Andersen, 'Knud Fæster 24. Juli 1887—30. Januar 1966', *Grønland* 6 (1966) 223.

the early eighteenth century. Swedish and Finnish architectural students also travelled on grants to study at renowned architectural schools abroad. The necessity to travel, study, and measure classical buildings on the spot was often emphasised by the professors of architecture in their lectures.[44] This area included the most common destinations for architects. The Mediterranean countries constituted an almost eternal source of inspiration for architects, but there were interesting objects in countries like France, Belgium, and the Netherlands too. Some of the most aesthetic destinations were in this region.

2.1 *Everything French*

France was one such destination and is the, by far, the most visited country in this region. Sweden's Ivar Tengbom[45] continued at the Academy of Arts to become an architect. He arrived in Paris in the spring of 1905 and was well prepared for his stay in France which he preferred over Italy. One reason was that he wanted to study modern methods, inspired by writings of architect Eugene Viollet-le-Duc. Viollet-le-Duc represented a practical and logical architecture in which an analysis of the relation between the building's task, material, and construction was central. These ideas were mediated by teachers at the Royal Institute of Technology and the Academy of Arts and exercised a strong influence on Tengbom. In Paris, he wanted to focus on ongoing construction projects and study the 'logical origin' of alien ideas. He also wanted to study how the French took care of and documented older buildings.[46] Boulevards, avenues, and monumental buildings stemming from Haussmann's mid-nineteenth-century re-building project, the first of its kind of a large city in Europe, were, of course, interesting study objects in Paris, as were the Eiffel Tower and many other steel constructions.

Most Nordic technicians headed for Paris regardless of whether they wanted to see the architecture or do something different. Around 60 per cent of those with known residences in France were in Paris. Swedish engineers worked for their countryman, industrialist Thorsten Nordenfeldt, who had immigrated to Britain in the 1860s and started the Nordenfeldt Guns and Ammunition Company. The company manufactured machine-guns invented by fellow Swede Helge Palmcrantz and had moved to Paris from London in 1890 after Nordenfeldt had experienced a personal bankruptcy. One employee was Nordenfeldt's nephew Per[47] who had followed his uncle from London. Nordic

---

44 Björklund, 'En sentida Grand Tour', 13.

45 CTI, civil, 1898.

46 Bergström, *Arkitekten Ivar Tengbom*, 24–26, 43–44.

47 KTH, mechanical, 1883.

technicians were employed to construct heating and water plants, owned their own architect's offices and engineering firms, headed and were employed at acetylene companies, electrotechnical industries, forcite industries, and motor and automobile factories. Two Swedish and one Danish technician worked for the watchmaker *Breuget*. The development in iron-concrete was another interesting field; a few Norwegian and Danish engineers worked for Parisian firms in this field and one of them also for a company in Marseille. French engineer Francois Hennebique founded a company in Brussels, which moved to Paris in 1898 and developed into the world's most prominent company in iron-concrete. Rudolf Christiani was in Paris before he returned to Denmark via Hennebique's subsidiary in Düsseldorf. He became a co-founder of one of the country's most successful iron-concrete companies specialising in bridges. A pioneer iron-concrete bridge in central Jutland, finished in 1905, was one of Christiani & Nielsen's first constructions. The company also went into other types of construction and utilised the so-called Hennebique principle.[48]

Christiani & Nielsen later had several engineers stationed in Paris and also one engineer involved in harbour construction in Cherbourg, and seems to have been the most represented Nordic-based company in France. However, Denmark's F. L. Smiths and Danalith as well as Sweden's AGA, ASEA, L. M. Ericsson, and SKF had a few representatives in Paris and the vicinity. France was also one of few countries where Norwegian engineers worked abroad for domestic companies other than Det Norske Veritas. During World War I, *Norsk Hydro* constructed a factory in Soulom at the foot of the Pyrenees, to produce ammonium nitrate and saltpetre for the French military. Kjetil Gjølme Andersen underlines that Norwegian engineers were sent to the southern French village to erect the factory, which started in 1916 but turned out to be a less than successful enterprise.[49] There were, however, at least four graduates from Kristiania that arrived in the late 1910s and early 1920s.

Myllyntaus stated that France's popularity as a student destination increased during the interwar years as the attraction of the German-speaking world started to decline.[50] This is not clearly reflected in these statistics, even if the migrant share to France was somewhat higher for the 1910s graduates. Many students took

48 Constructions according to the Hennibique principle; spinning-mill at a clothing factory near Copenhagen (1908) iron-concrete pier in Aalborg (1908). The spinning-mill, one of Denmark's first industrial buildings in iron-concrete, a material whose strength made possible the use of heavier machines and larger windows. Jespersen, *Biografiske Oplysninger*, 195; Grönberg, *Learning and Returning*, 87–89.

49 Andersen, *Flaggskip i fremmed eie*, 172.

50 Myllyntaus, 'Discovering Switzerland', 308.

their entire education in France. Also, several Nordic technicians also studied in France earlier. A few went to Paris to become painters and artists rather than continuing a career in architecture or engineering. Richard Birkeland[51] spent two years in Paris and followed the courses given by famous French mathematicians such as Henri Poincarè, Èmile Picard, and Èdouard Goursat.[52] Upon his return, Birkeland became the first professor of mathematics at the Norwegian Institute of Technology and held the very first lecture at the new technical university in Trondheim in September 1910. He also wrote a textbook in mathematical analysis, which was in use until the late 1930s.[53]

Paris's naval school formed a base for some successful maritime careers or as shipyard managers for some Swedish students of the 1880s and 1890s. Their countryman, John Luth,[54] went instead to the French navy's marine engineering school in Cherbourg. He continued to shipyards and workshops in Britain before he returned to be a co-founder of an electrotechnical company in Stockholm,[55]which was one of ASEA's major domestic competitors in the early years of the twentieth century. He probably knew August von Eckermann[56] who was a student in Cherbourg at the same time. He also went on to British shipyards and served as a superintending engineer for the construction of submarines in Turkey and Greece. Unlike Luth, von Eckermann made a career with the Swedish navy upon his return and participated in a mid-1910s investigation on the extension and modernisation of the navy shipyard in Karlskrona.[57]

France was a major aviation centre, and Jan Waernberg describes how Swedish pioneer Enoch Thulin acquainted himself with schools in Paris and its vicinity before he educated himself to be an aviator in northeast France and visited renowned motor factories.[58] Finland's Paul Hjelt[59] served in the military after he graduated in aeroplane construction in Paris around 1920; he was head of the Finnish air force's technical department before he started a diplomatic career in the late 1920s.[60] His Norwegian colleague Per Slinde[61] graduated

51 KTS, construction, 1899.
52 Birkeland also studied in Göttingen under the renowned German mathematicians David Hilbert and Felix Klein, but he spent more time in Paris.
53 Bent Birkeland, Richard Birkeland. In *Norsk biografisk leksikon*, https://nbl.snl.no/Richard_Birkeland 15 November 2017.
54 KTH, mechanical, 1880.
55 Indebetou and Hylander, *Svenska teknologföreningen*, 206.
56 KTH, mechanical, 1880.
57 Indebetou and Hylander, *Svenska teknologföreningen*, 228.
58 Jan Waernberg, *Enoch Thulin—forskare, flygare, företagare* (Lund 2004) 37–44.
59 STK, chemical, 1917.
60 'Sextioåringar', *Hufvudstadsbladet*. 24 November 1955.
61 BTS, mechanical, 1913.

from the same Parisian school at about the same time and returned to become head of drawing at the Kjeller aeroplane factory in Lillestrøm.[62] Students also had other objectives in France. The school for bridge and road construction in Paris was a foundation for careers in railway construction in Norway. Some Norwegian chemical engineering graduates also went to the Sorbonne after leaving school in the 1910s; some earned doctoral degrees. Others pursued similar studies in Strasbourg and Grenoble. The technical university in Grenoble gained an international reputation in the interwar years.[63] Erik Stephansen[64] studied there, was employed some years in France, and made a career with a refrigerator company in Oslo.[65] Margot Dorenfelt[66] represented the University of Kristiania at an international conference on flammable liquid in Paris in 1922 and stayed to study two semesters at Collège de France. Dorenfelt published scientific articles in French and German together with her husband Eugen Holtan and the famous radio-chemist Ellen Gleditsch, a co-worker of Marie Curie who became Norway's second female professor.

Cherbourg, Strasbourg, Grenoble, and the foot of the Pyrenees were not the only places hosting Nordic technicians outside Paris. Norwegian Harald Hansen[67] participated in the construction of the underground railway in Marseille while he was studying French in the city in the early 1920s. Also on the Riviera, Finnlander Eino Bergius[68] became the cohort's only visitor to Monaco as he went to motor exhibitions in the principality in the early 1910s.[69] An enamelling industry near Bordeaux attracted at least five Danish engineers, Nantes in the west received a few Nordic technicians to a paperworks and so did a cardboard plant in Honfleur in the northwest as well as shipyards and harbours in nearby Caen, Dunkirk, and Le Havre. The steel and iron industry in Le Creusot in the east was the workplace of at least three mechanical engineers from Sweden, while another worked for an electrotechnical company in Nancy in the northeast. The area in today's northeast France, Alsace-Lorraine, was German between 1871 and 1918 and some Nordic technicians worked in the 'floating' borderlands between Germany, France, Luxembourg, and Belgium. In Hayange,

62 Eskedal, *BTS-matrikkelen*, 56.

63 Andrè Grelon, 'The training and career structures of engineers in France, 1880–1939', in: Robert Fox and Anna Guagning (eds.), *Education, technology and industrial performance in Europe, 1850–1939* (Cambridge 1993) 51.

64 NTH, chemical, 1919.

65 Bassøe, *Ingeniørmatrikkelen*, 484.

66 NTH, chemical, 1919.

67 KTS, construction, 1896.

68 SPO, mechanical, 1907.

69 Heiniö, *Matrikel öfver Polytekniska institutets i Finland lärare och elever*, 277.

near today's French-Luxembourgian border, lay *Les Petits-Fils de François de Wendel,* which was one of Europe's major manufacturers of steel. The plant, adopting the Thomas process, employed at least five Swedish metallurgists in the 1910s. They worked for a French company in Germany. Torsten Tesch[70] continued to Carnegie Steel and North American industries and made a successful career as executive engineer at two of Sweden's major steel and ironworks upon his return.[71] Another traveller in the neighbourhood was David Waage,[72] who had been in mining in eastern and central Europe but was now engaged to tour French-Belgian-German borderland coal mines and study safety arrangements. He had prestigious employment in Geneva, at the International Labour Organisation's department of prevention of accidents and claimed that his education from the Norwegian Institute of Technology was valuable. He discovered that he was competitive regardless of country.[73] We have seen that this held true for many technicians educated in Scandinavia and Finland.

2.2 *Iron, Art Nouveau and Americans in Belgium and Luxembourg*

Belgium was the second largest 'other European' destination; 'visited' by two percent of all Nordic graduates and four percent of the transnationally mobile ones. Two-thirds of the visits were made by study travellers. Finns and architects had the highest percentages of total mobility. Victor Horta and his buildings were an attraction. Horta has been described as the major European Art Nouveau architect, and with Hotel Tassel in Brussels, he is sometimes credited for having introduced decorative arts to architecture.

Swedes and mining engineers and metallurgists also often went to Belgium, and Anduaga's Basque students were not the only foreigners at universities in Liege. A tradition to continue their study in Liege seems to have been established among students in Bergen; at least ten students followed this path. Most of them studied electro-technology as *Institut Electrotechnique Montefiore* was one of the pioneers in the field in Europe. This school provided one-year courses devoted entirely to electrical engineering and admitted only students who had completed a course in a traditional engineering branch.[74] Ole Houm[75] became technical and mercantile manager

70 KTH, metallurgist, 1900.

71 Indebetou and Hylander, *Svenska teknologföreningen*, 602.

72 NTH, mining, 1914.

73 Brochmann, *Vi fra NTH*, 53–54.

74 Jean C. Baudet, 'The training of engineers in Belgium, 1830–1940', in Robert Fox and Anna Guagning (eds.), *Education, technology and industrial performance in Europe, 1850–1939* (Cambridge 1993) 102–103.

75 BTS, mechanical, 1898.

of *Norsk Hydro,* as well as the company's manager upon his return. Because of these students, Liege and the surrounding region had a stronger appeal for Norwegians than other Nordic technicians. Nevertheless, there were colleagues from the other countries in the surroundings. Two Swedish engineers worked for the Liege Electric Company, and some of their countrymen were employed at the steel and ironworks Cockerill in Seraing, a pattern that also was revealed by Anduaga for the Basques. On the Luxembourgian side of the border, a few Swedes and Norwegians also worked in steel and iron, while three Finnish technicians arrived in Luxembourg to receive rail and other materials for the Finnish Railway Company. Their countryman Karl Verner Sulin[76] studied at the school for engineers and worked in the textile industry in nearby Verviers.

Many Nordic engineers in Belgium stayed in Brussels. The well-known British-American engineering firm Julian Kennedy, Sahlin & Company employed twelve Swedes in our cohort. Working together with their countryman, mechanical engineer, and part-owner, Axel Sahlin, was an attraction. He had spent several years at American steel and ironworks and was known as one of the most knowledgeable persons when it came to the production of pig iron and blast furnaces fuelled by coke. Brussels also hosted representatives of companies like ASEA, Electrolux and Luth & Rosèn, and the subsidiary of Julius Kahn's American iron-concrete company employed at least four Swedish technicians of whom one became associated with Kreuger & Toll upon return. Belgian match factories seem to have utilised Swedish knowledge. Two technicians were employed in Brussels, and Robert Sundström[77] assisted in the construction of Ghent's match factory with experience from similar establishments in Sweden.

Ghent attracted a few Nordic technicians, and some Swedes also worked for the dynamite factory in Balen, west of Antwerp. There were some important Nordic technicians connected to the city of Antwerp. Norwegian Trygve Olsen Herfeldt[78] led the construction of the Albert Canal between Antwerp and Liege and was later awarded an honorary doctorate in Aachen for this achievement.[79] Danish knowledge in cement manufacturing was utilised in Belgium; F. L. Smidth had a few representatives in the country, and the well-known factory Cannon Brand in Burcht, near Antwerp, employed a Dane as executive engineer and later manager in the late 1920s and early 1930s. General Motors had

76 STK, chemical, 1909.

77 CTI, mechanical, 1909.

78 BTS, mechanical, 1893.

79 Eskedal, *BTS-matrikkelen*, 13.

begun to construct Chevrolets in Antwerp in 1925; two years after a similar factory had opened in Copenhagen. The Danish factory was, from the beginning, organised after Taylorist principles including conveyor belts and methods engineering.[80] Aage Gram Blom[81] studied Industrial Management and Factory Management in Pittsburgh and underwent a special course at Ford's plants in Detroit. Upon return, he became an engineer at the factory in Copenhagen and later head of production in Antwerp.[82] Six Nordic engineers in the cohort were also employed with the Bell Telephone Company in Antwerp. Leif Brun[83] was first employed with Bell's department in Kristiania but relocated to Antwerp after one year. He left after two years to serve Bell in other countries but returned to the technical department in Antwerp in the mid-1920s.[84]

2.3 *Water Management and Architectural Modernism in the Netherlands*

Brun was rarely far away from Antwerp during his intermission. Outside Belgium, he mostly served Bell in the Netherlands. He was not the only Nordic technician serving a foreign telephone company, but the two Swedes that worked to establish Dutch telephone systems, switchboards and automatic stations worked for their own company L. M. Ericsson. Mechanical and electrical engineers and naval architects, however, visited the country to a lesser extent, while architects had the highest rate followed by civil and construction engineers and chemical engineers. Study trips constituted about five of six visits. Dutch modernist architect Hendrik Petrus Berlage, the man behind the commodity exchange in Amsterdam, was influential to many generations of future architects in the Netherlands as well as internationally. Dutch peat trade, mechanical and technical methods, ditching, cutting, storing, and so on, also attracted interest.[85] The struggle to master the water offered interesting objects to study. Large water management projects included dredging, the building of new levees, excavations of canals, dam construction, reservoirs, aqueducts, tanks, pumps, irrigation, and water pipes, that is, technical developments and artefacts interesting primarily to civil and construction engineers.[86]

---

80 Ole Hyldtoft, 'Perioden 1896–1930', 181.
81 KM, mechanical, 1916.
82 *Danske teknika,* 248.
83 NTH, electrical, 1919.
84 Brochmann, *Vi fra NTH*, 185.
85 Lundgren, 'Kunskap och resande', 88–89.
86 Harry Lintsen, 'Two Centuries of Central Water Management in the Netherlands', *Technology and culture: the international quarterly of the Society for the History of Technology* 43:3 (2002) 557–562; Martin Reuss, 'Learning from the Dutch. Technology, Management, and

Denmark was the Nordic country whose nature and topography reminded most of the Netherlands. Danish technicians served F. L. Smidth and Christiani & Nielsen, but also in places like a dredging company in The Hague, a cement factory in the southernmost province of Limburg, and a Chilean company that constructed a floating dock. Ejnar Hertzsprung[87] was appointed a professor in astrophysics in Leiden in 1920. The only Finnish migrant served at a sugar works in Amsterdam, while some Swedish colleagues served in the offices of Swedish companies in the same city. Mechanical industries in places like Breda, Deventer, Utrecht, and Hengelo also employed some Scandinavian technicians. Olaf Arneberg[88] worked at breweries in Copenhagen, Vienna, and the Netherlands before he returned to become a brewing master in southeast Norway and chemist and yeast physiologist at the country's largest brewery, Ringnes, in Kristiania.[89]

2.4 *Late Mobility to Newly Independent Eastern Europe*

Mobility to Estonia, Latvia, and Lithuania is very much the movement of technicians graduating in the 1910s, simply because moves to these areas were counted to the Russian Empire before independence in 1918. Norwegian technicians were totally absent. The only registered employment in Estonia was by a Danish engineer who served, hardly surprisingly, at a cement factory. An overwhelming majority, however, came from Finland as Tallinn was just a short boat trip away from Helsinki. Almost half the travellers were architects. Finland, study travellers, and architects also dominated in Latvia and Lithuania. A few Danish and Swedish technicians were, however, employed in construction in Riga in the 1920s building a dairy and as managers of superphosphate and match factories.[90]

Independent Poland was a larger destination. A major difference was that only a few architects found their way there. Danish and Swedish engineers often represented companies from their own countries but also other foreign companies. One Dane managed a French-owned petroleum mine in central Poland in the 1920s. F. L. Smidth had a few representatives in Poland, and a couple of Danes also served at Polish-owned cement industries. The machine

Water Resources Development', *Technology and culture: the international quarterly of the Society for the History of Technology* 43:3 (2002) 465–472.

87 PL, chemical, 1898.

88 KTS, chemical, 1894.

89 *Skrift ved 50 års jubileet for ingeniørene fra K.T.S. 1894* (Oslo 1944) 17.

90 *Danske teknika*, 186; Hannover, *Dansk Civilingeniørstat 1942*, 145, 318; Indebetou and Hylander, *Svenska teknologföreningen*, 468.

factory and aeroplane industry Nielsen & Winther was also represented, and one Danish engineer worked for Højgaard & Schultz during the building of the harbour at Gdynia from the mid-1920s. This was one of the largest tasks in the field in Europe to that time, and it led to several new tasks in Denmark and abroad. Swedish companies employed engineers as representatives in Warsaw; ASEA, SKF, and L.M. Ericsson. The Finn K. E. Ranko[91] served L. M. Ericsson in Russia before the revolution. In the early 1920s, Ranko travelled around in Poland for this company to evaluate local telephone nets. As a result, L. M. Ericsson and the Polish state took over the concessions to build nets in several cities. Ranko was responsible for the construction in Lodz between 1922 and 1924.[92]

In Czechoslovakia, Prague offered a blend of architectural styles to study including Roman, Gothic, Renaissance, Baroque and Art Nouveau. A few Norwegian and Danish engineers were also employed at ironworks like the earlier-mentioned Norwegian David Waage, who was executive engineer for a lead and silver industry in western Bohemia. Some machine factories in Prague employed Swedish engineers, and a few of them were also stationed in the Czechoslovak capital for domestic companies like AB Separator.

2.5 *To Hungary, the Balkans and Road and Railway Building in Turkey*

Budapest also offered architects a lot, displaying historicism and Art Nouveau and several buildings that took influences from Vienna, Germany, and many European countries. A couple of Danes also worked in the cement business for F. L. Smidth and to start a Hungarian factory. Swedes represented AB Separator and the Swedish Match Company, while a couple of countrymen and one Norwegian technician served the Ganz company. Started in the 1840s by Swiss-born Hungarian mechanical engineer Abraham Ganz, this company grew and became important. Its late nineteenth-century products furthered the expansion of alternating-current transmissions. The company also patented high-efficiency transformers, and employed some well-known engineers and electrical pioneers. Nikola Tesla's induction motor was patented in the United States in the late 1880s, but Tesla claimed that he got his idea while working for Ganz in the early 1880s. At least four Norwegians also worked for the Budapest-based contractors firm started by their countryman Gudbrand Gregersen. He became famous, and later ennobled, after he quickly built dams to protect Szeged after a disastrous flood in 1879. This success led to many tasks throughout

91 SPO, mechanical, 1902.

92 'Femtioåring. Ingeniör K. E. Ranko', *Hufvudstadsbladet*, 6 February 1928.

Hungary such as railway building and the modernisation of the harbour in Fiume (today Slovenian Rijeka). Over time, Gregersen's firm became one of Hungary's largest contractors.[93]

Single technicians also were in Szeged at a sugar works in Hatvan in northeast Hungary and in mining in today's Slovakia. One Norwegian cellulose engineer served at a factory in Turda. This city, turning Romanian in 1918, was also one of the places where Danish cement knowledge was utilised as P. F. Bagge[94] managed the local factory from the mid-1920s onwards. Bagge arrived from a similar establishment in Bulgaria.[95] Another Danish technician erected a Romanian glassworks, and a couple of Swedish and Finnish chemical engineers had positions at businesses in Bucharest, including an Austrian-American petroleum venture. Norway's Waage managed a mine near Plovdiv in Bulgaria, and was also involved in geological investigations in Serbia in the late 1910s. A few other Norwegian and Swedish engineers were also involved in mining investigations and metallurgic ventures in the Balkans. Swedish Orvar Sturzenbecker[96] served as works manager when Nuremberg-based Schukert & Co built a hydroelectric power station in Bosnia in the late 1890s, while his countryman Sven Springfeldt[97] constructed roads for the Turkish state in Kosovo in the early 1910s.[98]

Springfeldt was not the only one working for the Turkish government; a Danish colleague worked for the Turkish government with bridges and tunnels in Constantinople in the 1910s. A Swedish engineer built tramways in the same city for Germany's AEG around 1900. Most Nordic technicians served other foreign or domestic interests like Ture Söderlund,[99] who was Sweden's commercial attaché in Constantinople between 1908 and 1911.[100] Sweden's L. M. Ericsson and Denmark's Nielsen & Winther were also present in Turkey, but railway building was the main activity for Nordic technicians in Turkey. Two Norwegian engineers were involved with German firms in railway building in Turkish Asia Minor, but most Scandinavian engineers worked in the joint Danish-Swedish venture to project about 1,000 kilometres of new railway in the late 1920s and early 1930s. This was a cooperation between the Swedish

93 Jan Wiig (2015, 17. februar). Gudbrand Gregersen Saági. In *Norsk biografisk leksikon*, https://nbl.snl.no/Gudbrand_Gregersen_Sa%C3%A1gi 15 November 2017.

94 PL, chemical, 1909.

95 Hannover, *Dansk Civilingeniørstat 1942*, 197.

96 KTH, mechanical, 1893.

97 KTH, civil, 1905.

98 Indebetou and Hylander, *Svenska teknologföreningen*, 396, 701.

99 KTH, mechanical, 1892.

100 Indebetou and Hylander, *Svenska teknologföreningen*, 380.

locomotive workshop Nydquist & Holm and Denmark's Saabye & Lerche. The Swedes delivered locomotives, while the Danes built the railway. Per Kampmann[101] headed this project for six years.[102] At least twelve Scandinavian engineers, some also from Norway, were employed to construct railways from Ankara to Eregli, on the Black Sea coast, as well as from Fevzipaca to Diarbekir in eastern Turkey. According to Kampmann, the construction, managed from a joint office in the city that now was Istanbul, was difficult as it included the building of iron bridges and tunnels through rough mountainous areas and almost impenetrable forests.[103]

2.6 *Mining, Railways, and Architecture in the Mediterranean*

The Mediterranean countries received many Nordic technicians, of whom some worked for domestic or foreign companies from other countries. In Italy, Danish engineer Hermansen's thermo-engineering firm had an office in Genoa, and the Swedish companies ASEA, L. M. Ericsson and SKF had engineers in Milan and Naples. Emil Personne[104] was manager for ASEA's Spanish subsidiary in the late 1910s and at the same time Swedish vice consul in Madrid. At least ten Swedish engineers served ASEA in Madrid and Barcelona. Sweden's SKF, AGA and L. M. Ericsson, Denmark's Danalith and Norway's *Norsk Hydro* were other Scandinavian companies with representatives in Spain. As for Personne, he returned to Madrid in 1923 after an intermission in Stockholm. He was now representative of the vacuum cleaner and refrigerator producer Electrolux. Some year later, he became manager of this company's branch in Lisbon.[105] Nordic engineers also served at Germany's Siemens & Schuckert, America's Standard Electric, an American owned power supply company in Barcelona, and a German-Spanish mining company in Jaen in Andalusia. Another employer was Alfred Nobel, who spent the last years of his life in the 1890s in a villa in San Remo, had a laboratory in the neighbourhood, and initiated inventions that later were perfected by others. Ragnar Sohlman,[106] later known as

101 PL, construction, 1916.

102 P. H. Bendtsen and Povl Vinding, 'Per Kampmann', 2011, http://denstoredanske.dk/Dansk_Biografisk_Leksikon/Naturvidenskab_og_teknik/Ingeni%C3%B8r/Per_Kampmann?highlight=per%20kampmann, (15 November 2017).

103 Per Kampmann, 'Svensk-dansk jernbanebygning i Tyrkiet', *Teknisk Tidskrift. Allmänna avdelninge* 18 (1932) 177–179; Per Kampmann, 'Svensk-dansk jernbanebygning i Tyrkiet', *Teknisk Tidskrift. Allmänna avdelninge* 19 (1932) 189–192.

104 KTH, electrical, 1906.

105 Indebetou and Hylander, *Svenska teknologföreningen*, 697.

106 KTH, chemical, 1890.

executant of Nobel's will, worked in the laboratory to develop substitutes for caoutchouc and leather.[107]

Mining and railway construction also employed Nordic engineers around the Mediterranean Sea. Swedish Wilhelm Wahlman[108] worked in Greek mines already in the 1880s,[109] whereas his Norwegian colleague, A. W. Preus,[110] was involved in mining ventures and railway construction in Portugal and Morocco before he started his own business in Malaga in the mid-1920s.[111] Scandinavian knowledge was utilised in the Mediterranean, for example, in mining environments and metallurgic ventures in places like Narni in central Italy and Aosta in the northwest corner. Spain and Portugal utilised Danish competence in cement making, and one Dane was involved in this business in Bilbao. Italian cellulose factories and sugar works employed technicians from Scandinavia, and a rubber factory in Rome employed a Swedish engineer as technical director. Olaf Steen[112] projected, built, started up, and managed an electro-steelworks in northern Italy for the German firm Mannesmann in the early 1910s. From visits at industries in France and Torino, he concluded that the electric arc furnace invented by French Paul Heroult was the best solution. Steen described 'his' electro-steelworks as Europe's largest of its kind in the mid-1910s. Upon his return to Norway, Steen became manager of the electro steelworks in Stavanger, which he modernised, partly after his own plans. He described himself as the man who introduced Mannesmann to electric steel.[113]

More than anything else, however, the Mediterranean was a region where architects—and construction engineers who wanted to become architects—travelled. Italy was the classical architectural destination, but not the only one. The great Spanish architect Antoni Gaudi, whose magnum opus is Barcelona's *Sagrada Familia,* attracted colleagues from up north. Spain also hosted several other interesting objects such as Gothic Cathedrals.

The earlier-mentioned Hilding Ekelund was appointed lecturer in ancient and medieval architecture at the Finnish Institute of Technology in the mid-1920s, partly because of his experience from a journey to Italy together with his wife Eva Kuhlefeldt-Ekelund some years earlier (see below). His preparations

107 Göran Nilzén, 'Ragnar Sohlman', *Svenskt biografiskt lexikon. Bd 32, Sehlstedt-Sparre* (Stockholm 2003) 632.

108 KTH, mining, 1881.

109 Indebetou and Hylander, *Svenska teknologföreningen*, 228.

110 KTS, construction, 1901.

111 Bassøe, *Ingeniørmatrikkelen*, 405.

112 KTS, chemical, 1894.

113 *Skrift ved 50 års jubileet for ingeniørene fra K.T.S. 1894*, 92–101.

included, however, another Mediterranean journey, three and a half months to the eastern parts. His main object was to acquaint himself with classical Greek and Roman architecture, but he also was interested in Byzantine buildings and urban landscapes from any time. He spent some days in Istanbul where he was impressed with Aja Sofia; not primarily with the dimensions, but the ground plan, definite length and aisles providing rich views and scale to the space.[114] Ekelund continued to Smyrna and Sokia where he met up with some German archaeologists. He continued by ship to Pireus and from there to Athens. The Acropolis lived up to his expectations. Athens looked very much the same in the mid-1920s as in the nineteenth century.[115] Overlooking the Greek capital, he wrote in his diary:

> What makes the urban landscape of Athens so beautiful is the even white mass of buildings of the same height with their roofs all lying in the same direction, without dominant, heavy monumental buildings, towers or cupolas; even the churches are small and are lost in the mass of houses.[116]

Rome had long been one of the most important destinations for architects, and attention in the nineteenth century was directed towards urban landscapes in addition to the traditional monuments. However, interest in Rome's ancient monuments and public buildings began to decrease in the latter part of the nineteenth century, whereas architects started to pay attention to the ornaments and Pompeii's colourful villa architecture and the early renaissance architecture of northern Italy. A more diverse interest in Italian architecture was developed in the latter part of the nineteenth century; the classical period, the renaissance, medieval idioms, rococo and baroque were employed by architects after study tours in Italy and other countries. Early twentieth-century architects also focused on simpler architecture, with fewer decorations and forms.

Hilding Ekelund and Eva Kuhlefelt-Ekelund's study trip can exemplify journeys to Italy. They had read literature on classical Italian buildings and had heard about Italy from returning colleagues. Travelling by boat from Helsinki to Stettin, and then, by way of Berlin, Munich, and Innsbruck, the couple continued through the Brenner Pass and arrived in Trento late in October 1921.

---

114 Timo Tuomi 'The call of the Mediterranean: Hilding Ekelund's great journeys of the 1920s', in: Timo Tuomi, Kristina Paatero, and Eija Rauske (eds.). *Hilding Ekelund: (1893–1984): arkkitehi=arkitekt=architect* (Helsinki 1997), 78.

115 Tuomi 'The call of the Mediterranean', 78.

116 Cf. Tuomi, 'The call of the Mediterranean', 78.

Their main interest was early Renaissance architecture, but anonymous and simple buildings in smaller towns, countryside, and larger cities also attracted them. Interest in this *architettura minore,* minor architecture, was a typical trait among Nordic architect travellers to Italy during the time. These two architectural styles were also main inspirational sources for the so-called Nordic classicism. Parks and gardens were also interesting to the couple, and Eva had studied landscape gardening for a long period.[117] The couple's Italian route had roots in the Grand Tour tradition, continuing from the Alps to Venetia and cities like Verona and Venice and then through Emilia-Romagna with Bologna before the couple made a two-week stop in Firenze. They continued their journey through Tuscany, Umbria, and Latium and visited Siena before they arrived in Rome; the highlight of the tour. The couple stayed for a month and celebrated Christmas and New Year's together with other Scandinavian speakers. The tour continued southwards to Campania and to ancient Naples where they also visited famous ancient museums. They continued to investigate smaller towns in Sicily for about one month. On their way back, the couple spent one week doing sketches in Pompeii, where they also studied mural paintings. A second stay in Rome and its surroundings now also included tours to investigate the countryside near the capital. The couple also made another two-week stop in Firenze and visited Pisa, Genoa, Milan, and Verona and Venice a second time. Arriving in Helsinki at the beginning of June 1922, the couple had been away for eight months. This was a longer journey than for many other travellers who often lacked the time and resources to go south of Naples and Pompeii.[118]

Hilding later wrote the influential travel report *Italia la Bella* in an architectural journal and spurred several colleagues to visit Italy. Kim Björklund has identified some results of the couple's journey like Hilding's appointment at the Finnish Institute of Technology and his second-placed entry for a Parliament building from 1924. The drawing showed southern influences: a piazza in front of the main building with a fountain and a narrowing stair to a grand and sleek but stylistically pure façade. The 1927 church in the Helsinki neighbourhood Töölö had Italian reminiscences like medallions, the division of the ground floor into a gable tower, a church hall and a chapel and the even walls and their colouring. The art gallery in the same part of the city as well as cinema interiors also reminded of Italy with ancient decoration inspired by Pompeii. Eva's 1929 building for Helsinki's private Swedish-language girl's school had obvious impulses from the couple's journeys. The interior consists

117 Björklund, 'En sentida Grand Tour', 17–18; Tuomi 'The call of the Mediterranean', 78.

118 Björklund, 'En sentida Grand Tour', 16–17; Tuomi 'The call of the Mediterranean', 77.

of parts like a stone floor with geometric fields in different colours, columns, medallions, and arched niches for the taps. The school building contains many details reminding of the ones characterising Hilding's buildings, and Björklund underlines that the architect couple clearly was inspired by each other and had mutual opinions and views on architecture and art.[119]

3 The Old Industrial Empire Still Attracted

Britain did not exert the same attraction on architects as Italy. However, almost every fourth transnationally mobile architect visited the country; a higher share than the engineering specialisations. This was partly a result of their extensive study tours which often included many different countries. One Norwegian architect found his way to Dundee in Scotland, whereas everyone else focused on London. The earlier-mentioned Finnish female architect Salme Setälä worked with the famous architect and town-planner Raymond Unwin in 1917: Unwin was a pioneer in the garden city movement. Among the many architects inspired by garden cities, Tage William Olsson[120] suggested a garden city in his town plan for the north-Swedish mining community Boliden.[121] The garden cities were one reason for architects to cross the North Sea, but London also hosted many architect offices working in Art Nouveau-style. Gustaf Strengell[122] was employed with the well-known architect Charles Harrison Townsend; he became an advocate for a more functional and rational Finnish architecture upon his return. Norwegian architects working in London returned to rebuild Ålesund as one of Scandinavia's most distinct Art Nouveau towns after the fire in 1904.

Civil and construction engineers and chemical engineers had the lowest shares to Britain. According to Chandler and Hikino, Britain's chemical industry 'remained out of the game almost entirely'.[123] However, British food processing attracted some chemical engineers like the Danes who went to margarine factories around Liverpool. Techno-Chemical Laboratories in Clapham in south London was another destination. Erik Johan Wijkman[124] returned from there to become manager for Swedish margarine, oil, and fodder factories.[125]

119 Björklund, 'En sentida Grand Tour', 22–24.
120 TESÖ, 1903.
121 Grönberg, *Learning and Returning*, 209–210.
122 SPO, architect, 1902.
123 Chandler and Hikino, *Scale and scope,* 485.
124 KTH, chemical, 1910.
125 Indebetou and Hylander, *Svenska teknologföreningen*, 847–848.

F. L. Smidth and Christiani & Nielsen, who had some of their largest foreign branches in London, also employed Danish civil, construction and chemical engineers in Britain. Swedish engineers also worked for domestic companies, one as head of SKF's laboratories and one as manager for iron manufacturer Uddeholm's office in Birmingham

Two Swedes assisted when the Belgian zinc mining company *Vieille Montagne,* which also was involved in zinc mining in Sweden, established in Nenthead in Cumbria. Some studied mining at universities in London and Sheffield and some worked with electric furnaces in these two cities as well as in Luton.

Britain was the fifth destination according to the classification applied in table 3. Departing relatively early in the career and as a study traveller increased the likelihood that the destination was Britain. However, the strongest correlation is for graduates from the 1880s, that is, from the earliest period, when there were more remains of British industrial dominance. The United States and Germany gradually overtook Britain as industrial and technological models in the latter part of the nineteenth century. Every fifth Swedish engineer departing in the 1860s chose Britain; every sixth the United States, and every tenth chose Germany.[126] The British economy grew more slowly from the 1870s; the industrial decline was not absolute but relative, and the latter is to some extent important to understand the choices of the Nordic technicians. In the 1880s, the American and German industrial challenge had started, but Britain still held a relatively stronger position.

Nordic technicians were all over Britain in the 1880s; one Norwegian worked at the paper mill in Bury near Manchester, and a Swedish colleague participated in the building of the latter city's electricity works. A few Nordic technicians participated in railway building in southwest England and Wales; one was Finland's Carl-Erik Holmberg, who had studied in Charlottenburg and was mentioned here in the German chapter. Workshops in Liverpool and Ipswich are other examples of places where Nordic technicians served. London was, however, the main destination; Nordic engineers studied at some educational institutions, some were bridge builders and participated in the construction of an underground railway system. Swedish technicians opted more for London than colleagues from the other countries, and the main reason was the earlier-mentioned Thorsten Nordenfeldt who later moved to Paris. His company attracted at least twenty Swedish and single engineers from the other Nordic countries before 1890. Thorsten's nephew Per was the executive engineer and managed the construction office before joining his uncle to Paris.

---

126 Grönberg, *Learning and Returning,* 88–89.

Britain had a relatively strong appeal on the mechanical, electrical, and naval group also later. Bruce, Peebles & Co was an automated electrical workshop in Edinburgh and an experience for many engineers connected to ASEA. The two large American electrical manufacturers had British subsidiaries, and many engineers who were employed at these two companies were also at the parent companies in the United States. The Swedes went to General Electric's subsidiary British Thomson Houston in Rugby east of Birmingham and were often associated with ASEA. A few Danes and Norwegians were at British Westinghouse in Manchester: one of them was Olaf Eckbo[127] who later became manager for an electric factory in Sarpsborg and started some enterprises.[128] In shipbuilding, the British position remained relatively strong on to the twentieth century. True, adaptation to new technology was somewhat slower than in America and Germany, but British shipyards were still advanced in the early 1910s. Thirteen per cent of the Swedish naval architects went to Britain. This was a lower share compared to the United States and Germany, but a high share compared to other specialisations.[129] Glasgow's shipyards and its university's specialisation in naval architecture attracted and was as an experience for engineers graduating in Scandinavia, but British shipbuilding attracted in general. Some Norwegians made good careers upon return, like Reidar Darre Kaarbø,[130] who took over the family workshop in the northern town Harstad upon return from a shipyard in Grimsby. Holm Holmsen[131] was in Isle of Wight before he became manager for a workshop in Trondheim, during a golden age with profitable results.[132] British shipbuilding was especially relevant in Norway, and Andersen concludes that the British production ideal remained relatively strong on to World War II.[133] In this context, we may also mention that the British Isles were a major region of activity for *Det Norske Veritas*. Andersen and John Peter Collett have emphasised the offices in Glasgow, Newcastle-upon-Tyne, Sunderland, and Liverpool as major places of employment for Norwegian engineers.[134]

---

127 KTS, 1900.

128 Bassøe, *Ingeniørmatrikkelen*, 102.

129 Grönberg, *Learning and Returning*, Appendix 3:3; Olsson, 'To see how things were done in a big way', 434–435; Pollard and Robertson, *The British shipbuilding industry*, 49–51, 108–129, 231.

130 BTS, mechanical, 1898.

131 TTL, mechanical, 1892.

132 Inga Berntsen Rudi, 'Trondhjems Mekaniske Verksted',, http://www.kildenett.no/portal/artikler/2008/1203604964.07, (15 November 2017).

133 Andersen, *Fra det britiske til det amerikanske produksjonsideal*, 477–489.

134 Håkon With Andersen and John Peter Collett, *Anchor and balance: Det norske Veritas 1864–1989* (Oslo 1989) 61; Bassøe, *Ingeniørmatrikkelen*, 403.

Shipyards in Newcastle-upon-Tyne, Gateshead, Sunderland, and Middlesbrough were targeted. Hugo Hammar's workplace, Palmers Shipbuilding Company in Jarrow, on the south-side of the Tyne, was one of Britain's largest shipyards and hosted more Scandinavians. In Hammar's memoirs, he re-evaluated this experience and described it as more valuable than he had thought at the time.[135] The area around Glasgow was, however the major shipbuilding destination in Britain; Nordic technicians studied at the university and practised at some of the surrounding's numerous shipyards. Christian Stoltz Dekke[136] followed this pattern and later became a shipyard manager in Bergen, and his countryman Finn Christian Knudsen[137] became shipowner in Porsgrunn in southeast Norway. The shipyard founded by Danish-born Henry C. Lobnitz at the suburb of Renfrew was one of the biggest employers; Lobnitz specialised in dredging plants and rock-cutting machinery but also built many steamers for Danish and Swedish owners.

Scotland received a relatively large part of the Nordic technical mobility to Britain. Workshops in Edinburgh were already frequented, and in some cases, the sources say only 'employed in Scotland'. This may, of course, very well have been at one of the shipyards mentioned above. One Swedish engineer served at the papermills in Inverurie near Aberdeen. Rudolf Liljekvist[138] worked in bridge-building in France before he participated as a builder in the construction of the Forth Bridge over the Firth of Forth, connecting Edinburgh with Fife, between 1883 and 1890. This bridge was Britain's first in steel and was viewed as an engineering miracle. Liljekvist later founded and became manager of a factory to produce chlorine and alkali but is also known as one of the executors of Alfred Nobel's estate.

4 Among Nobel Employees and Finnish Technicians in Russia

The name Nobel is important also for Nordic—read Finnish and Swedish—technical mobility to Russia, which was the only European destination that was more of a migrant than a study-travel destination. Bengt Jangfeldt describes a lively late nineteenth- and early twentieth-century Swedish community in

135 See Alstad, *Trondhjemsteknikernes matrikel*, 89, 234; Bodman, *Chalmers tekniska institut*, 299; Eskedal, *BTS-matrikkelen*, 8, 21; *Danske Ingeniører fra Teknika*, 81–82, 340–341; *Danske teknika*, 345.

136 BTS, 1889.

137 TTL, mechanical, 1887.

138 KTH, civil, 1880.

Saint Petersburg, and among its members were the Nobel brothers Ludvig and Robert. They took over a mechanical workshop founded by their father Immanuel and developed it into one of the largest in Russia. Nobel was an important employer for the engineers.[139] Ludvig's son Carl[140] became part owner and managed the machine shop before he died young of diabetes. Anton Carlsund[141] was employed as executive engineer in the early 1890s and developed a smaller and more efficient diesel engine as well as the world's first reversible diesel engine.[142] The brothers also started naphtha production in Baku in the mid-1870s. The Russian oil industry was organised in the following ten years: The brothers set up laboratories and constructed the world's first tanker at a shipyard in Sweden. They also employed geologists[143] and—of course—Swedish engineers. In the cohort, about 25 Swedish engineers were employed by the Nobel brothers in Baku and some worked also for other companies around the city. The Nobel companies were also important for Finnish employment. The language provided no obstacles as most engineering graduates in Finland had Swedish native tongue. Around 30 Finnish graduates had one of their first employments with the Nobel brothers in Saint Petersburg, and there were also at least fifteen Finnish engineers employed in Baku. Jakob Estlander[144] later founded a factory for recycling waste from petroleum production; the method was his own.[145]

These engineers worked for a Swedish enterprise, even if the main office was on Russian soil. Companies based in Sweden were, however, also established in Russia at an early stage and favourable economic conditions attracted a company like the cement manufacturer *Skånska Cementgjuteriet*. A number of these companies remained after the revolution and Helmer Hesser[146] managed *Skånska Cementgjuteriet's* Soviet subsidiary before he became manager for brickworks in central Sweden in the 1920s. SKF, the steam turbine manufacturer De Laval, and telephone company L. M. Ericsson, which erected a large factory in Saint Petersburg and was granted concessions in several Russian cities, also had extensive activities in Russia.[147] ASEA established an early twentieth-century office in Saint Petersburg employing some Swedish engineers, and the

139 Jangfeldt, *Svenska vägar till S:t Petersburg*, 183–188.
140 KTH, mechanical, 1884.
141 KTH, mechanical, 1887.
142 Jangfeldt, *Svenska vägar till S:t Petersburg*, 190–191.
143 Jangfeldt, *Svenska vägar till S:t Petersburg*, 189.
144 SPO, chemical, 1883.
145 Engman, *Lejonet och dubbelörnen*, 152–157.
146 KTH, mechanical, 1905.
147 Jangfeldt, *Svenska vägar till S:t Petersburg*, 182.

company also began to erect a factory in Yaroslavl; a process interrupted by the revolution. At its resumption in 1927, civil engineer Axel Olsson[148] arrived to complete the factory, and at least another six Swedish engineers were in Yaroslavl in the late 1920s.[149] Danish engineers going to Russia often followed the pattern of employment for a Danish company. Russia was one of the main areas of operation for Christiani & Nielsen, Nyeboe & Nissen, and F. L. Smidth, and engineers travelled eastwards on the payrolls of these companies.

The pattern of going abroad for domestic employers was, as mentioned, much less significant in Norway and Finland than in Sweden and particularly Denmark. In the Finnish case, this was an activity that overwhelmingly was directed to Russia. One company with extensive activities in Saint Petersburg and Moscow before World War I was the stone industry *OY Granit*. Engman has counted eight employees in Russia, and one of them was the architect Ernst Gustaf Hedman.[150] He arrived in 1895, when a Finnish company erected a hotel in Nizhny Novgorod after his drawings. As an associate of Granit, Hedman worked in Saint Petersburg and Moscow, where he initiated a workshop specialising in softer stones. Crichton-Vulcan, a cornerstone in Finnish shipbuilding, dispatched engineers to its establishments in Saint Petersburg. However, most Finnish engineers in Russia worked for the Finnish National Railway Company, whose easternmost station was Saint Petersburg's *Finlyandsky* terminal until 1918. Roughly half of the Finnish railway engineers in Russia were civil servants of the Grand Duchy in the imperial capital.[151]

The Nobel family, Finnish companies, and the national railways were not the only contacts utilised for Russian employment, A few Finnish engineers chain migrated to the 1890s railway building between Moscow and Archangel. Nikolai Winogradoff[152] was head of construction and made later important contributions to railway building in Caucasia and Dagestan. The percentage choosing Russia was significantly higher for Finnish graduates compared

148 KTH, civil, 1914.

149 Indebetou and Hylander, *Svenska teknologföreningen*, 871–872, 941; Lasse Åsgård and Christer Ellgren, *Ericsson: historien om ett svenskt företag* (Stockholm 2000) 65–68; Therese Nordlund Edvinsson, *Att leda storföretag: en studie av social kompetens och entreprenörskap i näringslivet med fokus på Axel Ax:son Johnson och J. Sigfrid Edström, 1900–1950* (Stockholm 2005) 97, note 405.

150 SPO, architect, 1890.

151 Jonatan Reuter, 'Ernst Gustaf Hedman', in: Jonatan Reuter (ed.), *Finlandssvenska tekniker: biografiska anteckningar under medverkan av flera författare* (Helsinki 1925) 31–52; '60 år. Arkitekt E. G. Hedman', *Hufvudstadsbladet*, 19 December 1927; Engman, *Lejonet och dubbelörnen*, 152, 155; 'Femtio år', *Tammerfors Aftonblad*, 7 January 1928.

152 SPO, civil, 1880.

to Scandinavian, and this is not surprising considering Finland's status as a Grand Duchy and the geographical closeness. Hence, national philosopher J. V. Snellman was partly right when he argued that a Finnish Polytechnic Institute would educate for Russia. The tsarist empire played to some extent the same role in Finland as North America played elsewhere in Europe. Before 1860, more than 100,000 Finns went to Russia compared to 2,500 to the United States. Technicians were certainly spurred by some earlier migrants who had made successful careers, like Colonel Hugo Robert Standertskjöld. He made a military career in Russia and developed armament factories in Tula and Izjevsk into large industries before he returned and transformed a cotton-reel factory in Lappeenranta into a major cellulose industry. However, most migrants went to Saint Petersburg, for a long period Finland's 'second city'.[153] The imperial capital was a nearby place to go when Finnish technicians faced problems finding employment in Finland.

Many Finnish technicians also earned work experience elsewhere in Russia. Mining was one such field, and returnees often brought experiences and know-how back to Finland that should not be underestimated. Return migration from Russia to Finland after the revolution implied a boost in some technological fields. Radio engineering was one of the most important. Some technicians went significantly further away than Russian Karelia and the region around Saint Petersburg. Carl Jahn's[154] fascinating story shows the different implications of migration to Russia. Saint Petersburg was one of the closest international destinations, especially since Finland's eastern border ran just outside the city. Jahn travelled to another part of the world when he took the seaway through India to Vladivostok. He projected and constructed a sea canal before he was employed in harbour construction. Vladivostok hosted at least three Finnish technicians at the time: Jahn, a cement factory owner, and a manager and owner of gold-mines and gold-washing establishments around Amur River. Jahn did not even finish the harbour before he was engaged to project a narrow-gauge railway in Sakhalin. Its purpose was to exploit the island's rich carbon mines. Jahn was the only free man in Sakhalin; his workers consisted of deported prisoners. He also engaged in the building of the free port Dalnij, but this project was interrupted by the Russian-Japanese war.

153 Engman, *Lejonet och dubbelörnen*, 145–158, 164; Korkiasaari, *Suomalaiset maailmalla*, 12; Max Engman, 'The Finns in St. Petersburg', in: Max Engman (ed.), *Ethnic identity in urban Europe* (Aldershot 1992) 99–130; Jangfeldt, *Svenska vägar till S:t Petersburg*, 182; 'Standertskjöld, Hugo—Uppslagsverket Finland', http://www.uppslagsverket.fi/sv/sok/view-103684-StandertskjoeldHugo, 15 November 2017.

154 SPO, civil, 1896.

Jahn managed, however, to escape and returned to a career with the Finnish railways.[155]

5 Danish Domination in Asia and Oceania

Jahn and his two compatriots in Vladivostok were still some of the few Finnish technicians in Asia; the number in this cohort is nine, which can be compared to 32 Danish technicians in Siam alone. Nordic technical mobility to Asia and Oceania was significantly stronger from Denmark than from the other countries. East Asian ties were traditionally stronger in Denmark than elsewhere in the Nordic area. The only major exception from the Danish pattern in the region was Japan, which was more of a 'Swedish' destination. The distribution between these two nationalities was also relatively equal in British India. Civil and construction engineers went to this region to a greater extent; the only major exceptions were Japan and Australia. It was a late rather than an early destination and attracted experienced technicians who had worked some years before they departed. It was also late in the sense that the attraction was stronger among the 1910s graduates, with Australia and New Zealand as exceptions.

5.1 *The Danish Colony in Siam*

No other destination in this region was thus as Danish as Siam. Swedish chemical engineer Emanuel Nilsson[156] was the only non-Dane but was still employed by the Danish-owned Siam Cement Company. Steen Sehested[157] travelled to Bangkok early in 1914 to lead this company. One and a half year later, however, he joined the English engineer John Hunter Swanson in the engineering firm Swanson & Sehested. This firm was active all over Siam and opened a branch in Singapore in 1917, whereto the entire business was moved a few years later. The firm also employed Danish engineers in Siam and later also in Singapore. Nilsson and Sehested were not alone at the Siam Cement Company; at least seven of the Danish engineers had the same employer.

When Hans Niels Andersen founded The Danish East Asiatic Company in 1884 and began operating passenger and freight traffic between Copenhagen, Bangkok, and other destinations in the Far East, it opened possibilities

155 Engman, *Lejonet och dubbelörnen*, 161; 'Direktör Carl Jahn', *Hufvudstadsbladet*, 31 March 1937.

156 TESM, chemical, 1912.

157 PTI, construction, 1910.

for technicians in Siam. A few engineers worked for this company managing their Siamese mines and windmills. Siamese migration can be connected to Andersen and Andreas du Plessis de Richelieu. He arrived in 1875 and made a military career. Over time, he became head of the navy with a shipyard as well as all coastal stations and sea fortresses and started, in addition, a power station and the country's first tram and railways. De Richelieu strengthened Siamese-Danish ties and recruited many countrymen to civilian and military positions before he returned due to health reasons in 1902. Siamese railways were described in *Ingeniøren* in 1920, but the author also claimed that Siam was the only country in the world where Danes constituted one of the largest foreign nationalities. Siam Electric Company, Siam Cement Company, and the first and only private railways in the country were founded and managed by Danes. Aage Westenholz[158] served as director of a brick works, which soon recruited another two Danish engineers. Later, he became manager of Bangkok's first tramway and initiated the switch from horse power to electricity. The electrification was finished in 1894, which was much earlier than any city east of the Suez Canal. In 1898, Westenholz founded the Siam Electric Company and became responsible for the process to introduce electric light in Bangkok's streets. Siam Electric Company later took over the running of tramways and railways and employed many Danish engineers in these activities.[159] Hugo Zachariae[160] arrived in 1908 and served as executive engineer for Siamese tramways and railways until he returned to Denmark in 1932.[161] People from Denmark had always been warmly welcomed by the Siamese administration and contributed to Siamese progress. Danish medical doctors and military officers started to arrive in Siam; a few of the arriving engineers made military careers and were civil servants, working with bridge and road construction. Danish engineers continued to travel to Siam in the twentieth century and the Siamese royal family's trust and friendship with de Richelieu—who continuously maintained his contacts with Siam—opened many doors for other Danes. One of the engineers became, for example, a professor of chemistry at Bangkok's university.

---

158 PL, civil, 1884.

159 Hans Christian Bjerg, 'Aage Westenholz—Dansk Biografisk Leksikon', http://denstoredanske.dk/Dansk_Biografisk_Leksikon/Naturvidenskab_og_teknik/Ingeni%C3%B8r/Aage_Westenholz, (15 November 2017).

160 PL, mechanical. 1907.

161 Povl Engelstoft and Svend Dahl, eds., *Dansk biografisk leksikon. XIX Quaade—Rongsted* (København: Schultz, 1940) 493–494; Kaarsted, *Admiralen*, 16, 35–36, 165–174; Hugo Zachariae, 'Siams Jærnbanevæsen', *Ingeniøren* 54 (1920).

5.2 *Danish Connections in East Asia and Oceania*

China was the largest destination in Asia; receiving a total of 80 technicians in the cohort, of whom 41 were Danes. Technicians from Norway were less numerous and Finland contributed only with Werner Paavola,[162] who worked with railway construction from 1902 to 1904.[163] Some of the Norwegian engineers also worked with railway construction, but Ole Waldemar Aga[164] practised as an architect in Shanghai from 1901 to 1906. He later opened his own architect's office in San Francisco.[165]

Some Danish technicians worked at so-called Conservancy Boards to undertake hydrographic investigations of the rivers, but this was more of a Swedish pattern. Swedish engineers are noted for several activities in China; a few served in railway and bridge construction. Others were employed by Swedish companies such as ASEA, Bofors, and L. M. Ericsson, whereas Danish engineers often worked for F. L. Smidth, Nielsen & Winther, and the Aarhus Oil Factory. These companies were established in Shanghai, a city that hosted a large early twentieth-century Danish community. This was even more due to the presence of The Danish East Asiatic Company and the Great Northern Telephone Company,[166] two companies that also employed several Danish engineers. Some also started their own enterprises. Shanghai was one of, basically, five places where Danish engineers worked; the others were Beijing, Hong Kong, Tientsin in the northeast, and Hankow in eastern central China. They served at cement factories and also in municipal bodies such as waterworks and harbours. One engineer was an advisor for the Chinese government, another owned a prospector enterprise and dug for gold in China and Siberia. One was employed to investigate the Yellow River, while another served as a missionary rather than an engineer. Carl Gimbel[167] arrived in China in 1908 and projected and headed the setup of the Imperial Waterworks in Beijing and was later a professor of mathematics and statistics at The Imperial University. In the mid-1910s, he was appointed district superintendent in the Chinese salt-administration, and served in different parts of China as well as in the central administration until the administration was dissolved by the Nationalist government in the late 1920s. Gimbel retired and returned to Denmark.[168]

162 SPO, civil, 1902.
163 *Suomen Insinöörejä ja arkkitehtejä*, 654.
164 BTS, mechanical, 1899.
165 Eskedal, *BTS-matrikkelen*, 22.
166 Bramsen, *Open doors*.
167 PL, construction, 1904.
168 Jespersen, *Biografiske oplysninger*, 236–237; Povl Vinding, 'Carl Gimbel—Dansk Biografisk Leksikon', http://denstoredanske.dk/Dansk_Biografisk_Leksikon/Naturvidenskab_og_teknik/Ingeni%C3%B8r/Carl_Gimbel (15 November 2017).

Denmark had long traditions of connections with these parts of the world and relatively large shares of the crews sailing for the Dutch East India Company in the seventeenth century came, for example, from the Danish Kingdom, from Denmark proper, and from Schleswig and Norway.[169] In 1911, *Ingeniøren* wrote about the Dutch government's problem recruiting their own engineers to their colony in today's Indonesia. Engineers from the Netherlands were described as reluctant to leave the native country, but more important was the comparably low wage. The working conditions may imply a lot of adventure; the article claimed that an engineer travelling to Dutch East India had to combine the ability to adjust with an energetic personality. There was little to build on from Danish education and domestic work experiences, but construction, mechanical, and electrical engineers still had good career possibilities. The author claimed that it was only an 'elite' among engineers who had a chance for success.[170]

This somewhat cautionary tone did not keep Danish engineers away from the Dutch colony. Occasionally, *Ingeniøren* reported on Danish employment. In 1920, there was a note stating that the colony primarily needed construction engineers, to some extent electrical and mechanical engineers with experience in telephone or telegraph technology, and a smaller number of chemical engineers. Conditions were described as good; contracts were signed for five years with a yearly wage increase; a first-class journey from Copenhagen to Batavia (Djakarta) via The Hague was paid for by the Dutch state and included family members for married engineers. Even return journeys for engineers who had to quit due to circumstances beyond their control were covered.[171] Such conditions must have spurred Danish engineers to do a few years in Dutch East India. It was, however, not uncommon to stay for long periods, sometimes even for life. One settler was Niels Langkilde Thiele.[172] Upon arrival in 1912 and until 1917, Thiele projected and built a hydroelectric power station in a river in Java. For some years, he worked in Bangkok, Singapore, and Kuala Lumpur and returned briefly to Denmark. In 1924, he settled in Batavia, first to become head of the local branch of a Dutch engineering firm. From 1926 onwards, he had his own business, specialising in iron-concrete.

---

169 Sindre W. Aarsbog, Med Mars og Merkur: en analyse av norsk deltakelse i VOC basert på skipssoldbøker 1633–1794 (Trondheim 2003) Master Thesis Norwegian University of Science and Technology.

170 'Bør danske Ingeniører søge til hollandsk Indien?', *Ingeniøren*, no. 50 (1911) 506–508.

171 'Danske ingeniører til nederlandsk Indien', *Ingeniøren*, no. 3 (1912) 40; 'Sporvejsingeniør til Sumatra', *Ingeniøren*, no. 8 (1913) 60; 'Elektroingeniører til nederlandsk Indien', *Ingeniøren*, no. 8 (1914) 39; 'Ingeniører til Nederlandsk Indien', *Ingeniøren*, no. 66 (1920) 493–494.

172 PL, construction, 1906.

Some Nordic technicians worked with railways, roads, and as city engineers in Sumatra, Borneo, Celebes, and even in Timor. Two Norwegian brothers constituted the only mining engineers and served as geologists for the Dutch government. One of the few Swedish technicians in Dutch East India headed a rubber factory in Sumatra, whereas one of the Norwegians served as a harbour engineer in the same island. Danish Axel Obelitz[173] served several years as city engineer in the port city of Makassar in southwest Celebes. However, most technicians served on the main island of Java where the capital city also was located. Asger Smitt[174] arrived in 1912 and came to serve in Batavia's municipal bodies. Smitt planned and erected the waterworks, and became its manager before he returned to Denmark in 1923.[175] Most engineers served in municipal and governmental bodies, for example, with different aspects of water; waterworks, water power, and irrigation. Construction of railways, tramways, and to some extent roads also were among the most common activities. There were also Nordic engineer-migrants who were privately employed, for example, at cement, sugar, tobacco, and petroleum factories.

Denmark also dominated in the most remote destinations—Australia, New Zealand, and some Pacific islands—but not as markedly as in East and South East Asia. On New Zealand's North Island, Swedish Yngve Övergaard[176] installed the electric equipment for the Arapuni Power Station in 1929 as an employee of ASEA,[177] whereas Norwegian Kolbjørn Thorvaldsen Jenssen[178] was responsible for the harbour in Wellington from 1904 and onwards. Three technicians served Danish cement industries in New Zealand, while Poul Sylow[179] worked for Sydney-based Colonial Sugar Refining Company in Auckland. Sylow also served this company in Sydney and Brisbane as well as in the Fiji Islands.[180] He was accompanied by Frantz Albert Velschou,[181] who worked with irrigation in the mid-1880s.[182]

---

173 PL, construction, 1911.

174 PL, construction, 1908.

175 Jespersen, *Biografiske oplysninger*, 289.

176 KTH, electrical, 1919.

177 Alstad, *Trondhjemsteknikernes matrikel*, 200; Alstad, *Tillegg til Trondhjemsteknikernes Matrikkel*, 56; Bassøe, *Ingeniørmatrikkelen*, 249; Talvitie, *Suomalaisten teknikkojen*, 191; *Suomen Insinöörejä ja arkkitehtejä*, 653; Indebetou and Hylander, *Svenska teknologföreningen*, 1116.

178 TTL, mechanical, 1903.

179 PL, chemical, 1883.

180 Jespersen, *Biografiske oplysninger*, 103.

181 PL, civil, 1880.

182 Jespersen, *Biografiske oplysninger*, 86.

Velschou and Sylow were also in Australia, the largest destination in Oceania. Denmark dominated, but a few Norwegian technicians worked around Perth in Western Australia—in the governmental laboratory, for example—and some Swedes at SKF's office in Melbourne. Swedish Bror Laurell[183] constructed a smaller sulphite factory for The Queensland Pine Company in the early 1910s. In 1918, *Ingeniøren* published a report on Queensland as a working field for young engineers, and the author claimed that the conditions in Brisbane and elsewhere in the state probably were valid for all Australia. Among the foreigners down under no group was more appreciated than the Danes, wrote the author. The Danes were looked upon as reliable and calm. The climate was good, especially in the southern part of Queensland, and Australia was doubtlessly the most democratic country on earth. On the other hand, the war was an obstacle for Danish technical immigration and the author believed that Australia needed between six and eighteen months to make provision for the country's returning soldiers after peace had been established. It was, however, only a matter of time before Australia was ready to offer good employment conditions for young Danish technicians, primarily for the government, and to some extent in sugar refining and mining. The author advised that the Danish Engineer Association should create an information bureau about Australian conditions.[184] Most Danish engineers worked in the sugar refining and cement industries, but around Sydney and not in Queensland.

5.3 *Exceptions to the Danish Rule in Japan and British India*

British India and Japan were the only two major destinations in Asia and Oceania where Denmark was not the Nordic country contributing most engineers. The number of Swedes was higher, especially in Japan, where they constituted two-thirds. This was basically due to the offices of Gadelius & Co in Tokyo, Kobe, and Osaka, which employed almost all of them. Gothenburg merchant Knut Gadelius initially founded a trading company to sell iron and steel as well as pulp and paper products to the Far East. Later, he represented several Swedish engineering firms in Japan. Most Nordic engineers in Japan worked for foreign companies; the only Finnish graduate represented an American pulp and paper company, one of the Norwegians a British steel corporation, and some Danes were in Japan for the Copenhagen shipyard Burmeister & Wain and a London-based cement manufacturer.

183 TESÖ, chemical, 1903.

184 K. L. Rahbek, 'Queensland som Arbejdsfelt for unge danske Ingeniører', *Ingeniøren* 66 (1918) 459–460.

Danish engineers in India worked primarily in the cement business, partly for Danalith and F. L. Smidth. Swedish engineers were important for the iron industry. Olof Sahlin[185] managed the Mysore Ironworks in southern India in the mid-1920s,[186] and his colleague Gustaf Fisk[187] participated in the setup of India's first ironworks, Tata Iron & Steel Company, in Jamshedpur in the northeast.[188] However, there were single Nordic technicians in several branches in India; a few Swedes served at electricity works and power stations; another was in the match industry, and one served as a construction engineer for the Indian government. Olle Andersin[189] from Finland headed a mechanical workshop as well as an industrial school in central India for ten years. At the same time, he served as a missionary.[190] Thus, India did attract some interest among technicians in the Nordic countries. Norwegian Engebret Rua[191] went on a business trip to investigate the possibility of starting up wood processing.[192] Finland's Väinö Vähäkallio[193] became one of few study travellers to Asia and the only architect going to India in the cohort when he included the country on a 1911 study trip that also went to Turkey and Egypt.[194]

5.4 *Study Travelling, Railway Building, and Mining in Africa*

Visiting Egypt was, together with countries like Algeria, Tunisia, and Morocco, often a part of the Mediterranean study tours. Hilding Ekelund was at least in Tunisia and Algeria during his Mediterranean tours and described how he visited Algerian Timgad:

> This becomes a remarkable experience, a town almost as preserved as Pompeii, with handsome regularly arranged and beautiful sites. I have never seen so many columns all at once, some of the streets lined with covered columned or pillared walkways, a well-preserved triumphal arch the most monumental of all.[195]

---

185 KTH, mining, 1916.
186 Indebetou and Hylander, *Svenska teknologföreningen*, 980.
187 TESÖ, mechanical, 1905.
188 Forsberg and Adlers, *Tekniska föreningen i Örebro*, 334.
189 SPO, chemical, 1908.
190 '50-åring', *Åbo Underrättelser*, 8 September 1934.
191 KTS, mechanical, 1898.
192 'Ingeniør E. Rua', *Ringerikets blad*, 22 April 1956.
193 STK, architect, 1909.
194 'Sjuttioåringar', *Hufvudstadsbladet*, 16 June 1956.
195 Cf. Tuomi, 'The call of the mediterreanen', 79.

Tunisia was visited by two Swedish study travellers, one architect and one construction engineer. Countries like Algeria and Morocco were also study-trip destinations for architects: Finnish female architect Elsi Borg was, as mentioned, the only woman in the cohort outside Europe and North America when she travelled in Morocco in the 1920s.

Railway building and surveying involved a few Nordic technicians in the Gold Coast and the Niger territories as well as Mozambique, Southwest Africa (Namibia), and Senegal. Two technicians owned and headed coffee plantations in Kenya, and one served as city engineer in a Rhodesian city. Mining engineers and metallurgists found their way to several African countries. Gold mining in West Africa attracted a few. Norwegian Wilhelm Koren[196] served in Congo and managed a gold mine in Tanganyika in the late 1920s and 1930s, and a few of his countrymen went as geologists to investigate mines in Ethiopia and Sudan. Two Swedish colleagues worked in Madagascar; one of them was Sven Schwartz,[197] who later became vice manager of the north-Swedish mining company Boliden and chairman of the Swedish Employer's Association (SAF).

5.5 *Mining and Construction in South Africa*

The most frequent African destination was the one furthest away. South Africa had a stronger appeal in Norway for mining engineers, civil and construction engineers, and technicians who left school before 1900. Kjetil Kvist concludes that some of the graduates from Trondheim went almost at the same time, but did not necessarily influence each other. He notes that most graduates departing in the late 1890s were architects, but in our cohort, civil and construction engineers dominated the pattern from both Norway and Denmark.

Almost all graduates in Kvist's study went to Cape Town. Martin Bergmann Torstenson[198] managed well in South Africa, according to Kvist. He opened an architect office after the Boer War and was later appointed a teacher in physics and chemistry at two high schools in Cape Town. His diploma from the school in Trondheim was evaluated as on a par with a university degree when he was applying for employment in South Africa. Torstenson settled in South Africa for good.[199] Swedish Gösta Emanuelsson,[200] who served as a builder on the

196 UiK, mining, 1911.
197 KTH, mining, 1915.
198 TTL, architect, 1896.
199 Kjetil Kvist, Arbeidslyst—eventyrlyst: Trondhjems Tekniske Læreanstalts-elevers utferdstrang i 1890-årene (Trondheim 2005) Master thesis Norwegian University of Science and Technology, 79–80.
200 TESÖ, mechanical, 1916.

electrification of the railway between Glencoe and Pietermaritzburg,[201] also settled in South Africa.

Among the Swedish technicians in South Africa, we can also find the earlier-mentioned Emil Lundqvist and Ivar Kreuger. Lundqvist was sent to South Africa to sell German electrician Bergmann's machines, but spent some years as an electrician and works engineer in the diamond mines in Kimberley. This may reflect the desire to apply American methods in South African mines, a pattern that, as mentioned, was revealed by Teisch and Higginson.[202] Lundqvist was clearly inspired by American principles. Kreuger had established contacts with a London-based engineering firm to participate in the building of a large hotel in Johannesburg. He worked in London with drawings and planning and visited a steel plant in Saarbrücken to control the material before he departed for South Africa. After finishing the hotel, Kreuger unsuccessfully looked for other employment in Johannesburg's construction business. To earn money, Kreuger and his friend opened a downtown restaurant and sold it at a good profit after some months. Kreuger now left for India but was soon present in Paris for language studies. One year later, he was back in New York to obtain the knowledge that later made him a Swedish pioneer in reinforced concrete.[203]

6 Infrastructure, Mining, and Sugar Plantations in Latin America and the Caribbean

Kreuger was a globetrotter and seemed always to be on his way somewhere. During his first years abroad, he worked in three cities in the United States, but also in Havana, Cuba, and Vera Cruz, Mexico.[204] He was one of about 400 Nordic technicians working in the Americas south of the United States. There were certainly Nordic technical migrants to remote places in this part of the world. One Norwegian settled for good in the Galapagos Islands and started a coffee plantation there in the 1920s.[205] One Finnish and one Norwegian technician participated in railway building in Costa Rica and Honduras respectively and were the only visitors to these two countries.[206] Synnøve Ones Rosales identifies a few Norwegian engineers participating in the modernisation of

201 Forsberg and Adlers, *Tekniska föreningen i Örebro*, 323.
202 Higginson, 'Privileging the Machines', 1–34; Teisch, 'Home is not so far away', 139–160.
203 Thunholm, *Ivar Kreuger*, 24–25.
204 Indebetou and Hylander, *Svenska teknologföreningen*, 625.
205 Brochmann, *Vi fra NTH*, 76; Bassøe, *Ingeniørmatrikkelen*, 223.
206 *Matrikel öfver Tekniska realskolans*, 269.

Guatemalan infrastructure, but her examples graduated after 1919 and are not included in our dataset.[207] Panama was a common study-trip destination. Some arrived to see the building of and the finished Panama Canal, and the employed ones usually also served at the canal.[208] However, Argentina was, by far, the most common country to go to and accounted for almost half the visits in the region. Brazil, Chile, Mexico, and the Danish West Indies were other larger destinations.

Danes went to this region more often than the other nationalities, but the differences were smaller compared to Asia and Oceania. This reflected general migration patterns. Argentina, for example, received about 13,000 Danish immigrants before World War II, compared to around 4,500 Swedish immigrants. Norwegians also went here relatively often. Steinar A. Sæther estimates that engineers represented between five and ten per cent of the nineteenth and early twentieth century Norwegian population in Latin America, although many exaggerated their titles, raising the share of engineers.[209] Sæther emphasises the importance of Norwegian shipping for the general emigration to Latin America: hundreds of Norwegian vessels sailed between European and Latin American port and thousands of young Norwegian men worked on them.[210] This linkage to Latin America had no counterpart in Sweden. Regarding technicians, however, the real number of Swedes was almost equally high, but this is not surprising when we consider that the Swedish technical schools educated more than twice as many as the Norwegian ones. Finns were very few, but—apart from in Argentina—single technicians were in Brazil, Costa Rica, Cuba, Mexico, and Uruguay. Latin America and the Caribbean primarily attracted civil and construction engineers and mining engineers and metallurgists. The former group accounted for the highest numbers in most countries; Mexico and Colombia were major exceptions. However, mining engineers and metallurgists accounted for higher shares to the countries in the Andes, that is, Chile, Bolivia, Peru, but also in Brazil. Chemical engineers often went to the Caribbean. Study travelling was very uncommon, but some journeys to Panama were undertaken by graduates who already lived in the region.

207 Synnøve Ones Rosales, 'Opportunities for the few and select: Norwegians in Guatemala (1900–1940)', in: Steinar A. Sæther (ed.), *Expectations unfulfilled: Norwegian migrants in Latin America, 1820–1940* (Leiden 2015) 148–151.

208 http://www.hemneslekt.net/getperson.php?personID=I75458&tree=Hemne, 15 November 2017.

209 Steinar A. Sæther, 'Making sense of a minor migrant stream', in: Steinar A. Sæther (ed.), *Expectations unfulfilled: Norwegian migrants in Latin America, 1820–1940* (Leiden 2015) 42–43.

210 Steinar A. Sæther, 'Introduction', in: Steinar A. Sæther (ed.), *Expectations unfulfilled: Norwegian migrants in Latin America, 1820–1940* (Leiden 2015) 10.

The region was a less likely destination for graduates of the 1890s, possibly because of the earlier-mentioned Baring crisis.

6.1 *Infrastructure Building in Eastern and Southeast South America*

Technicians went to Argentina because they loved the adventure, claimed a Danish engineer in a 1912 lecture. An adventurer would, however, he continued, be disappointed in the land of the Pampas.[211] Sæther states that Norwegian engineers were employed in 'port contruction, in the building of dams, in hydrological surveys, and in the petroleum industry, both off-shore in Comodoro Rivadavia and in the inland exploration and production in the northern provinces'.[212] Argentina's first Norwegian engineers were hired by the head of the country's border commission to explore Patagonia in the 1890s. The purpose was to establish the exact border between Argentina and Chile. Later, a small army of Norwegian engineers was established to defend Argentine interests.[213] An overwhelming majority of the Nordic engineers served, however, the railways with different tasks such as bridge building. Danish engineer Christian Clausen[214] was employed at the *Pacifo* railway and the underground rail in Buenos Aires until 1913. Later, Clausen travelled around in the world as the representative of an oil factory in Aarhus. Danish engineers also worked at Argentine cement factories, some for F. L. Smidth and Christiani & Nielsen. Some served in drainage, primarily in Buenos Aires.[215] Thomas Segelcke[216] was, however, second manager and responsible for operations as well as the laboratory at Argentina's largest distillery in the early 1910s, but returned to Denmark because of blindness. Norway's Alf Storm Stig[217] was a chemist with the railways and later with two butcheries before he started his own enterprise dealing with citrus fruits. Stig also made several inventions. Some Nordic engineers worked in hydroelectrics. *Compania Hidro-Electrica de Tucuman* employed at least four Swedish engineers in water-power construction in northwest Argentina in the early 1910s.[218]

211 Lauritz Nielsen, 'Om den underjordiske Bybane i Buenos Aires samt om Forholdene for Ingeniører i Argentina', *Ingeniøren*, no. 104 (1912) 807; Hannover, *Dansk Civilingeniørstat 1942*, 192.
212 Sæther, 'Making sense of a minor migrant stream', 50.
213 Sæther, 'Making sense of a minor migrant stream', 50.
214 PL, construction, 1911.
215 Harnow, *Den danske ingeniørs historie*, 233; Hannover, *Dansk Civilingeniørstat 1942*, 222.
216 PL, chemical, 1888.
217 KTS, chemical, 1912.
218 Indebetou and Hylander, *Svenska teknologföreningen*, 603.

A few technicians also proceeded to south-Atlantic islands. Hans Hindrum[219] set up a whaling station in South Georgia and was one of three Scandinavian technicians on this island. Migration to Uruguay and Paraguay was also connected to moves to Argentina. Thirteen of a total of seventeen technicians in Uruguay were in both countries, and the same was true for five of seven in Paraguay. Employment patterns in these two countries resembled the Argentine; Nordic engineers participated in the construction of railways and tramways, and Montevideo's waterworks and Asuncion's harbour are other examples of employers. One Danish engineer in Uruguay worked in the cement business; a countryman constructed a wireless telegraph for the government, and another was a chemist at the agricultural ministry. A Swedish colleague served in Paraguay's marine and war ministry.

Brazil differed from Argentina and Uruguay as the country—relative to Danes and Norwegians—attracted more Swedes, but employment in railways and railway building was a mutual pattern. Brazil was also a frequent destination among mining engineers and metallurgists. There were single employments in harbour construction, mining, electrical engineering, and one graduate from Trondheim served as an architect in Sao Paolo.[220] Swedish and Danish technicians also served companies like Christiani & Nielsen, AGA, ASEA, and SKF. Viktor Theodor Nyquist[221] was employed in a drawing office in Rio de Janeiro and later as a deputy for a mapping expedition to Patagonia. Nyquist became city engineer in La Franca, started an enterprise for bridge building, sewing, and so forth together with a Spanish colleague and was appointed as manager for one of Brazil's most distinguished railway builders before he fell ill. Nyquist passed away in the Canary Islands on his way back to Sweden.[222]

6.2 *Mining in the Andes*

Railway building was also a major activity in western South America, for example, the construction of a railway all through the elongated Chile in the 1910s. Norwegian engineers worked on the construction of the Transandine railway.[223] Swedish technicians also served a Bolivian oil company and a Peruvian rubber company. A few Swedes participated in the re-building of the harbour

219 KTS, construction, 1901.

220 Kjartan Fløgstad, *Eld og vatn: nordmenn i Sør-Amerika* (Oslo 1999) 161.

221 TESM, mechanical, 1882.

222 *Malmö teknologförbund*, 368.

223 Cecilia Alvstad, 'Migrants on skis: Norwegian-Latin American return migration in the 1890s', in: Steinar A. Sæther (ed), *Expectations unfulfilled: Norwegian migrants in Latin America, 1820–1940* (Leiden 2015) 82.

in Peru's coastal city Callao, and some were representatives of domestic companies and other Swedish interests in Lima. In Chile, Swedish technicians also represented companies such as AGA and Swedish Match, whereas Danish technicians worked in the cement business, often for F. L. Smidth and subsidiaries.

Mining was, however, the most frequent activity in these mountainous regions. In 1924, the Norwegian journal *Teknisk Ukeblad* published a note directing the attention of mining engineers to Chile. The country was the third destination in the region and attracted many Norwegians, chemical engineers, mining engineers, and metallurgists. 'Mining engineering in the southern Andes developed into a specific migrant niche for Norwegian engineers between 1890 and 1930', writes Sæther.[224] As mentioned in chapter 3, the Guggenheim family had formed an exploration company to search for and buy profitable mines, and their pit copper mine in north-Chilean Chuquicamata developed into the world's largest. The writer in *Teknisk Ukeblad* reported that the mine and copper works were looking for people. A Norwegian engineer who had arrived one year earlier described good opportunities for those who knew English, could make themselves understood in Spanish, and had practised in British and American mining. Chile offered good wages, interesting work, and was one of the world's most beautiful and pleasant countries.[225] Norwegian Elias Anton Cappelen Smith[226] arrived earlier, in the 1910s, and developed a method to extract ore poor of copper with brochantite and was awarded a gold medal from the American Institute of Mining and Metallurgy.[227] Chuquicamata, together with the copper mining environments in Potrerillos, also in the north, and in Rancagua in central Chile employed more than fifteen technicians of whom a large majority were Norwegians. The patterns in Bolivia and Peru resembled the Chilean. Nordic engineers worked in mining, and some Norwegians served the American Anaconda Mining Company in Bolivia.[228]

6.3 *Building A Telephone Net in Mexico*

A few Nordic technicians served in mining also in Mexico, but railway construction was a more common feature there. A chemical engineer from Denmark

224 Sæther, 'Making sense of a minor migrant stream', 50.

225 C. K., 'Gode utsikter for bergingeniører i Chile', *Teknisk ugebla.d*, no. 9 (1924) 92.

226 TTL, chemical, 1893.

227 Fløgstad, *Eld og vatn*, 161; Arne Espelund (2009, 13. februar). Elias Anton Cappelen Smith. In: *Norsk biografisk leksikon*, https://nbl.snl.no/Elias_Anton_Cappelen_Smith, 15 November 2017.

228 Fløgstad, *Eld og vatn*, 160.

managed a distillery; a Swedish colleague owned a metallurgical-chemical laboratory in the capital, while a Norwegian managed Mexico City's construction of waterworks and streets.

However, the Swedish telephone company L. M. Ericson was an important factor in the mobility to Mexico and contributes to a relatively strong Swedish and mechanical, electrical, and naval pattern. The company obtained the concession for the telephone net around Mexico City in 1905. Mexico's telephone net was, from the beginning, built by Americans, but the country obtained knowledge to develop it from Sweden. Hungarian-born American José Sitzenstatter had been Bell's representative in Stockholm, knew the Swedish language and the country's telephone industry. He contacted L. M. Ericsson in London to ask for an offer on a telephone station but could not arrange the financing. Sitzenstatter asked L. M. Ericsson's office in Stockholm if they were interested in taking over his concession. A foothold in Mexico was a good starting point to establish the company in Latin America, and Ericsson cooperated with another Swedish telephone company to operate the net.[229] Mexico became one of Ericsson's most important markets, and several Swedish engineers worked for Ericsson in Mexico at the same time. Erik Östlund[230] superintended the telephone net and settled in Mexico forever, whereas Helge Rost[231] served twenty years as executive engineer.

6.4 *Railways and Sugar Making in the Caribbean*

Juhani Wiiste[232] was the only Finn and one of two Nordic architects in Mexico. He later continued to Cuba, and his travels also embraced North America and some European countries, before he returned to a position as county architect in Viipuri. In Cuba, too, he was the only Finn and the only architect. Nordic architects were uncommon in the Americas.

In general, migration to Cuba was described as an 'insane' idea by an early twentieth century Norwegian writer. Engineers constituted the only group with a chance to find employment.[233] Some Nordic technicians in Cuba worked in railway construction, including Norwegian Waldemar Michelsen,[234] who also invented a centrifugal machine while he was residing in Jatibonico in

229 Åsgård and Ellgren, *Ericsson*, 70, 73–75.

230 KTH, chemical, 1897.

231 KTH, electrical, 1907.

232 STK, architecture, 1915.

233 Mieke Neyens, 'The good, the bad and the rational: desirable and undesirable migration to Cuba and Mexico', in: Steinar A. Sæther (ed.), *Expectations unfulfilled: Norwegian migrants in Latin America, 1820–1940* (Leiden 2015) 112.

234 KTS, mechanical, 1902.

the central part of the country in the early 1910s. Knud Sadolin,[235] who became part-owner of a well-known colour and lacquer company upon return to Denmark, also projected a Cuban railway station and served a railway company. Norwegian Arne Kolbjørnsen[236] became railway manager in Haiti.[237] Other Caribbean activities included mining in Cuba, serving a British oil company as a geologist in Trinidad and Tobago, a single Dane in the cement business, and Swedish representatives of ASEA in Havana.

Sugar making and distillation in Cuba, the Dominican Republic, Jamaica, Puerto Rico, and Haiti involved a few Swedes, but primarily Danes. The largest Caribbean destination was, however, the Danish West Indies, the islands of St Croix, St. Thomas, and St. John, an area that attracted technicians only from Denmark. Danish engineers projected railways and harbours, performed inspections and surveys of land and buildings, and founded and managed telephone companies, mostly on St. Croix. The sugar business was a major employer also in the Danish West Indies, and refineries and plantations drew their managers from among technicians educated in Copenhagen. Folmer Andersen[238] was the only one still living in the islands in the early 1930s. Another omnipresent engineer was Carl Jens Gudik Sørensen,[239] who constructed a far-reaching irrigation plant which was viewed as a solution to St. Croix's dry spell. Sørensen created, as *Ingeniøren* expressed it in 1909, 'more rain for Sancta Cruz', one wish that Caribbean plantation owners in the old days used to propose a toast for. Plantation manager Gunnar Hagemann[240] had following the journal created another field for able engineers when he employed Sørensen.[241]

7 Summary

This chapter has dealt with transnational mobility to destinations outside the two 'main' ones; German-speaking Europe and North America, a pattern that embraced about one-third of the Nordic technical school graduates and two-thirds of the ones who are registered for going abroad. It was most common among Finnish graduates and least among their Norwegian colleagues.

---

235 PL, construction, 1904.
236 TTL, mechanical, 1882.
237 Fløgstad, *Eld og vatn*, 160.
238 PL, construction, 1914.
239 PL, construction, 1904.
240 PL, chemical, 1902.
241 Valdemar Nielsen, 'Irrigationsanlæget paa Plantagen La Grange. Brev fra St. Croix', *Ingeniøren*, no. 32 (1909) 299–301.

Mobility between the Nordic countries was the most common pattern, besides going to the two main areas. Sweden was an interesting place to study industrial development and trends in architecture and more of a study-trip destination. Danes, however, went to Sweden to manage brickworks and participate in activities such as harbour, bridge, and railway building for Danish companies. Denmark offered interesting study objects in architecture, brewing, and food processing. The study-trip character of the mobility to Sweden and Denmark made the countries Finnish destinations. Norway and Finland received many native-born technicians who came back from studies in Sweden, and they were dispersed to different fields such as shipbuilding and shipping as well as mineral and textile industries. Norway and Finland were also subjects to Swedish and—to a smaller extent—Danish expertise migration. Some technicians represented Swedish and Danish companies in electro-technology, construction, and cement making. Swedish technicians also served in mining, steel and iron, and hydroelectric power. Technicians going to Iceland were often native Icelanders educated in Denmark. Infrastructure building such as road and harbour construction was the focus. Greenland also received technicians from Denmark who worked primarily in telecommunications and cryolite mining.

Mobility to other European countries, that is, Europe outside the Nordic countries, Britain, and Russia, constituted the fourth largest mobility pattern and was characterised by extensive study travelling; it was the only region in which the real number of study travellers was higher than the number of migrants. France was, by far, the most frequented country, but Belgium, Italy, the Netherlands, Spain, Hungary, Poland, and Czechoslovakia also received relatively large numbers. Architects were most likely to go to this region, which included some of their most classical destinations. France offered interesting objects such as the Eifel Tower, other steel constructions, and Parisian monumental buildings and streets. Modern methods and the practical and logical French architecture were also inspirational sources. Belgium as a centre for Art Nouveau, Dutch modernist architecture, Spain's Gaudi and Gothic buildings, a mixture of architectural styles in Budapest and Prague, Greek, Roman, and Byzantine architecture in the eastern Mediterranean were all attractions that made this into a pilgrimage region for Nordic architects. The Ekelund couple's study trip in the early 1920s illustrates the attractions of Italy; parks, gardens, mural paintings and ornaments, early Renaissance and *architettura minore*.

There were other interesting objects in continental Europe. France offered possibilities to deepen the knowledge of reinforced concrete as the home of pioneer Francois Hennebique. His principles were essential to the foundation of the Danish construction company Christiani & Nielsen and used in many of its constructions. The company later brought many Danish engineers to France by setting up an office in Paris. France was also a targeted for studies in fields such as marine engineering, aeronautics, and aviation, bridge and road construction, chemistry, and for those who wanted to change track and go into fields like mathematics or painting. The country offered employment possibilities in a variety of different fields. In Belgium, the region around Liege offered possibilities both for university studies and employment in fields such as mining and metallurgy, electrical engineering, and textile technology. Dutch water management and peat trade also attracted interest among Nordic engineers. Several returnees from these activities later played important roles in Nordic industries, implying technology transfer. Employment with the Budapest-based Ganz Company—wherefrom Tesla claimed he had gotten the idea for the induction motor—was also a means to learn about efficient electrotechnical constructions from some pioneers in the field.

The employment patterns partly point to technology transfer through the Nordic technicians to the adopted countries, but not necessarily transfer from the Nordic countries. Many technicians represented American and German companies in different countries. The American companies General Motors and Bell Telephone employed some Nordic engineers at their large European branches in Antwerp, and Germany's AEG had, for example, a Danish technician assigned to the tramways in Constantinople. Some technicians worked for businesses founded and owned by fellow countrymen but based in the adopted country. The engineering firm Julian Kennedy, Sahlin & Company in Brussels, the arms factory Nordenfelt in Paris, and Alfred Nobel in San Remo are Swedish examples, while the Budapest entrepreneur Gregersen constitutes a Norwegian one. Technology, or knowledge, transfer from the Nordic countries was in some cases obvious. Belgian, Polish, Romanian, and Spanish cement industries utilised, for example, Danish technicians, whereas Belgian and Baltic match factories employed Swedish colleagues. Czechoslovakian, Hungarian, and Italian mining ventures, steel and iron industries, rubber factories, and cellulose industries as well as, for example, Turkish and Portuguese road, railway, tunnel, and bridge construction involved Nordic technicians. Many technicians did, of course, mediate knowledge through employment with domestic companies: F. L. Smidth was represented in most European countries, engineers working for the Swedish telephone company L. M. Ericson played major roles in the build-up of telephone systems in the Netherlands and Poland after her independence, and *Norsk Hydro*'s establishments in the Pyrenees imply transfer from Norway.

Britain was the fifth destination according to our division but would have been the third on country level in numbers, and the fourth in percentages.[242] Finnish study travellers, Danes—whose multinational companies had large offices in London—architects and, not least, early graduates went to Britain. We are studying a period when a relative, although not absolute, British industrial decline had started, but the 1880s can be viewed as a period when the British world position remained comparably strong. The 1880s saw Nordic technicians in railway building in England and Wales, English and Scottish mechanical workshops, the construction of London's underground system, and, not least, employed by Swedish entrepreneur Nordenfeldt's London-based machine-gun factory before it moved to Paris in the early 1890s. The mechanical, electrical, and naval group found their way to advanced naval architect schools and shipyards in Glasgow and in northeast England already in the 1880s, and this pattern continued all through the studied period. Automated electric workshops like Bruce and Peebles in Edinburgh and the British subsidiaries of the large American electrotechnical companies were other targets. Mining engineers and metallurgists studied in Sheffield and London and served in mining and the steel and iron industry in, for example, Cumbria. London was also a marked destination among architects who went to Britain to study among prominent colleagues in town planning. The garden city movement had its attraction, so had many Art-Nouveau-inspired architects. Civil, construction, and chemical engineers were less likely to choose Britain but had employment in bridge and railway building, food processing, and the pulp and paper industry. The Swede Rudolf Liljekvist participated in the construction of the so-called engineering miracle, the bridge connecting Edinburgh and Fife.

Liljekvist later became known as one of the executors of Alfred Nobel's will, which provides—at least indirectly—a connection to Russia, where the Nobel brothers and their establishments in Saint Petersburg and Baku functioned as a major employer for Swedish and Finland-Swedish technicians before the revolution. Russia was the only European destination where study travellers were underrepresented compared to migrants. It was the sixth destination on the pan-Nordic level; very rare among Norwegians, but more frequent among Danes and Swedes who often went there as representatives of domestic companies. Some of the activities of these companies were resumed after the revolution, for example, ASEA's erection of a factory in Yaroslavl. Geographic proximity, political ties, and Russia proper's function as a substitute America

242 Sweden attracted a lower number, but a higher percentage, since we are only counting technicians from Denmark, Finland, and Norway going there.

made mobility over the intra-empire border a very common trait among technicians in the Grand Duchy of Finland. Some technicians served in mining. In Finland, this field, as well as radio engineering, later utilised returnee knowhow. Railway constructions like the one between Moscow and Archangel was another field. Employment for domestic companies was very uncommon among Finnish technicians, but Russia was an exception. Some served Finnish granite industries, but most worked for the Finnish national railway company whose railway net stretched to Saint Petersburg. Finnish mobility to Russia had different implications; it was mostly very short-distance as the border ran just north of Saint Petersburg. However, it was also not only European, as reflected in Carl Jahn's ventures in Vladivostok and Sakhalin.

As one of very few Finnish technicians in Asia, Jahn was unusual. Nordic technical mobility—read migration—to Asia and Oceania was a Danish trait. The most marked Danish destination was Siam, where a Swede working for a Danish company was the only non-Dane. The Danish pattern is connected to Asian ties in general and the Siamese colony in particular. The cement business also contributes, Danish technicians represented F. L. Smidth and were involved in domestic cement ventures in Asian countries as well as in Australia and New Zealand. Infrastructure building also contributed. Civil and construction and electrical engineers participated in road and railway building as well as in the electrification of Bangkok's tramways and in similar activities in the Dutch East India where Danish technicians functioned, to some extent, as a substitute, as the Dutch government sometimes had problems recruiting the country's own engineers to the colony. Some served in municipal bodies in the Dutch colony and managed, for example, waterworks and harbours, a pattern that also was implemented in places like Beijing, Shanghai, Hong Kong, and New Zealand's capital city Wellington. Other Nordic technicians were present in most Asian countries and Australia and New Zealand, but British India and Japan were the only destinations where a non-Danish nationality accounted for a higher number. Swedes served at Indian steel and ironworks, electrical companies, and in the match industry. In Japan, two-thirds of the Nordic technicians were Swedish, and one major reason was that Gothenburg-born merchant Gadelius's trading company brought many Swedish technicians to the country.

Northern Africa was sometimes part of the Mediterranean study tours, which, of course, also included the European destinations we have described above and thereby closely connected to European travel. The experiences and views could, however, be equally remarkable, as Hilding Ekelund's observations from Algerian Timgad reveals. Working in Africa south of the Maghreb countries and Egypt involved, for example, railway building and surveying in the Gold Coast, and also in the southern parts of Africa, as well as mining in West and East Africa and, too, in a country like Madagascar. Mining, railway building, and different kinds of construction were also major activities in the largest destination on the African continent, South Africa. The country was also visited by two Swedish engineers who later were described as important for industrial development. Emil Lundqvist served in gold mines in Kimberely, whereas Ivar Kreuger was engaged by a British firm to participate in the setup of a large hotel in Johannesburg. However, their pioneering in rationalisation and reinforced concrete upon return to Sweden is usually connected to their American experiences.

Kreuger also spent time in the Americas south of the United States. Like in Asia and Oceania, Danes dominated, but not as markedly. Civil and construction engineers shared Latin America and the Caribbean with mining engineers and metallurgists. Mobility to the area fell in the light of the Baring Crisis in the 1890s but increased again after the turn of the century. Like in the other remote areas, study travelling was rare, but a few went to see the building and the finished Panama Canal. Argentina was the overwhelmingly largest destination, not only in this region but of all countries outside Europe and North America. Brazil, Chile, Mexico, and the Danish West Indies constituted other relatively major destinations. Technicians working in Argentina, Brazil, Uruguay, and Paraguay often served in infrastructure building: railway and tramway construction was important, so was managing of waterworks. Mining was the major field in Chile, Bolivia, and Peru, where, not least, the Guggenheim venture in Chuquicamata in northern Chile played an important role for employment. Railway construction was, however, also a field of work in the Andes countries. In Mexico, too, mining and railway building played a role for employment, but the mobility there is characterised by the Swedish company L.M. Ericsson's concession to build-up the early twentieth-century telephone net. Employment in the Caribbean, finally, was partly secured by railway building ventures, but more often at sugar plantations. Sugar refining in the Danish West Indies was dependent on recruiting engineers, often chemical, from the technical institutes in Denmark.

In the early 1930s, Folmer Andersen still lived on the islands that used to be the Danish West Indies before they were sold to the United States to become the American Virgin Islands in 1917. He was the only one in our cohort remaining. Some of the others had died, but many had also returned to Denmark. They were not the only Nordic engineers and architects going back. In the next chapter, we will join our technicians on their homeward journeys.

CHAPTER 7

Burning No Bridges Behind Them

> Unlike a large portion of those who left the country districts of Norway to take up land in the Middle West, the engineers burned no bridges behind them; in fact a majority had every intention of returning to the homeland after acquiring experience, perhaps a fortune, and possibly, too, a great reputation. They had no farms to sell and no families to care for. A ticket for the voyage to America, a few dollars to keep them going until they found a job, some articles of clothing—these with exceptions were all that they carried with them. In a short time they would return to visit parents and friends in Europe; a few years more and they would return to take over engineering posts in Norway.[1]

Bjork has, as mentioned, noted how Norwegian engineers diverged from ordinary emigrants as they often intended to return. The observation at the 1901 meeting in Sweden has also been mentioned, and we have exemplified with several returned engineers and architects.[2] In figure 18, study trips are excluded as they always implied a return. Once a study traveller decides to stay abroad, he/she turns into a migrant. Figure 17 confirms the main target migration or placements abroad pattern: return rates for the Scandinavian countries lay roughly at 70 per cent; Finland's at 90 per cent.

Kristian Hvidt, Keijo Virtanen, Ingrid Semmingsen, and Lars-Göran Tedebrand have shown that return rates among transatlantic emigrants in general rarely exceeded 20 per cent on a yearly basis.[3] Here, we should keep in mind that we include all destinations and four of five graduates to destinations

1 Bjork, *Saga in Steel and Concrete,* 35–36.

2 'Svensk-amerikanska ingeniörers och arkitekters möte i Gefle', 209.

3 Kristian Hvidt, *Flugten til Amerika eller drivkræfter i masseudvandringen fra Danmark 1868–1914* (Aarhus 1971) 326; Semmingsen, *Veien Mot Vest. 2,* 460; Lars-Göran Tedebrand, 'Remigration from America to Sweden', in *Historia och demografi: valda texter* (Umeå 1999) 201; Virtanen, *Settlement or Return,* 66.

 | DOI:10.1163/9789004385207_008

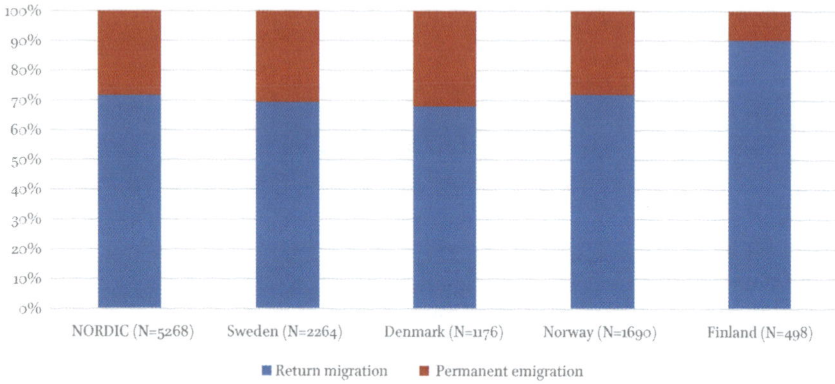

FIGURE 17 Pre-1930 return migration and permanent migration in numbers and percentages of Nordic engineers and architects educated 1880–1919 (migrants only)
SOURCES: SEE FIGURE 1.

outside North America returned. This rate almost reached 95 per cent in Finland. Nevertheless, engineers and architects also showed much higher return rates from North America. The pan-Nordic rate was 58 per cent, and the Scandinavians rate lay close to this average, whereas almost four of five graduates returned from America to Finland.

Interest in technical development, a will to raise the status of the profession as well as the individual position were, following Henrik Björck, reasons for engineers' will to introduce and work with new technology.[4] A fourth criterion is the will to contribute to the native country's technical, economic, and industrial development. These return rates indicate interactions between these criteria. Development nationalism included the transfer of primarily German and American ideas and to make use of experiences gained by returned emigrants.[5] These returning graduates were very well suited for a region on the edge of industrial society, including two countries that were about to reach full independence. To leave the native country almost became, paradoxically or not, a patriotic act in this context.[6] The reason was, after all, to return and be ready to contribute with successful work.

Table 6 shows that the median duration on the pan-Nordic level was three years and the average duration was four and a half years. The Swedes generally

4 Björck, 'Bilder av maskiner', 298.
5 Runeby, 'Americanism, Taylorism and Social Integration', 21–46.
6 Grönberg, 'The Engineer Paradox', 96–105.

TABLE 6 Median and average duration of stay (first intermission) abroad among engineers and architects educated in the Nordic countries 1880–1919, who returned before 1930

Duration	NORDIC	Sweden	Denmark	Norway	Finland
Median duration	3 years	3 years	3 years	2 years	2 years
Average duration	4,5 years	4,9 years	4,4 years	4,3 years	3,3 years

SOURCES: see figure 1.

spent a somewhat longer time abroad, while the Finns were away for shorter periods.

About 25 per cent of the returns occurred the same year or the year after the migration, whereas more than 50 per cent had returned within three years. It was rare to stay abroad for more than ten years, which occurred only for every tenth graduate. The propensity to return decreased with increasing time spent in the adopted country. Studies have estimated that the adopted place, region, or country asserts more influence than earlier places of residence after five years.[7]

The countries diverged, and figure 18 confirms the patterns indicated in table 6: Migrating Finnish graduates were more in a hurry to return than the Scandinavians, and this reveals a closer relationship between migration and study travelling than in the other countries. Considering the markedly higher share of study travellers, we can conclude that transnational mobility of Finnish technicians had more of a study-trip character than in Scandinavia, even if target migration was the dominating pattern in all countries. The general observation here is that Finns stayed out shorter and Swedes somewhat longer.

1 What Technicians Returned?

Finland's high percentage of returnees can be connected to a will to contribute as the country moved towards and gained full independence. Myllyntaus argues that the growing demand for technical experts made Finnish students

7 Anders Bränström, Jan Sundin, and Lars-Göran Tedebrand, 'Two Cities: Urban Migration and Settlement in Nineteenth-century Sweden', *The History of the Family*, 5:4 (2000) 415–429.

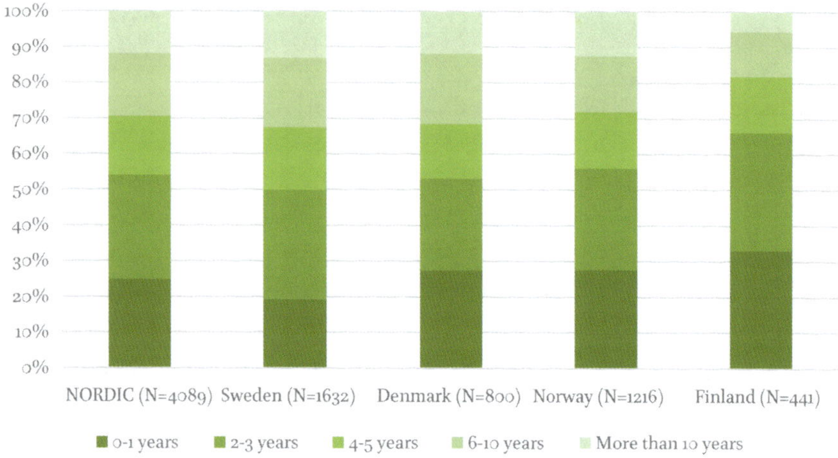

FIGURE 18 Duration of stay abroad (first intermission) among engineers and architects educated in the Nordic countries 1880–1919, who returned before 1930
SOURCES: SEE FIGURE 1.

abroad return more than colleagues from Norway and Eastern Europe. The connection between foreign experience and prestige was present in all countries, but Myllyntaus states that engineers rose to key positions in Finland after 1917.[8]

As opposed to Norway, also reaching full independence during the period studied, the transnational mobility of Finnish technicians had a comparably weak connection to general transatlantic emigration. There was a dominance of relatively nearby destinations among migrating Finnish graduates. Some studies have concluded that the likelihood of return migration decreases with increasing migration distance,[9] and this is a pattern that can be observed for the Nordic cohort and all individual countries. North America constituted a smaller share of the migration of graduates from Finland, whereas moves to other overseas destinations almost were absent. The target migration patterns dominated in all four countries, but nowhere else in this study was it as dominant as in Finland. The extraordinarily strong study travel pattern further underlines that Finnish graduates more often performed mobility for learning purposes.

---

8 Myllyntaus, '"The Best Way to Pick Up a Trade"', 138–163.

9 Sune Åkerman, 'Theories and Methods of Migration Research', in: Harald Runblom and Hans Norman (eds.), *From Sweden to America: a History of the Migration* (Uppsala 1976) 75; Virtanen, *Settlement or Return*, 201.

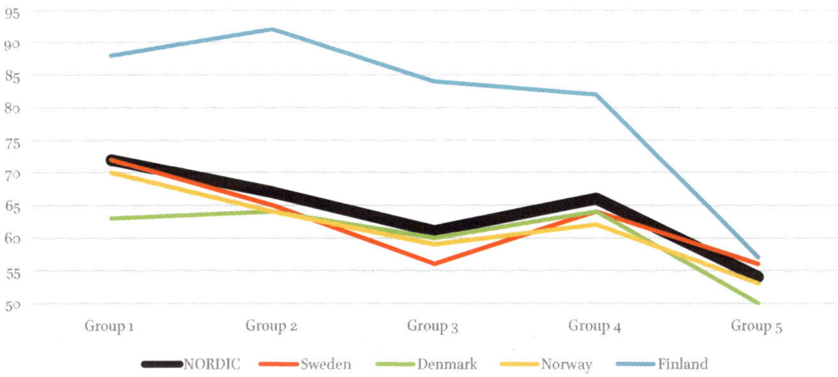

FIGURE 19 Pre-1930 return migration per social class in percentage of transnational migration Nordic engineers and architects educated 1880–1919
SOURCES: SEE FIGURE 1.

1.1 *Children of the Upper and Upper-Middle Classes Saw More Possibilities Back Home*

Nine of the ten migrating female graduates returned. Nevertheless, earlier studies have described return migration as a male rather than a female phenomenon, and we may assume that the overwhelmingly male dominance among these graduates contributed to the return rate. Return migration was integrated into the plans of young male industrialists and businessmen, and temporary intermissions abroad were used as a part of a maturity process, especially for upper-class men.[10]

This is reflected in the social origin pattern, which basically shows a decreasing trend with lower social origin, even if graduates with origin in the working class returned more frequently than those whose fathers had middle-class occupations. We may presume that few foreign employers asked for fathers' positions and family ties, while such bonds—Pierre Bourdieu's social capital if you like—were important assets in the native country. Those originating from the upper class got several important family ties to economic and political elites for free, whereas those from a lower social origin had to work harder to establish contacts. Higher shares of graduates from the lower ranks estimated that the chances of a good career were better abroad, although there are reasons to point out that majorities also returned among graduates born into other societal layers.

10 Lindqvist, *Herrar i näringslivet.*

1.2 *A Foothold in the Capital Cities Increased Return Migration*

First, we can mention that the return rates increased with the size of a graduate's birthplace, that is, capital-born returned to the greatest extent, and rural-born to the lowest, but differences were small. Whereas birthplace had little to say for return migration, we can note that it mattered for return migration in which city the graduate had studied. In Sweden, 75 per cent of the students from Stockholm returned, compared to 55 per cent of the students from Gothenburg, Malmö, and Örebro. In Norway, the corresponding shares were 70 per cent for Kristiania and 63 per cent for Trondheim and Bergen. Denmark noted 60 per cent for Copenhagen and 55 per cent for the schools in Funen and Jutland. We cannot discuss Finland here since all graduates in the cohort studied in Helsinki.

Engineers and architects that had studied in the capital cities were more likely to return. Having studied in Stockholm seems especially important for return of Swedish graduates. One explanation is that the Royal Institute of Technology was a more prestigious school than Chalmers in Gothenburg and especially the upper-secondary schools in Malmö and Örebro. Björck has argued that Chalmers, even if it is in practised functioned as technical university already around 1900, really had a position between the Royal Institute of Technology and the upper-secondary schools.[11] A degree from the Royal Institute of Technology was, therefore, more valuable than degrees from elsewhere.

In the Scandinavian neighbour countries, the pattern was less clear. Copenhagen hosted an intermediate school besides the Polytechnic Institute, and the number of transnationally migrating students from the schools in Odense, Aarhus, and Horsens was also low. Norway's most prestigious school was in Trondheim rather than in Kristiania, a relation that was formalised when the Norwegian Institute of Technology was inaugurated. These factors contributed to somewhat weaker results than for Sweden, but it was nevertheless important to have a foothold in the capitals also in Denmark and Norway. To even higher degrees than Stockholm, Copenhagen and Kristiania played important roles for Danish and Norwegian industrial development. Thus, employment opportunities back home often occurred in capitals and their surroundings.

1.3 *Graduating Young and Departing Shortly After Graduation Implied Return*

One hardly surprising trait is related to graduation age. Just as the probability for transnational migration was higher for a younger graduate, so was the

11 Henrik Björck, *Staten, Chalmers och vetenskapen.*

TABLE 7 Pre-1930 return migration per age of graduation in percentage of transnational migration among Nordic engineers and architects educated 1880–1919

Graduation age	NORDIC	Sweden	Denmark	Norway	Finland
Younger than 25 years	67	65	61	66	87
25 to 29 years	66	59	62	59	89
30 years and older, unknown	45	50	33	36	85

SOURCES: see figure 1.

likelihood of a return. The pattern was, as table 7 shows, more marked in Norway and Sweden, but the oldest age group returned to the lowest extent in all countries. Younger graduates not only had more time to go abroad, they also had more time to do a career back home upon return.

The same is true for graduates who departed shortly after graduation, as figure 20 shows. Return rates increase for those who departed two to three years after graduation, compared to graduates who went abroad the same year as they left school, or the year after. It seems like return rates decrease gradually for departures from the fourth year after graduation.[12] We can assume that graduates departing shortly after leaving school often were target migrants, whereas those who waited for a longer time to a greater extent were ordinary or labour-market migrants. Stang writes the following about engineers: 'They travel to acquire competence: an engineer fresh from school is still considered an unfinished product'.[13]

1.4 *Being Native-Born Was Positively Correlated with Return*

Table 8 shows that graduates born in the country of education generally returned much more than foreign-born, which is not surprising.

The basic explanation to at least the pan-Nordic, Danish, and Swedish patterns is simple and governed by our choice to use country of education rather than country of birth as the point of departure; many of these graduates had really returned when they moved back to their birth countries, that is, this was mostly the Finnish- and Norwegian-born students at

12 Finland differs as the return rates increase again for graduates who departed after more than ten years, but this rate is calculated on a small number (ten graduates).

13 Stang, 'A measure of relative development?', 93.

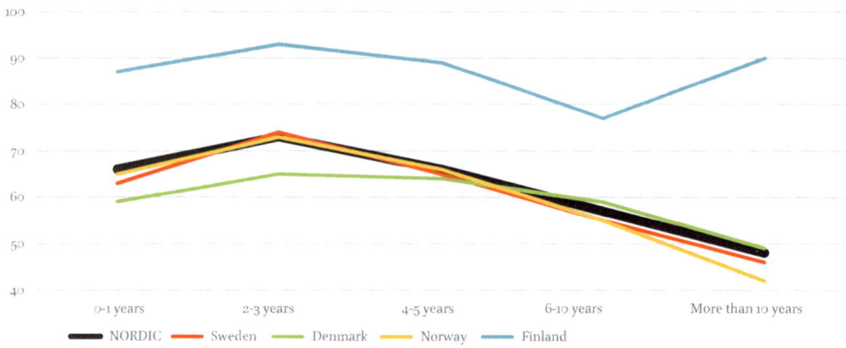

FIGURE 20 Pre-1930 return migration per TIME BETWEEN GRADUATION AND MIGRATION in percentage of transnational migration of Nordic engineers and architects educated 1880–1919
SOURCES: SEE FIGURE 1.

TABLE 8 Pre-1930 return migration per country of birth in percentage of transnational migration among Nordic engineers and architects educated 1880–1919

Country of birth	NORDIC	Sweden	Denmark	Norway	Finland
Native	67	67	61	65	87
Foreign	41	26	47	51	100

SOURCES: see figure 1.

Chalmers and the Icelanders in Copenhagen. We get the largest difference for Sweden, as the country attracted, as mentioned, more foreign-born students than Denmark. In Norway, most foreign-born graduates were Norwegian-Americans who possibly carried a positive picture of the United States. Finland shows a completely different pattern, although the number of foreign-born is low. Nevertheless, all 22 migrating graduates born outside of Finland returned. This is probably a random result, but it would probably have looked different if they had not been a born abroad by Finnish parents.

1.5 *Architects Returned More Than Engineers, but the Countries Diverged*

Return rates were, as figure 21 shows, generally high for all specialisations; the highest ones are noted for architects. In Finland, only two architects stayed

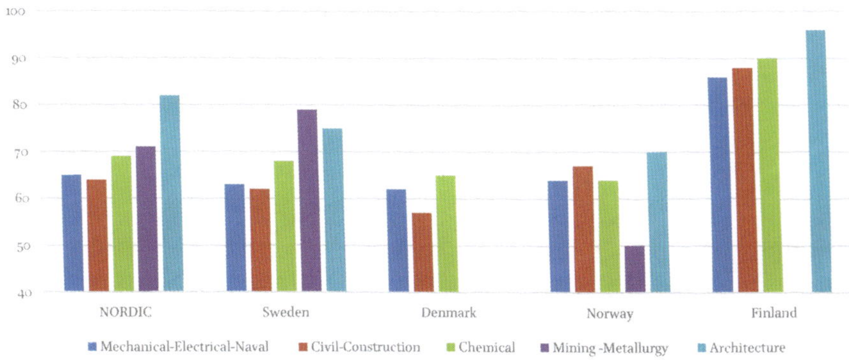

FIGURE 21 Pre-1930 return migration per SPECIALISATION in percentage of transnational migration of Nordic engineers and architects educated 1880–1919
SOURCES: SEE FIGURE 1.

abroad of whom one was the earlier-mentioned American city-building and planning pioneer Eliel Saarinen.

There are some differences between the countries; one is that civil and construction engineers generally note lower return rates than the other specialisations, but this does not apply to Norway. The category mining engineers and metallurgists notes the highest return rate of all specialisations in Sweden, whereas their Norwegian colleagues had the lowest return rate. Chemical engineers have—relative to other specialisations—higher rates in Denmark and Finland than in Sweden and Norway. These differences are most likely explained by differences in the education and variations of the labour markets.

1.6 *Departures between 1890 and 1919 Implied Return, but Denmark Diverged*

Figure 22 shows that return rates were higher if the graduates departed between 1890 and 1919, and lower before and after. This is valid for the pan-Nordic level as well as for Sweden, Norway, and Finland. The low rates for the 1920s are, of course, influenced by 1930 as the stoppage year.

Sweden and Norway lay closest to the pan-Nordic pattern, whereas Denmark diverged with considerably higher return rates for graduates.

1.7 *Return Migration Increased with Shorter Migration Distances*

Figure 23 shows that migrants who had their first destination in Europe generally returned more. There is a correlation between shorter distances and return migration, a pattern that has been noted in earlier migration studies.

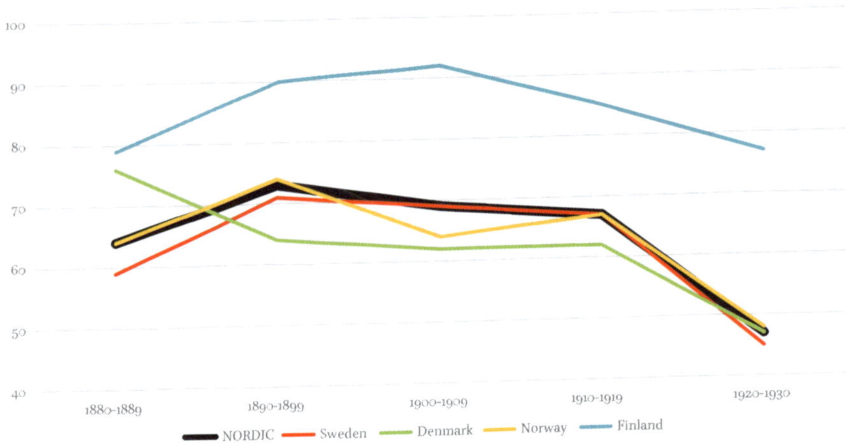

FIGURE 22 Pre-1930 return migration per DECADE OF DEPARTURE in percentage of transnational migration of Nordic engineers and architects educated 1880–1919
SOURCES: SEE FIGURE 1.

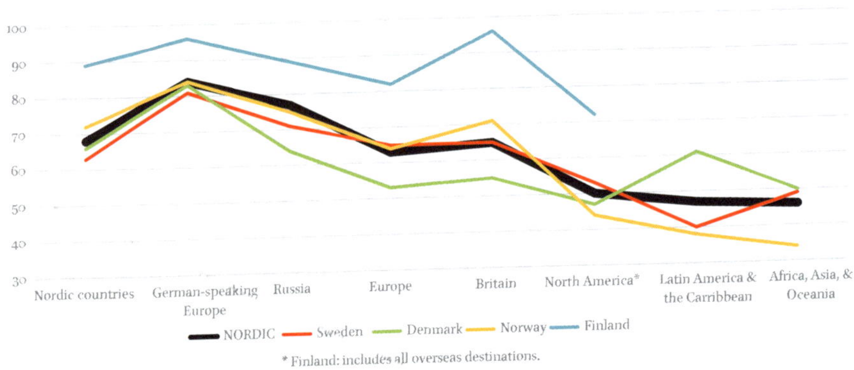

FIGURE 23 Pre-1930 return migration per FIRST DESTINATION in percentage of transnational migration of Nordic engineers and architects educated 1880–1919
SOURCES: SEE FIGURE 1.

Keijo Virtanen noted, for example, that Finns settling on the American East Coast were likelier to return compared to those who lived on the West Coast.[14]

However, the return from other Nordic countries lay lower than the return from German-speaking countries, which thereby stand out as the prime target migration destinations. The German-speaking countries constituted an area from which nearly everyone aimed to return, whereas the duality of migration

14 Virtanen, *Settlement or return*.

to North America—target migration and traditional transatlantic emigration—implied lower return rates. The return rates were even lower for graduates who chose a Latin American, Caribbean, African, Asian, or Oceanic destination as their first migration target, even if Denmark diverged with a return rate from Latin America and the Caribbean that was on a par or higher than from all other destinations except the German-speaking countries. All these rates were distinctly higher than the total return rates of emigrants from North America, but our main purpose here is to compare different destinations for technicians. Finland notes the highest return rates regardless of destination, whereas return migration to the Scandinavian countries diverges by immigration area.

1.8 *Foreign Experience Facilitated the Careers Upon Return*

What was the importance of foreign experience for the graduates' careers? We have, as mentioned, used the international HISCLASS system, mainly developed by Marco van Leuven, Ineke Maas, and Andrew Miles. This system is constructed for historical contexts, and classifications are based on data from several different countries including Scandinavia.[15] When engineers and architects leave school, they are, according to this scheme, in group two, that is, 'higher professionals'. There is only one group above, 'higher managers'. Upward occupational mobility is, thus, based on graduates stepping from group two to group one. Clearly, higher shares of technicians with foreign experiences had made this step, as table 9 shows.

Generally, occupational mobility was more common among those who had foreign experience;[16] the only exception in table 9 was in Norway in 1920 when graduates without foreign experiences had been somewhat more successful than study travellers. However, the differences between migrants and study travellers were not large. In Denmark, study travellers were more successful than migrants at all three points of observation. Denmark was the country where study travelling was least common, whereas table 9 shows that it was

15 van Leeuwen, Maas, and Miles, *HISCO*. See chapter one for a discussion on the advantages and shortcomings of the HISCLASS system.

16 This upward mobility was not only due to foreign experience. Urban origin was more valuable than rural. Age and experience in general, that is, how long a graduate had been in the profession, were also valuable assets in this context. Mining engineers and metallurgists, as well as chemical engineers, were more likely to climb up the ranks, whereas architects were less likely to do so. Danish and Finnish graduates had better chances than Norwegian and Swedish. A technician who was ready to work in a smaller urban or rural-industrial setting outside the capitals and other larger cities were also more likely to experience upward occupational mobility. The competition for positions was probably less severe in smaller geographical settings.

TABLE 9 Percentages of migrants, study travellers, and graduates with no foreign experience who had experienced upward occupational mobility among Nordic engineers and architects in 1900, 1910, and 1920

		1900	1910	1920
Nordic	Migrant	12	14	18
	Study traveller	14	13	16
	No mobility	7	8	9
Sweden	Migrant	12	14	19
	Study traveller	12	10	14
	No mobility	6	8	9
Denmark	Migrant	13	13	15
	Study traveller	19	21	23
	No mobility	10	9	9
Norway	Migrant	9	11	16
	Study traveller	7	12	7
	No mobility	6	6	9
Finland	Migrant	18	18	25
	Study traveller	14	12	18
	No mobility	12	10	12

SOURCES: see figure 1.

most rewarding to be a migrant in Finland, that is, the country where study travelling was an institution. This is perhaps still explained by the study-trip-like character of Finnish migration. More than the Scandinavian graduates, the Finns travelled to acquire knowledge to use upon a return. They were also to a greater extent expected to perform such mobility. Therefore, longer intermissions might have implied more, to use a Bourdieuan term, 'symbolical capital', that is, they were expected to be more knowledgeable. In Denmark, the study trips possibly symbolised that the engineers really had been abroad to learn. Many Danish migrants had, as mentioned, been travelling on payrolls of Danish companies, implying technology and knowledge transfer from, rather than to, Denmark.

One striking pattern is that destination seems to have mattered relatively little. Not on the pan-Nordic level, nor in any Nordic country was it—at least in this context—especially rewarding to go to the German-speaking countries and/or North America.[17] It was, rather, foreign experience, regardless of destination, that counted. The question is: how large were the shares of technicians with such experiences?

2 More Than Every Third Nordic Technician Had Foreign Experience

One idea in this study is that significant shares of foreign experiences among technicians provide critical masses for technical change and impact. There is, following Bovenkerk, a correspondence between shares and numbers of returnees and their likelihood of making an impact. One of the conclusions in the Norwegian returned emigrant project from the 1980s was that many returnees acted in accordance with American practise over time and in many places and their ideas had a significant impact as they also convinced their neighbours to evaluate old patterns of behaviour. Their conclusion is for Norway as a whole, albeit the returnee shares of the total population hardly exceeded 1 per cent.

For distinct parts of society—technology, industry, engineering, and architecture—such a conclusion seems even more reasonable. Figure 24 shows that shares of technicians with foreign experience in the Nordic countries lay around 40 per cent in the earliest decades of the twentieth century. Experiences from migration lay approximately 10 per cent lower.

Norway and especially Finland hosted higher shares of technicians with foreign experience than Sweden and Denmark. In Finland, between two-thirds and three-fourths of the technician corps had foreign experience, either through study travelling or migration. We can, however, conclude that quantities indicate that the shares of returnee technicians were significant in all Nordic countries. In addition, there were both foreign-born technicians and native-born technicians who had taken the entire education abroad present in the Nordic countries.

A calculation based on the 1910 census shows that somewhat less than 1 per cent of the population in Norway were returned emigrants,[18] and corresponding shares were hardly higher in Sweden and the low-emigration countries Denmark and Finland. We can compare to our lowest observed share of returned technicians: 17 per cent in Denmark at the turn of the century. There are

---

17 Upward occupational mobility in per cent per destination on the pan-Nordic level: Britain 16, Russia 16, other Nordic countries 15, Europe 14, Latin America and the Caribbean 14, German-speaking countries 14, North America 12, and Africa, Asia and Oceania 11.

18 Hans Storhaug, 'The 1910 Norway Census and return migration: The case of Nedstrand', *AEMI—Journal*, 2 (2004) 4.

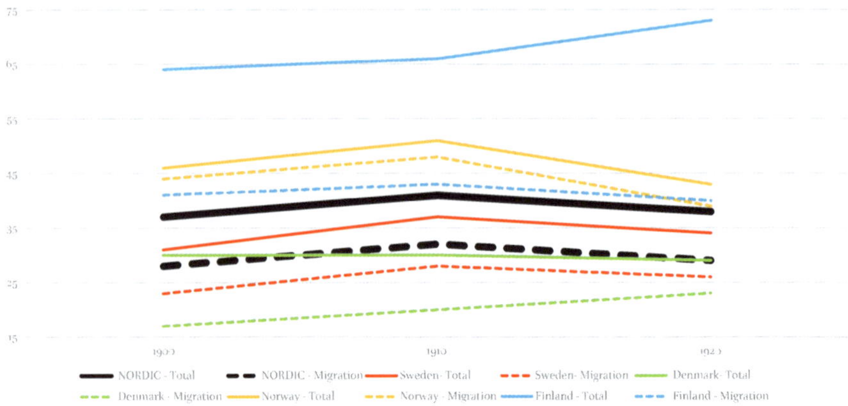

FIGURE 24 Percentages of engineers and architects with foreign experience in the Nordic countries 1900, 1910, and 1920. Total experience (Tot) and experience from migration (Mig)
SOURCES: SEE FIGURE 1.

reasons to believe that most Nordic technicians had either worked or studied abroad for a significant period or knew a colleague who had. Even Lange has calculated that about 40 per cent of the members in the Norwegian Engineer's Association in the first half of the twentieth century had foreign experience.[19] Lange's calculations are in line with the findings of this study.

Sweden and Denmark had, as we can see, still less accumulated foreign experiences within their communities of technicians than Norway and Finland. We may assume that the lack of higher technical education in the two 'less' industrialised countries to some extent was compensated by these high frequencies of foreign-experienced technicians. Norway had the highest shares of foreign-experienced engineers and architects if we look only at those with experiences from placements abroad, but its share dropped, as we can see, somewhat after the establishment of the technical university in Trondheim in 1910. The Finnish community of technicians had, however, accumulated markedly higher amounts of foreign experience when we add study trips. These journeys gave Finland the significantly highest share of foreign-experienced technicians in the Nordic area through the period. Myllyntaus writes:

> At the turn of the century, almost two generations of Finnish engineers had been educated in a cosmopolitan spirit. For them, studying abroad

19 Lange, 'Norske ingeniører i Amerika', 8–9.

> and going on 'educational business trips' was not just a fashion of the time. For some devoted professionals, studying at renowned foreign universities of technology and later regularly travelling abroad turned into an obsession when they claimed that without this, an engineer's competence was questionable. In any case, these two generations had managed to create a popular image that engineers were knowledgeable persons versed in languages and internationally oriented.[20]

The differences in shares of technicians with foreign experience between Norway and Finland on the one hand, and Sweden and Denmark on the other hand, may even be underrated in this calculation. A considerable number of Finns and Norwegians took, as mentioned, their entire education abroad. Bassøe has calculated that 386 Norwegian engineers were educated only at foreign institutes between 1901 and 1920, and of those, four out of five made their careers in Norway. The same happened in Finland, but not to the same extent in Sweden and Denmark. The earlier-mentioned Sam Eyde is one example: He searched for a university education from the very bottom. Eyde's father was a ship owner, and Grimnes argues that there was no natural step between the secondary school and a Norwegian technical school for an upper-class son.[21] Carl-Selim Westman, who lectured on Henry Ford in Finland, took all his education in Darmstadt.

As far as the author knows, no previous studies on countries outside the Nordic area have tried to estimate the presence of technicians with foreign experience in percentages of the communities of engineers/technicians at certain points in time. We can, nevertheless, ascertain that this was not only a Nordic pattern. Countries like Portugal, Greece, Czechoslovakia, and Japan employed many foreign technicians and sent many students abroad. Several hundred Japanese students returned from Europe.[22]

2.1 *National Differences in Foreign Experience among the Specialisations*

Specialisation is a good point of departure to look at foreign experiences within different fields of engineering and architecture. Considering the difficulties of measuring technological influence through statistics, the shares of returnee technicians still indicate that transnational mobility of technicians was

20 Myllyntaus, 'The Best Way', 160.

21 Grimnes, *Sam Eyde*, 50.

22 Pauer, 'Techologietransfer', 40–43.

important in the entire Nordic area. However, were there differences based on specialisation? We can study this in table 10.

The pan-Nordic pattern reveals that civil and construction engineers were least internationally experienced. This was also the case in Sweden and Finland, whereas Danish mechanical, electrical, and naval engineers and Norwegian architects lay on roughly the same shares, and mining engineers and metallurgists had the least experiences in Norway. Architects had somewhat higher shares of foreign experiences than engineers but had often acquired experiences through study trips and less frequently through placements abroad. In 1910 and 1920, they had the least experiences of migration but the highest shares of total foreign experience. Norway was an exception, as architects also had the highest shares of migration experiences. The explanation can be connected to Norwegian architects having a stronger identity as technicians than their colleagues from Sweden and Finland who probably viewed themselves more as artists. Somewhat generalised: the technician was more prone to migrate, whereas the artist study-travelled.

If we put most weight on placements abroad, mechanical and electrical engineers together with naval architects and chemical engineers had most migration experiences in the Nordic area. This applies largely to Finland, but also to the other countries for the year 1900, whereas the pattern is more

TABLE 10 Percentages of technical school graduates with foreign experience (total) in the Nordic countries, 1900–1910–1920. Total experience above, placements abroad below

Specialisation	Nordic	Sweden	Denmark	Norway	Finland
Mechanical,	40-43-40	36-42-38	29-26-25	52-49-46	73-72-77
electrical & naval	32-36-32	26-33-30	24-21-23	50-46-43	64-55-49
Civil &	30-34-34	19-22-21	24-29-31	40-51-45	46-51-60
construction	21-25-26	14-13-13	13-18-24	38-49-42	16-21-23
Chemical	44-46-40	30-36-33	51-46-33	56-51-42	87-86-91
	31-34-30	22-26-25	18-26-22	51-48-35	74-72-62
Mining &	36-37-41	37-40-42	xxxx	29-20-27	xxxx
metallurgy	18-28-32	17-29-35		21-20-21	
Architecture	46-46-53	35-30-32	xxxx	15-56-71	59-57-64
	25-22-22	16-10-9		15-52-55	32-24-20

SOURCES: see figure 1.

ambiguous in 1910 and 1920. Nine of ten chemical engineers in Finland had foreign experience, by far the highest rates in the Nordic area, and the only group that came close to this was their countrymen in mechanical engineering and their two sub-groups. This implies that the Finnish pulp and paperworks, the country's dyeing industry, mechanical and electrical workshops, and shipyards were among the most frequent receivers of foreign impulses in Nordic industry. Chemical engineers also belonged, however, to the most experienced groups in Denmark and Norway. Mechanical engineers and their sub-groups were Sweden's most experienced specialisations before 1920 when they were overtaken by mining engineers and metallurgists. Swedish mining was increasingly mechanised, modernised, and automatized by the help of returned engineers in the early twentieth century. Mining engineers and metallurgists in Norway did not experience a similar increase in international experience. Nevertheless, naval architects would have been Sweden's most experienced specialisation by the dawn of the 1920s if we had separated them from mechanical and electrical engineers. Olsson has described the foreign influence on Swedish shipyards, mainly when it came to rationalisation and corporate welfare. Swedish electrical engineers were also above average in experience, and Fridlund has revealed how American, German, and Swiss impulses were important for the development of the country's electrical industry.

2.2 *The German Domination Was Most Striking in Norway and Finland*

The first striking thing when we are looking at experiences from different regions is the dominance of the German-speaking countries. In figure 25, we can study the experience shares in the Nordic area as well as the individual countries in 1900.

Experience from Germany, Switzerland, and Austria dominated significantly in the first decades of the twentieth century, whereas experience from Britain was as common as experience from North America by the turn of the century. The share with experiences from Russia increased a little over time, and especially between 1910 and 1920. This might be a result of the October revolution. The increasing shares for North America over time reflects the growing interest in American technology. The developments in transport technology also contributed to an increase for the United States and Canada. It became, over time, both easier and cheaper to go to North America and thereby also to extend more or less pure study trips to another continent. However, in 1920, there were still twice as many Nordic technicians with experiences from the German-speaking countries. The reasons mentioned are many students, placements abroad, and study trips, because of Germany's industrial leadership and its renowned technical universities. This made the German-speaking countries

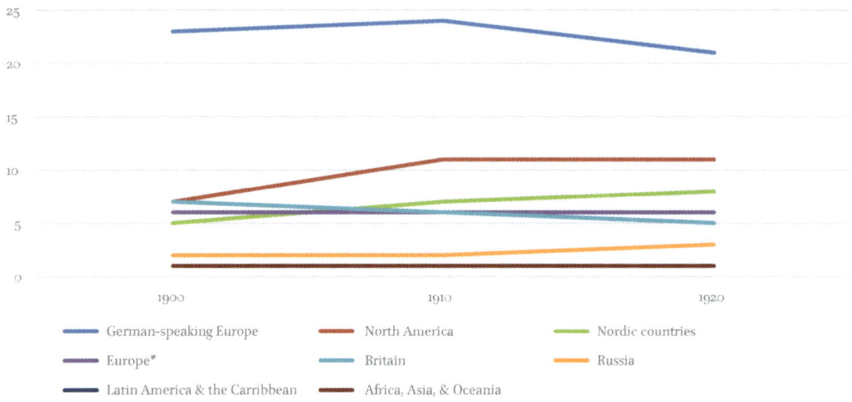

FIGURE 25 Percentages of foreign experiences per 'area of experience' among engineers and architects in the Nordic countries 1900, 1910, and 1920
SOURCES: SEE FIGURE 1.

the major target mobility destination. The geographical closeness contributed too. The countries were easily accessible, and it was relatively easy to get back home once the Nordic technicians were there. In figure 26, we can study the patterns for Sweden.

The German-speaking countries accounted for a somewhat lower share in Sweden compared to the pan-Nordic level, whereas North America lay a little higher. The other areas lay roughly on a par with the pan-Nordic level. Figure 27 shows the Danish patterns and reveals both similarities and differences between the pan-Nordic and Swedish ones.

In Denmark, too, German-speaking Europe accounted for lower experience shares compared to the pan-Nordic patterns, but North America also lay lower and did not dominate over other areas as in the pan-Nordic and—even more—Swedish patterns. Britain and continental Europe partly score higher because of the payroll migration for Danish companies. Figure 28 shows that Norway differed more from the pan-Nordic and the patterns in the two more industrialised countries.

The German-speaking countries dominated much more in Norway, mainly because they were destinations for graduates who wanted to complete their technical studies. North America was on a par with the pan-Nordic pattern, whereas the other destinations lay lower. Figure 29 reveals a Finnish pattern that diverged more than the patterns in Scandinavia.

Germany, Switzerland, and Austria dominated even more in Finland, whereas North America basically lay on the same levels as in the Nordic area and Norway. However, the other European areas accounted for many more experiences

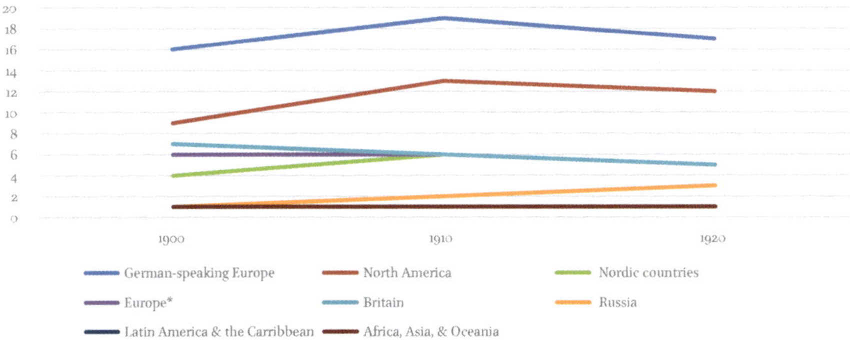

FIGURE 26 Percentages of foreign experiences per area of experience among engineers and architects in Sweden 1900, 1910, and 1920
SOURCES: SEE SWEDISH SOURCES, FIGURE 1.

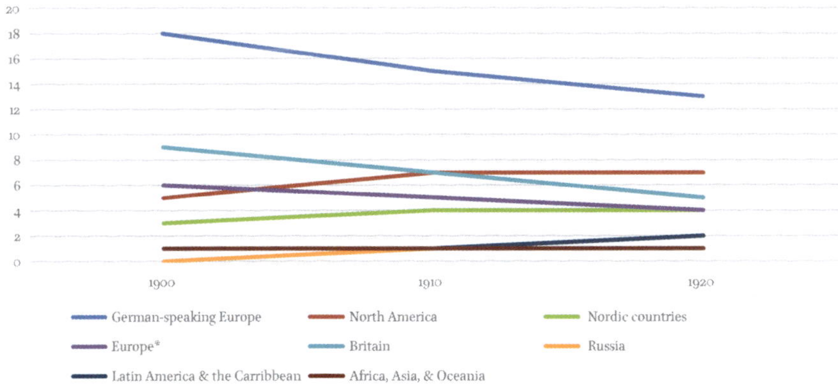

FIGURE 27 Percentages of foreign experiences per 'area of experience' among engineers in Denmark 1900, 1910, and 1920
SOURCES: SEE DANISH SOURCES, FIGURE 1.

in Finland. Therefore, North America was only the sixth experience area in 1900 and the fifth in 1920. The October revolution also brought back many technicians with experience from Russia. Michelsen has argued that Finland imported almost all production technology from the Western industrialised countries despite being a Russian Grand Duchy. This pattern may, however, have implied transfer of originally Western technologies that had reached the tsarist empire through foreign investments and the arrival of engineers from, for example, Germany.

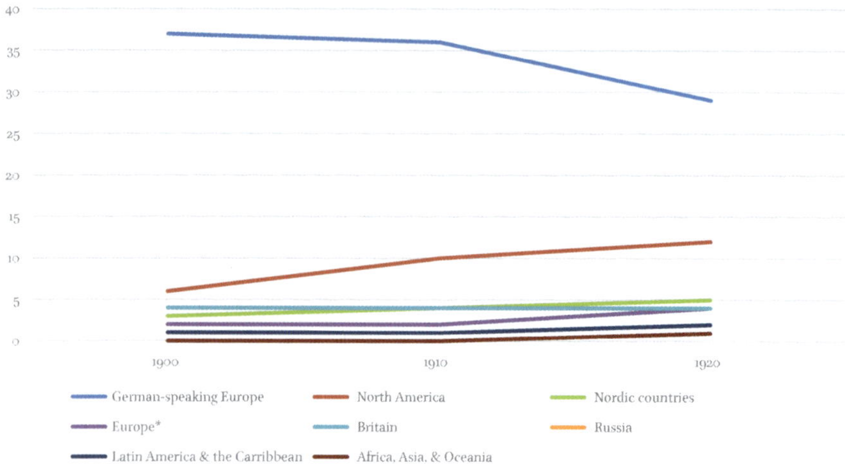

FIGURE 28 Percentages of foreign experiences per area of experience among engineers and architects in Norway 1900, 1910, and 1920
SOURCES: SEE NORWEGIAN SOURCES, FIGURE 1.

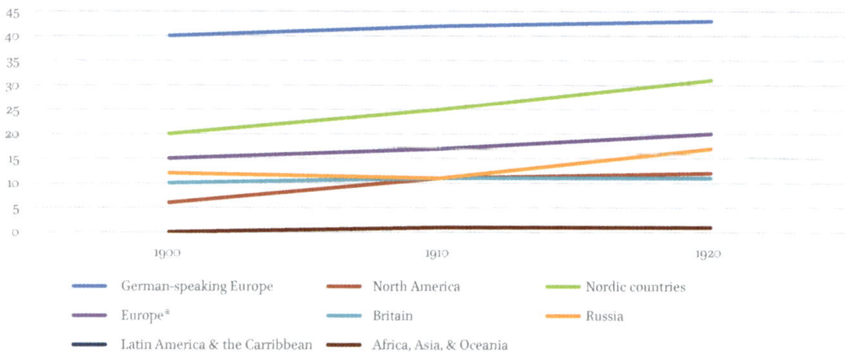

FIGURE 29 Percentages of foreign experiences per 'area of experience' among engineers and architects in Finland 1900, 1910, and 1920
SOURCES: SEE FINNISH SOURCES, FIGURE 1.

The German-speaking countries thus dominated in all four countries, but more markedly in Norway and Finland. The Norwegian pattern can be explained by the fact that many graduates had studied there before the inauguration of the Norwegian Institute of Technology in 1910. The Finnish pattern can also be explained by the lower level technical education before the 1908 transformation of the Polytechnic Institute in Helsinki to a technical university. However, the extensive study travelling that often had the German-speaking countries as its target contributed with an extra dimension that was not present in Scandinavia. This is related to the higher percentage of architects in the Finnish cohort, which also explains why so many Finns had experienced other Nordic countries, Britain and non-German continental Europe. Russia was, as indicated, a special case because of the Grand Duchy's political ties and geographical closeness.

Large shares of Nordic technicians had experiences from the leading industrial and educational countries in German-speaking Europe and North America. The pattern reflects the German-American overtaking of Britain's role as the leading industrial nation and major source of inspiration.[23] British shares were higher back in time; the country was the most common destination among Swedish technicians graduating in the 1860s, and many technicians took employment and studied technology there in the early nineteenth as well as back in the eighteenth century.[24] We can note a decline for Britain also in this context, but there are reasons to believe that this downward tendency started earlier and was more pronounced before the turn of the century.

Bovenkerk's criterion about the quality of the training and the acquired skills is relevant in this context. It was at German-language technical universities like the ones in Berlin-Charlottenburg, Dresden, and Zurich that the students could acquire the qualitatively best education. German and American industries offered experiences that were valuable, even if Stang underlines that an engineer that 'got stuck as an assistant draftsman in Westinghouse'[25] hardly could count on his experiences automatically being transformed to the 'kind of prestige which secured better opportunities at home'. We must, however, regard the German-speaking countries and North America as the regions where technicians could acquire the most valuable knowledge and skills. Scholars have described the development in these countries as extraordinary. Olsson has argued that Swedish naval architects more easily could get responsible positions in the United States, and that is one of several reasons why they oriented themselves away from Britain, despite its world leadership.[26] From a general point of departure, therefore, we may interpret the dominance of these two countries as a prerequisite for return migration of Nordic technicians to be influential. If they had experienced environments in countries on a similar level of development or perhaps what Stang calls 'work in municipal bodies in an

23 Hyldtoft and Johansen, *Teknologiske forandringer i dansk industri 1896–1972*, 1.

24 Grönberg, *Learning and Returning*, 88–89; Gårdlund, *Industrialismens samhälle*, 243; Sven Rydberg, *Svenska studieresor till England under frihetstiden* (Uppsala 1951) 139–202.

25 Stang, 'A Measure of Relative Development'?, 93,

26 Olsson, To see, 454.

obscure South American town'[27] they could possibly have acquired the same knowledge at home. We should underline that this discussion is on an aggregate level; individual technicians may have acquired equally valuable knowledge and experience also in other countries.

Using a German-American dichotomy, the overall tendency is that the relative importance of North American experience grew over time. This is mostly a result of American growth from 1900 to 1910 and German decline between 1910 and 1920. Higher shares of Nordic engineers and architects brought back experiences from the United States and, to a smaller extent, Canada and this was valid for all the individual countries.

Sweden had the highest shares of 'Americans' within her corps of engineers and architects all through the period, and the difference between the two major areas was smaller than in the other countries. It can be argued the shares of 'American' technicians in Norway and Finland were almost equally high, but they were still significantly outnumbered by engineers and architects who had experienced the German-speaking countries. In addition, Finland's corps consisted, as mentioned, of technicians with higher shares of foreign experiences than most countries, and almost all European regions accounted for more than North America throughout the period. American impulses have been viewed as important for Sweden's development,[28] but this has been done without applying a comparable perspective with neighbours and other countries. These shares point in a direction where the United States had a relatively stronger role in Sweden than in other Nordic countries. One explanation is, perhaps, the fact that the Swedish companies generally were larger and more directed towards an American style mass production compared to the industry in the other Nordic countries.[29]

We may, however, call our period a primarily German one when it came to technological impact. Technicians from all Nordic countries made most travels to the German-speaking world. The Germanised world overtook from a British era some decade before our starting point and was replaced by an American after our ending point, but primarily after World War II. Once again, we may underline the difficulties of discussing technical influence based on statistics. German patterns were more obvious in Norway and

27 Stang, A Meaure, 93.

28 Kurt Samuelsson, *From Great Power to Welfare State: 300 Years of Swedish Social Development* (London 1968) 265.

29 See for example, Bo Stråth, *Union och demokrati: de förenade rikena Sverige och Norge 1814–1905* (Nora 2005); Francis Sejersted, *Socialdemokratins tidsålder: Sverige och Norge under 1900-talet* (Nora 2005).

Finland, and these quantities can be interpreted in the light of technological and educational impact as, for instance, Andersen and Nerheim have noticed for Norway and Myllyntaus for Finland. Quantitatively speaking, the shares of technicians with total experiences from Germany, Switzerland, and Austria were always highest in Finland. However, when we look at technicians on placements abroad in one of these countries, the pattern becomes somewhat different as table 11 shows.

Migration in the German-speaking world accounted, as we can see, for more in Norway than in Finland at all three years. It is, of course, difficult to ascertain to what extent there were different implications of these somewhat dissimilar mobility experiences. A temporary residence including employment and/or enrolment at a technical university implies, on the one hand, longer exposure time and a higher probability of discovering something interesting. A study visit may only have lasted for a day or two, but we may, on the other hand, assume that many Finnish technicians labelled relatively long intermissions abroad as study trips. Mobility to the German-speaking countries had, however, an almost certainly stronger implication for technical development in Finland and Norway than in Sweden and Denmark.

National differences are probably underestimated as many engineers from the two countries lacking universities of technology often took the entire education in some German-language university. The German shares in Finland and Norway decreased a little after the transformation of the polytechnic institute in Helsinki's Kamppi area and the inauguration of the Norwegian Institute of Technology at Gløshaugen in Trondheim. At least the formal need to complete the education abroad was no longer present. Table 11 indicates that the

TABLE 11 Experience from migration in German-speaking Europe in numbers and percentages among engineers and architects in the Nordic countries in 1900, 1910, and 1920

COUNTRY	NORDIC		Sweden		Denmark		Norway		Finland	
Year	Number	%	Number	%	Number	%	Number	%	Number	%
1900	432	16	127	9	45	9	193	35	67	24
1910	961	17	341	12	98	9	392	34	130	24
1920	1470	15	494	11	207	9	600	27	169	20

SOURCES: see figure 1.

shares having experienced placements abroad had decreased more in Norway and Finland than in Sweden and Denmark between 1910 and 1920.

The Finnish and Norwegian institutes of technology applied many ideas from the *Hochschulen* in their curriculum, but the university establishments may also have meant the beginning of a decline for the German technological influence in these two countries. Many technicians worked, as mentioned, some years in German industry before returning, but as the migration for studies diminished, so did the number of engineers with experience from enrolment and employment at German industrial works. In Finland, however, it seems that more people began to travel for shorter visits to Germany.

2.3 *The German-Speaking Countries Dominated in All Fields Except Mining*

Table 12 shows that on the pan-Nordic level, the German-speaking countries constituted the most common experience for technicians of all specialisations, except mining engineers and metallurgists, where it was a small American majority.

Architecture and chemical engineering were the most German groups, but the region of European countries outside the German-speaking and Nordic ones, Britain, and Russia were almost equally large among architects. The German domination was clearest among chemical engineers since the superiority was very marked. Germany's world leadership within the chemical industry is clearly mirrored in this pattern.

More than every second chemical engineer in Finland had, for example, experienced a German-speaking country. The German shares were somewhat higher in Norway than in Denmark and Sweden, but the domination is clear among all chemical engineers in the Nordic area. Nordic neighbouring countries accounted for most experiences besides Germany on the pan-Nordic level and in all countries except Sweden where other European countries and North America accounted for more. However, the North American rates were still much higher in Finland than in Scandinavia. This is basically the return of engineers from New England, most prominently pulp and paper technicians from Holyoke, Massachusetts, and textile technicians from Lowell in the same state. This is also the group where experiences from Russia were most frequent, but the pattern for the eastern empire was even, and this applies to Finland and Denmark but not as much to Sweden. Finally, Chandler's observation that Britain's chemical industry remained out of the game is also reflected here. Few chemical engineers had experience from there; only overseas destinations outside of North America were less represented.

TABLE 12 Foreign experience per specialisation among graduates from technical schools in the Nordic countries working in Sweden, Denmark, Norway, and Finland in the years 1900, 1910, and 1920

Experience	TOTAL	Mechanical, electrical & naval	Civil & construction	Chemical	Mining & metallurgy	Architecture
German-speaking countries	27	27	23	33	18	37
Sweden	21	23	13	26	19	22
Denmark	18	16	16	29	16	33
Norway	37	37	38	37		50
Finland	53	56	44	69		
North					20	6
America	15	19	10	14	23	9
Sweden	17	22	7	13	9	8
Denmark	9	8	10	9		3
Norway	16	20	14	15		
Finland	15	24	4	28		
Europe*	10	8	9	11	13	34
Sweden	9	8	8	9	12	24
Denmark	8	6	11	8	16	15
Norway	6	6	5	10		48
Finland	30	24	24	30		
OTHER Nordic countries	10	9	8	14	12	24
Sweden	7	6	6	9	14	9
Denmark	6	4	6	13	7	22
Norway	7	8	3	15		37
Finland	41	44	40	43		
Britain	8	10	6	8	6	13
Sweden	8	10	4	6	7	8
Denmark	8	7	9	7	3	17
Norway	7	9	4	8		15
Finland	17	22	10	17		

(continued)

TABLE 12 Continued

Experience	TOTAL	Mechanical, electrical & naval	Civil & construction	Chemical	Mining & metallurgy	Architecture
Russia	4	4	3	4	2	4
Sweden	3	3	2	3	2	1
Denmark	2	2	2	2	2	1
Norway	1	1	1	0		8
Finland	18	20	19	25		
Latin America & the						
Caribbean	2	1	3	2	2	1
Sweden	1	1	2	0	2	0
Denmark	3	1	3	5	1	1
Norway	2	1	4	3		1
Finland	1	0	0	1		
Africa, Asia						
& Oceania	2	1	1	1	3	3
Sweden	2	1	3	1	1	3
Denmark	3	2	1	1	8	2
Norway	1	1	0	0		2
Finland	1	0	1	0		

*Except German-speaking and Nordic countries, Britain, and Russia.

Architects constituted another group with comparably high shares of experiences from the German-speaking countries. Many students returning from architectural studies, primarily of Art Nouveau, in, for example, Berlin, Dresden, and Munich contributed. The pattern was, however, more even; many architects had been in countries such as France and Italy. Experiences from other European countries was far more frequent among architects compared to any other group. The Ekelund-Kuhlefeldt couple and their admiration for Italy is one example, and they can also illustrate the relative usualness of experiences from the Nordic neighbouring countries as they worked with the Swedish classicists in Stockholm before they returned to establish their office in Helsinki. British experiences, for example, the study of garden cities, were also relatively common. One striking pattern is—with an exception for Sweden—the low frequency of experiences from North America; only about every twentieth Nordic architect had been there. Surprisingly enough, the share also decreased from 1900 to 1920, but this is mostly because a few Swedish architects had remained for some years in the United States after the 1893 Chicago fair. Swedish architects generally had more American experiences than their Nordic colleagues and more migration experiences from the United States than from German-speaking areas. Adding study trips, however, also makes Swedish architects more European. Silja Laine has shown that it was very rare among Finnish architects to go to the United States before 1930. American architecture was often viewed as only rational, whereas European counterparts also were interpreted as esthetical. Continental Europe was often viewed as the prime destination, as it also was argued in Borgstedt's travel account. This view was reflected in the experiences architects brought back to the Nordic area.

Several construction engineers continued their studies to become architects, and this is one part of the explanation of why the German-speaking countries also dominated the experience pattern in this group. The German shares were however somewhat lower, but the differences from other areas were relatively significant since this was the group with the least foreign experiences in general. About every fifth Nordic engineer had, however, experienced bridges, railway building, and other aspects of what Rudolf Christiani called the 'great genius of German engineering'. Germany, Switzerland, and Austria dominated in all the Nordic countries. North America, other Nordic countries, and other European countries were other common experiences, but the Nordic countries diverged. In Norway, it was clear that the United States and Canada constituted the main alternative experiences, and there also were relatively many engineers who had been in Latin America. Civil and construction engineers seem to have been Denmark's most Americanised engineers, whereas they almost were the ones with the least such experiences in Sweden. It was also very uncommon that Finnish engineers with these specialisations had been in North America; they had, instead, often travelled to other Nordic countries and somewhat more often to Russia compared to graduates specialising in mechanical and electrical engineering as well as naval architecture.

These graduates also had somewhat fewer experiences from German-speaking Europe, but the area was nevertheless the most common experience in the four countries and especially in Norway and Finland. Mechanical engineers could specialise in electrical engineering or naval architecture at a renowned technical university in, for example, Zurich or Berlin. Many had experienced Berlin's electrical industry and the German Baltic Sea coast shipyards and, thereby, modern, rational, and technologically advanced industrial settings. Nevertheless, North America had a relatively stronger position within this group. Except for mining engineers and metallurgists, this group was the most Americanised one, and many returnees had worked at General Electric and Westinghouse and experienced large-scale manufacturing there as well as at mechanical workshops like Baldwin and Sellers. A significant number of Nordic engineers had, thus, experienced Taylor's and Ford's country or, as Edström expressed it, 'the best place in the world to learn to organise and run a workshop cheaply'. Mechanical and electrical engineers and naval architects in Sweden, Norway, and Finland had more experience of North America than their Danish colleagues; it was still more common that Danish mechanical engineers had been in Britain around 1900. Together with architecture, this group was also more British when it came to experiences, which is not very surprising as it includes naval architects. On to the twentieth century, Britain was still dominating world shipbuilding, and many Nordic engineers had, for example, worked at shipyards around Newcastle-upon-Tyne and Glasgow as well as studied naval architecture at the university in the latter city. British experiences were more common in 1900 than later; this is partly a reminiscence of the times when the island was the main industrial model and was especially marked in Denmark. The shares of British experiences in this group were highest in Finland, and the country shows, in addition, a reversed trend: Britain accounted for more in 1920 than in 1900. Nevertheless, this is because Finnish shares for all areas generally were higher. Britain was still the fourth or fifth most common experience in Finland, whereas it was the third in Scandinavia.

Mining engineers and metallurgists were the least German group in this context. We should underline that these specialisations include only Swedish and Norwegian graduates, which might have contributed to giving the group an American touch as these two countries were major countries of transatlantic emigration in general. Nevertheless, experiences from the German-speaking countries—for example from mining in Silesia or studies at the *Bergakademie* in the Austrian town of Leoben—were more frequent among mining engineers and metallurgists on the Scandinavian peninsula in 1900. North America overtook as the main experience in 1910 and 1920, and several engineers had now experienced the mechanised mines in the Adirondack Mountains in upstate New York as well as in Michigan and Minnesota. This Nordic change is, however, only because of the Swedes, and Swedish mining communities were sometimes described as 'miniature Americas' in the earliest decades of the twentieth century. German-speaking Europe continued, however, to dominate the pattern of experiences in Norway, even if American methods also were used in Norwegian mining.

3 Geographical Dispersion of Returnee Technicians

Were there certain areas in the Nordic countries where foreign-experienced technicians were more common, and could it thereby be relevant to speak of a geographical impact pattern? The problem of doing a study like this is, of course, that smaller communities account for very small numbers, and the results may be dependent on random circumstances. To use larger areas is not completely appropriate since the experienced technicians might have lived far away from each other. Nevertheless, we can reveal some interesting patterns.

One is that more industrialised areas generally had somewhat higher shares of foreign-experienced technicians, possibly with Norway as a minor exception which we will return to. We can assume that a more heavily industrialised area was more interesting to an engineer, regardless of whether he or she had foreign experience, but we may hypothesise with the idea that this was even more relevant for technicians with experience from leading industrial countries who possibly had observed and learned about more advanced technology and means of production.

3.1 *Swedish Concentrations in Larger Cities and Industrial Areas*

The largest concentrations of returnee technicians in Sweden were around the large cities, the counties of Stockholm, Gothenburg and Bohuslän and Malmöhus. This is, however, a result of these regions hosting the highest numbers of technicians in general. We can still conclude that the percentages of foreign-experienced technicians over time were higher in some counties around Lake Mälaren. Stockholm's share increased from 1900 to 1920. Västmanland, hosting ASEA in the provincial capital of Västerås and some of Sweden's major ironworks, and Södermanland, hosting some of the country's largest mechanical workshops in Eskilstuna, also belonged to the Mälar counties with higher shares of returnee technicians.

Östergötland, Jönköping, Kalmar, and the island of Gotland in the southeast also hosted higher shares, be it that Gotland's percentage is calculated at a very low number. Östergötland and Jönköping also hosted some major mechanical workshops. The big-city counties in the southwest, Malmöhus, and Gothenburg and Bohuslän also had relatively high percentages, as well as Halland in the southwest and Värmland in western central Sweden. Finally, there were also relatively many returnee technicians in two counties along the southern part of the north Swedish coast: Gävleborg, hosting one of Sweden's largest ironworks, and Västernorrland, a centre for the sawmill and pulp and paper industries.

The lowest shares are noted for the three northernmost counties, Norrbotten, Västerbotten, and Jämtland, of whom at least the latter two were sparsely

TABLE 13 Foreign experience per region among graduates from technical schools in Sweden 1880–1919, working in Sweden in the years 1900, 1910, and 1920

County	1900	1910	1920
Stockholm	133 (31%)	332 (38%)	558 (37%)
Uppsala	4 (24%)	14 (39%)	15 (23%)
Södermanland	6 (38%)	14 (36%)	25 (35%)
Östergötland	9 (25%)	26 (46%)	50 (34%)
Jönköping	9 (41%)	18 (39%)	32 (40%)
Kronoberg	0 (0%)	3 (16%)	3 (12%)
Kalmar	3 (33%)	7 (33%)	22 (42%)
Gotland	0 (0%)	3 (75%)	5 (56%)
Blekinge	4 (20%)	12 (30%)	20 (31%)
Kristianstad	3 (20%)	9 (39%)	11 (31%)
Malmöhus	54 (36%)	125 (40%)	162 (30%)
Halland	4 (36%)	11 (46%)	13 (45%)
Älvsborg	11 (39%)	32 (33%)	37 (30%)
Gothenburg and Bohuslän	50 (40%)	104 (38%)	172 (34%)
Skaraborg	3 (18%)	12 (36%)	9 (24%)
Värmland	14 (50%)	18 (35%)	32 (30%)
Örebro	15 (31%)	32 (32%)	46 (29%)
Västmanland	15 (31%)	48 (40%)	81 (41%)
Dalarna	10 (20%)	50 (40%)	50 (31%)
Gävleborg	9 (24%)	24 (46%)	43 (36%)
Västernorrland	7 (35%)	26 (43%)	28 (30%)
Jämtland	3 (27%)	10 (38%)	5 (17%)
Västerbotten	1 (100%)	4 (24%)	4 (25%)
Norrbotten	7 (19%)	15 (25%)	17 (22%)
UNKNOWN PLACE	10 (19%)	17 (23%)	28 (29%)
SWEDEN	**1234 (31%)**	**2591 (37%)**	**4331 (34%)**

SOURCES: SEE SWEDISH SOURCES, FIGURE 1.

industrialised. Foreign-experienced technicians were overrepresented in the eastern parts of Sweden at all three observation points, whereas we can note a constant underrepresentation in the four northern counties. Some counties in the south and southeast also hosted relatively few returnee technicians: Blekinge, Kronoberg, and Kristianstad as did Älvsborg and Skaraborg in the southwest and Örebro, Dalarna, and Uppsala in central Sweden.

Many returnee technicians in the Swedish capital had public employment. The National Railway Company and Board as well as the Water Power Board and the Swedish Patent and Registration Office were common employers, and many served the military as naval engineers. The Royal Institute of Technology also belonged to one of the most frequent employers; at least six returned engineers became professors of different subjects. Edvard Hubendick,[30] who studied in Zurich and became one of Sweden's leading experts on internal combustion engines.[31] Municipal bodies were also major employers; the manager of Stockholm's gas and electricity works, Robert Dahlander,[32] had experience from Switzerland and Germany and performed extensions, modernisations, and technical improvements.[33]

Many architects in Stockholm were publicly employed but also had their own offices. One was the Jugendstil architect and former student in Munich Albin Brag. Some of Sweden's most famous architects also belonged to this group; Ragnar Östberg[34] travelled to the Chicago fair in 1893 and Germany, Britain, France, Italy, Spain, and Greece in the late 1890s.[35] He is known as the architect of Stockholm City Hall, which is believed to be influenced by the Doge's Palace in Venice. Östberg was far from alone; names like Gunnar Asplund,[36] Torben Grut,[37] and Ivar Tengbom[38] are found among the internationally experienced architects who were active in early twentieth-century Stockholm. Kreuger, with experience of reinforced concrete, also had Stockholm as a base and his and Toll's company employed other returnee engineers.

One of the most frequent private employers of returnee engineers in Stockholm was the electrotechnical company Luth & Rosèn, one of ASEA's smaller competitors on the national scale. Two of its managers had been abroad, and among them was the earlier-mentioned founder John Luth, who had continued his studies in Cherbourg, been employed in Britain, and study-travelled in the United States. Among other larger employers of internationally experienced technicians in Stockholm, we found the Bolinders mechanical workshops. Erik

30 KTH, mining, 1900.

31 https://www.ne.se/uppslagsverk/encyklopedi/l%C3%A5ng/edvard-hubendick, (16 November 2017).

32 KTH, civil, 1890.

33 Johan Axel Almquist, Bertil Boëthius, and Bengt Hildebrand, *Svenskt Biografiskt Lexikon. Bd 9, Cornell-Dal* (Stockholm 1931) 596.

34 KTH, architecture, 1888.

35 Indebetou and Hylander, *Svenska Teknologföreningen*, 316.

36 KTH, architecture, 1909.

37 KTH, architecture, 1894.

38 CTI, construction, 1898.

August Bolinder[39] returned in the late 1880s to overtake this family business, one of Sweden's largest workshops, after four years in Philadelphia. The workshops bought American milling machines and introduced interchangeable parts.[40] 'The firm was entirely re-organised, new buildings were erected and equipped with the most modern machine tools and appliances which could be found,'[41]wrote one of his friends admiringly. Telephone manufacturer L. M. Ericsson also employed returned engineers: Gottlieb Piltz[42] practised in the United States and studied the rational telephone system, with a switch and a central battery. Piltz headed reconstructions and extensions of the telephone systems in Moscow and Warsaw and introduced the system in Stockholm. He had a crucial role for automatic telephoning in Sweden.[43] Lighthouse inventor Dalèn's AGA also frequently employed engineers who had been abroad, and so did two companies connected to inventor Gustaf de Laval, one manufacturing steam turbines and another producing a separator for skimmed milk and churns as well as other apparatuses for dairying. AB Separator was one of Sweden's fastest growing companies, and its executive engineer was Erik August Forsberg.[44] He is described as Sweden's most prominent Taylorist and introduced an organisation based on time and motion studies. His foreign experiences were, however, not from the United States but from an electromechanical workshop in Saint Petersburg.[45]

There were smaller communities with technicians with foreign experience. Sweden had a geographically dispersed industry tending to locate in areas outside the larger cities. The reason was, following Jan Glete, twofold: an inclination to build up industrial enterprises around natural resources and transport systems and the relatively strong support from local and regional institutions whose period of greatness was the latter half of the nineteenth century.[46] Some industrial centres were still located relatively close to the capital. Södertälje, some 30 kilometres to the southwest of Stockholm, was one. This town hosted several larger mechanical workshops and was also the home of the car manufacturer Scania-Vabis. In 1910, two out of three observed technicians in

39 CTI, 1884.
40 Grönberg, *Learning and Returning*, 226–227.
41 Pollock, *The Bolinder Book*, 5.
42 TESÖ, mechanical, 1892.
43 Jan Kuuse, 'Gottlieb M T Piltz', in: Göran Nilzén, et. al. (red.), *Svenskt Biografiskt Lexikon. Bd 29, Pegelow-Rettig* (Stockholm 1997) 313.
44 KTH, mechanical, 1896.
45 Alf Johansson, *Arbetarrörelsen och Taylorismen*, 95. Erik A Forsberg, https://sok.riksarkivet.se/sbl/Presentation.aspx?id=14336, (15 November 2017).
46 Glete, *Ägande och industriell omvandling*, 71–73.

Södertälje had foreign experience, and four managers and one executive engineer at the town's mechanical workshops had been abroad. To the south lay Norrköping, the country's fourth-largest city. Relatively few returnee technicians worked in the city, but the picture might have looked different if the graduates of Norrköping's technical secondary school had been included. It was, as the pet name 'Sweden's Manchester' indicates, dominated by the textile industry and a few returned engineers worked within this branch—one as manager of a wool-factory—as well as in municipal bodies. Gustaf Ambjörn[47] had studied naval architecture at the Massachusetts Institute of Technology and became manager for the shipyard in Norrköping before he became a professor at Chalmers.[48]

The industrial town Eskilstuna on the south shore of Lake Mälaren hosted foreign-experienced technicians: a manager for a company manufacturing chains, and the executive engineers for two major mechanical workshops, both known for their modern methods of production. ASEA's hometown Västerås was another concentration around Lake Mälaren. We have mentioned J. Sigfrid Edström's reorganisation of the company after American principles, Karl-Erik Eriksson's establishment of a laboratory like General Electric's, how Edström's friend and roommate from Chalmers and Pittsburgh, Emil Lundqvist, introduced mass production in the ASEA workshop and made it reminiscent of Westinghouse's counterpart, how Carl Silvander participated in the erection of a workshop inspired by AEG in Berlin and in manufacturing the largest capacity generators ever after having studied General Electric's production of generators. There were many more ASEA examples.[49]

In the 1905 English language brochure 'What we can do', ASEA wrote: 'The location of our works at Westerås is excellent, in the close proximity to the best swedish (sic) steel, iron and copper works, possessing good railway and steamship communications and cheap electric power from our own waterfalls'.[50] The nearby mining and steel and iron district, Bergslagen, hosted a number of industries where returnee technicians served, often in leading positions. ASEA had a major branch in Ludvika in Dalarna, where relatively many returnees also worked. The region's steel and iron industry had several returnees in responsible positions. Hugo Carlsson[51] took over as executive engineer and later

47 KTH, naval architecture, 1910.

48 Olsson, *Technology Carriers*, 149–150.

49 On returned engineers at ASEA, see: Fridlund, *Den gemensamma utvecklingen;* Glete, ASEA *under hundra år;* Grönberg, *Learning and Returning*.

50 Cf. Grönberg, *Learning and Returning*, 153.

51 TESÖ, chemical, 1904.

as manager in Fagersta after sixteen years in the United States and Canada and managed the company during a very profitable period. The steel and iron mills in, for example, Avesta, Borlänge, Hofors, Kolsva, and Filipstad also had returned engineers among their staff. We have mentioned Esselius's and Magnusson's rationalisations and modernisations of the rolling-mills upon return from the United States. The armament industry, Bofors, near Karlskoga, constituted another Bergslagen concentration of returned technicians. Ragnar Sohlman, whom we mentioned worked in Alfred Nobel's laboratory in San Remo, had also worked at gas-light companies in Chicago and for the Swedish delegation at the 1893 Chicago fair. He became the long-time chief executive officer for Bofors Nobelkrut but is, as stated, known more as one of the main organisers of the Nobel Prize. Mining in the region also utilised foreign experience, like Bror Andersson's example in Grängesberg, the 'piece of America' in Sweden, shows. A few returnee technicians worked in the copper-mining town of Falun. Some were mining engineers, but two of the managers and leading engineers at the town's locomotive and wagon manufacturer had practised at locomotive workshops in the United States and Germany. The long-time town architect had practised at an architect office in Helsinki.

Town architects, town engineers and other managers of municipal services such as harbours were also often returnees. This was the case in Örebro, a town hosting relatively many foreign-experienced technicians; the others managed a paperworks and a brewery, owned a mill, a furniture store, and a firm within heating and sanitary engineering, headed a chemical factory, served the national railway company, and taught at the technical secondary school. Gävle, on the east coast, hosted a shipyard manager and a construction company owner who also had experienced the United States. The town also had a manager of the gas and electricity works with experience from General Electric. This was also the case in many other towns such as Luleå, Boden, Skellefteå, and Umeå in the north.[52]

Northern Sweden lagged when it came to the employment of foreign-experienced technicians, but there were exceptions. One was the power station and the village in Porjus in northern Lapland where the works engineer had been in Shawinigan Falls in Quebec. Porjus's Klondike character resembled the development around the Canadian power station. Electricity from the power station was used for the railway that was to transport iron ore from the mines of Malmberget and Kiruna to the ports in Luleå and the Norwegian coastal town Narvik. Returnee engineers participated in the processes. Körner, mentioned in the very beginning of this book, was crucial for the electrification of

52 Grönberg, 'Tillbaka till Framtidslandet', 49–51.

the railway, whereas Fagerberg participated in the modernisation and mechanisation of the iron ore mines.[53] The manager of the LKAB mines in Kiruna, Hjalmar Lundbohm, also studied rationalisation in North America in the early 1890s.[54] In the late 1920s and early 1930s, some returnee technicians—Falkman, Palén, and William-Olsson—also participated in the development of another north-Swedish mining enterprise, Boliden and the adjacent smelting plant.

Sawmills and pulp and paperworks in the northern and north-central parts of Sweden also employed a few returnee engineers as managers and in other leading positions at, for example, Hissmofors near Östersund where two brothers from Scania with experience from Germany arrived. Establishments along the southern parts of the north-Swedish coast, near Söderhamn, Örnsköldsvik, and Sundsvall also followed this pattern. Sundsvall was a major centre for the forest industry, but also a town where a man who had been enrolled in Zurich and studied harbours during a trip to the world fair in San Francisco in 1915 was town engineer.[55] Returnee technicians were town engineers and town architects in some smaller urban settings around Sweden: Karlstad in western central Sweden, the Stockholm suburb of Djursholm, Linköping in the southeast, Visby on the island of Gotland, Skövde and Strömstad in the west, are some examples. Not far from these two towns, returnees were involved in weaving and spinning mills and at chemical industries and stone industries.

In Gothenburg, Sweden' second city and centre for the western parts of the country, some returnee engineers made careers as teachers at Chalmers and at least five of them became professors. It was also common that returned technicians worked in municipal bodies: tramway managers had experience from the United States, Germany, and Switzerland. J. Sigfrid Edström, for example, served as tramway manager in the city before he was recruited to ASEA in Västerås. Gothenburg also had a manager of the gas and electricity works who had acquired experience abroad and returnees on the harbour board and at the city engineering office.

The Swedish ball-bearing factory, SKF, was one Gothenburg-based company with many returnees. Manager Uno Forsberg[56] had, as mentioned, been at the

53 Hansson, *Porjus*, 174–182; Gösta Malm, *I min krafts dagar* (Stockholm 1963) 55; Grönberg, 'Tillbaka till Framtidslandet', 51–54, 56–57; Eriksson, *Gruva och arbete*, 103.

54 Curt Persson, *Hjalmar Lundbohm—En studie av ledarskap inom LKAB 1898–1921* (Luleå 2015) 42–44.

55 Lars-Göran Tedebrand, 'Politik, väljare och valda efter 1888', in: Lars-Göran Tedebrand (ed.), *Sundsvalls Historia*, vol. D.2. (Sundsvall 1997) 259.

56 KTH, mining, 1907.

company's American factories. He used his experience from the United States to implement rationalisation of manufacturing and workshop organisation, establishing a planning department and introducing time and motion studies.[57] Some other mechanical and electrical workshops around the city also had returnee managers, like the Davy Robertson machine factory, and the electrical manufacturer Boye & Thoresen. Another electrical manufacturer, Eck in the suburb of Partille, was also a large employer of returnee engineers. The founder, Carl Eck, was a returnee from America who had managed an electrical factory in Belleville, New Jersey.[58]

The Götaverken shipyard was one of the most common employers of returned technicians in Sweden's second city and so were two other Gothenburg shipbuilding establishments, Lindholmen and Eriksberg. The latter had a manager with long-time foreign experience, mostly from the United States but also from Germany and Britain.[59] Götaverken's manager, the earlier-mentioned Hugo Hammar, had experienced American warship production and was inspired by welfarism. His colleague Ernst A. Hedèn[60] became deputy manager and was inspired by modern electric machines and introduced a template system, based on what he had observed in Camden near Philadelphia.[61] This was a solution to the labour shortage and helps to explain the rapid shipyard expansion in the 1910s.[62] Lars O. Olsson assumes that shipyards in Gothenburg looked more over the Atlantic for inspiration, whereas shipyard in southernmost Sweden looked instead to Germany.[63]

Trollhättan was another cluster of returnee technicians in the southwest. Many of them served at the Nydqvist & Holm workshop, a company that, for example, manufactured locomotives. Trollhättan was, however, most known at that time for its power station, and at least fifteen returnee engineers served there; many had been at General Electric and the large electrical firms in Germany. According to a 1907 letter from ASEA's manager Edström to a colleague, Trollhättan's power station was planned after the Ontario power station at

57 Martin Fritz and Birgit Karlsson, *SKF—a global story: 1907–2007* (Stockholm 2006) 52.

58 http://www.hestories.info/company-history-and-historical-facts-concerning-makers-of-elec.html?page=2, (17 November 2017).

59 On *Eriksberg's* manager, Gunnar Engberg, see Olsson, *Technology Carriers,* 67, 99, 101, 105–106.

60 CTI, naval architecture, 1898.

61 Olsson, *Engineers as System Builders,* 50–51.

62 Bodman, *Chalmers tekniska institute,* 110; Olsson, 'To See How Things Were Done in a Big Way', 455.

63 Olsson, 'Amerikaemigrationen', 243.

Niagara Falls. Mats Fridlund has described Trollhättan as an 'America in Sweden' or a 'Swedish Niagara'. Some of the engineers had experienced General Electric's manufacture of the world's largest capacity generators for use in Canada's development of Niagara Falls.[64]

The dual towns of Jönköping and Huskvarna on the south shore of Lake Vättern hosted many returnee technicians. They served at match factories, major mechanical workshops and at one of the country's largest armament industries. John Sandwall[65] worked a few years in Sunderland before he returned to become manager of the family workshop in the 1890s. We found returnees as engineers and builders at the Karlskrona naval station in the southeast, in leading positions at cement factories on the Baltic islands of Gotland and Öland, and at some glassworks in Småland.

There were a few more clusters of returnee technicians in southern Sweden. The region of Halland, located along the coast between Gothenburg and Malmö, belonged to the counties where the returnee share among technicians was among the highest; some had been in Germany and the United States and managed sugar works, sulphite factories, and mechanical workshops in and near the provincial capital of Halmstad. This town, as well as neighbouring Falkenberg and Varberg, also had town engineers sharing experiences from these two countries. Rudolf Sunström[66] spent five years in Germany, Switzerland, and the United States and returned to become engineer in charge at the *Gunnebo* ironworks branch in Varberg.

On the other side of the provincial border, in the southernmost region of Scania, returnee engineers from Germany and America were important to the development of coal mining around Höganäs. Landskrona's shipyard employed engineers with experiences from Germany and the United States, and Danish engineers assisted in constructing the largest dry dock in the Nordic countries. Trelleborg's port was, as mentioned, also finished by Danish engineers using their competence in constructing ferry berths. Landskrona's town architect had studied in Germany, and the returnee contingent also included managers of a foundry, a mechanical workshop, and an oil-factory. Some leading technicians at motor and machine factories in the same town had also worked in Russia. Jakob Beskow[67] managed a superphosphate and sulphuric

64 Fridlund, *Den gemensamma utvecklingen*, 63–67; Grönberg, *Learning and Returning*, 129, 157, note 87; 'Some Important Events in the History of the General Electric Company', *Schenectady Works News*, 6 April 1923.

65 KTH, mechanical, 1889.

66 CTI, mechanical, 1905.

67 KTH, chemical, 1896.

acid factory in another Scanian town, Helsingborg. He worked in a chemical laboratory in Paris and studied distillation of coal tar in France and England as well as the manufacture of superphosphate and sulphuric acid in Germany. Beskow was an important inventor within superphosphate technology who also constructed furnaces and manufactured red phosphorus.

Most returnee engineers in Scania served in Malmö, the provincial capital of Malmöhus and Sweden's third city. The manager of the Kockums shipyard, Georg Ahlrot,[68] had worked ten years in German shipbuilding.[69] This points to relevance in Olsson's assumption that shipyards in Scania were oriented towards Germany.[70] Johan Hallgren[71] managed a cement factory in the suburb of Limhamn and had practised in mechanical workshops in Bristol and Berlin and spent one year in Philadelphia in the late 1880s.[72] After a study trip in Germany around 1905, he got the idea to suggest the introduction of revolving furnaces.[73] Some returned technicians worked in Malmö's municipal bodies, for the national railway company and there were several sugar works in the countryside outside the city managed by engineers who had been in Germany. The architects Nils August Ewe[74] and Carl Oskar Melin[75] had also been in Germany, and their mutual business employed many architects and construction engineers with similar experiences. Fellow architect Frans Fredriksson spent eight years studying at the technical university in Charlottenburg and at Berlin's academy of arts in the late 1880s and early 1890s. He designed many Art Nouveau buildings in Malmö and the neighbouring university town of Lund. The latter town also hosted two Swedish-American managers of mechanical workshops.

3.2 *Copenhagen's Dominance Echoed in the Danish Dispersion of Returnee Technicians*

Today's Malmö is connected to Copenhagen by a bridge, but the historical ties between the former Danish province of Scania and the capital on the other side of the Öresund Strait are strong. In the 1850s, when the Swedish four-estate parliament discussed the establishment of the technical secondary school in Malmö, one of the arguments was that the school could prevent

68 CTI, naval architecture, 1889.
69 Olsson, *Technology Carriers*, 100, 105–107.
70 Olsson, 'Amerikaemigrationen', 243.
71 KTH, mechanical, 1880.
72 Indebetou and Hylander, *Svenska teknologföreningen*, 218.
73 Karin Kock, *Skånska Cement Aktiebolaget 1871–1931: Minnesskrift* (Uppsala 1932) 99.
74 TESM, construction, 1898.
75 KTH, architecture, 1902.

TABLE 14 Foreign experience per region among graduates from technical schools in Denmark 1880–1919, working in Denmark in the years 1900, 1910, and 1920

County	1900	1910	1920
Copenhagen	104 (32%)	233 (31%)	424 (31%)
Holbæk	0 (0%)	1 (33%)	7 (28%)
Frederiksborg	3 (38%)	3 (30%)	4 (18%)
Roskilde	1 (33%)	2 (25%)	5 (45%)
Sorø	2 (33%)	4 (31%)	8 (28%)
Præstø	1 (17%)	4 (40%)	3 (16%)
Maribo	3 (25%)	7 (27%)	10 (24%)
Bornholm	2 (100%)	3 (100%)	4 (50%)
Svendborg	2 (29%)	4 (25%)	5 (23%)
Odense	0 (0%)	1 (3%)	25 (35%)
Southern Jutland*			3 (50%)
Vejle	6 (67%)	5 (22%)	10 (21%)
Ribe	1 (33%)	1 (11%)	8 (44%)
Ringkøbing	1 (17%)	2 (14%)	7 (22%)
Skanderborg	3 (60%)	3 (17%)	6 (30%)
Aarhus	9 (35%)	13 (34%)	24 (28%)
Randers	6 (67%)	6 (55%)	9 (41%)
Viborg	0 (0%)	1 (13%)	3 (11%)
Thisted	0 (0%)	2 (29%)	0 (0%)
Aalborg	4 (22%)	7 (32%)	10 (22%)
Hjørring	0 (0%)	2 (22%)	3 (15%)
UnkNown place	2 (12%)	14 (30%)	17 (15%)
DENMARK	**146(30%)**	**318 (30%)**	**595 (29%)**

*The counties Tønder, Haderslev, Sønderborg and Aabenraa in 1920.

SOURCES: SEE DANISH SOURCES, FIGURE 1.

Scanian youngsters from going to technical studies in Denmark. We, however, are now crossing the strait and arrive in Copenhagen.

Table 14 shows the concentration of returnee technicians in Copenhagen as somewhat more than 70 per cent of the registered positions in 1900, 1910, and 1920. Aarhus is the only county outside Copenhagen that is noted for more than ten returnee engineers in 1910, and, together with Odense, is noted for more than twenty in 1920. The latest year, Maribo, Vejle, and Aalborg, count for ten returnee engineers each. The low number makes it rather useless to discuss shares for more than the counties of Copenhagen, Aarhus, and Odense. Copenhagen lay somewhat above average at all three observation points, whereas Aarhus was above in 1910, and Odense in 1920. We can see that the northernmost counties, Thisted, Aalborg, and Hjørring, attracted fewer returnee technicians.

Hyldtoft writes that the Danish capital 'differed from other capitals in its degree of industrial dominance over the rest of the country',[76] and this is reflected in this distribution of returnee technicians. We must remember, however, that Copenhagen graduates are overrepresented in our cohort compared to the 'real' distribution between the Danish schools (see table 1). Even if all those assigned to Unknown Place above were outside Copenhagen, it would not change the distribution significantly.

Many returned engineers in Copenhagen served in the city bodies. Aage Rørbye Angelo had as mentioned studied in Zurich before he returned and became a pioneer in alternating-current. Poul Gerlow was employed by the municipal electricity company and had studied, and was impressed by, its counterpart in Berlin. The Polytechnic Institute and the University of Copenhagen employed teachers and professors who were former migrants and study travellers. Erland Thaulow, a professor in mechanical technology, had studied his subject in Dresden and London. Alfred Lehmann, who was the manager of the psychophysical laboratory at the university, besides having his own laboratory, had, as mentioned, studied with Professor Wundt in Leipzig.

F. L. Smidth was, by far, the most frequent private employer of returnee engineers around Copenhagen. One important returnee was Poul Sehested Larsen.[77] He had been a builder in Scotland and became a partner upon return. During a study trip in the United States, Larsen noticed new and large rotary furnaces. Later, they were installed at a cement factory near Aalborg as the first ones in Europe.[78] The mechanical workshop and iron foundry Titan also employed returnee engineers. Frode Kjems[79] studied and worked in Pittsburgh, was inspired by Taylorist ideas, and performed rationalisations upon his return.[80] The Laur. Knudsen workshops was another Copenhagen establishment that became known for rationality; its manager, Harald

76 Hyldtoft, *Københavns industrialisering*, 418.
77 PL, mechanical, 1882.
78 Hyldtoft, 'Perioden 1896–1930', 190. http://www.denstoredanske.dk/Dansk_Biografisk_Leksikon/Naturvidenskab_og_teknik/Ingeni%C3%B8r/Poul_Larsen,(17 November 2017).
79 KM, mechanical, 1908.
80 Nørregaard, 'Ingeniørernes indsats', 195; Hyldtoft, 'Perioden 1896–1930', 180.

Helweg-Larsen,[81] had studied modern work methods in the United States. Hans Blache[82] became executive engineer and machine director for the shipyard Burmeister & Wain after he returned from employment in Stuttgart and London and was important for the development of diesel engines.[83] Rudolf Christiani, who had experienced the Hennibique iron-concrete method in Paris and Düsseldorf, started a building contracting company together with Aage Nielsen and was among the major employers of internationally experienced engineers in the Danish capital. The workshop and iron foundry Atlas introduced serial production in the mid-1910s under one of Denmark's Taylor pioneers. Icelandic-born naval architect Erik Zimsen had experience from shipyards in the United States and Russia as well as the Bell Telephone Company in Antwerp.[84] The manager of the Smith, Mygind & Hüttemeier workshop had been in America. The executive engineer at the same workshop had been in Germany. This workshop also introduced methods inspired by Taylor and employed foreign-experienced engineers. The same can be said about Nyeboe & Nissen—specialising in heating and brewing technology—whose founders both had been in the United States and employed a few other transatlantic returnees. The J. G. A. Eickhoff factory for printing machines, N. C. Monberg, machine and aviation manufacturer Nielsen & Winther, tool maker V. Løwener, the engineering firm Saabye & Lerche—specialising in bridge, railway, and harbour construction, and the Danish sulphuric acid and superphosphate factory were other Copenhagen-based companies with returnee engineers among their staff.

Returnee technicians outside Copenhagen were scattered, but present in all parts of Denmark. North of Copenhagen, in Elsinore, Edgar Madsen[85] modernised a brewery based on his experiences from Bavaria, Austria, and France.[86] In Roskilde, west of the capital, a few returnee technicians were employed in the municipal engineering offices and the regional road administration. A shipyard in Køge, south of the capital, was managed around 1920 by a returnee with experience in shipbuilding from Glasgow, Bremen, and Stavanger.[87] The limestone factory in Stenlille on central Zealand was owned in the 1920s and 1930s by a returnee who had studied in Karlsruhe,[88] and the town engineer in Holbæk in

81 KM, mechanical, 1904.
82 PL, mechanical, 1896.
83 http://www.denstoredanske.dk/Dansk_Biografisk_Leksikon/Naturvidenskab_og_teknik/Ingeni%C3%B8r/Hans_Blache, (17 November 2017).
84 Hyldtoft and Johansen, *Teknologiske Forandringer i Dansk Industri 1896–1972*, 179.
85 PL, chemical, 1888.
86 http://www.rosekamp.dk/DBL_All/DBL_15_text.pdf, 16 November 2017.
87 Hannover, *Dansk Civilingeniørstat 1942*, 188.
88 Hannover, *Dansk Civilingeniørstat 1942*, 186.

the 1910s and early 1920s had participated in the construction of the pillars to the Saint Erik bridge in Stockholm.[89] Thorvald Kerstens[90] visited municipal engineering plants during a study trip in the Netherlands and Germany in 1898. Later, he became the long-time town engineer in Slagelse.[91] In the same town, Henrik Rasmussen[92] started an iron-foundry upon his return from employment in Lancaster in England in 1898.[93] Around 1904, Rasmussen's son Ludvig travelled to Germany and brought back a modern moulding machine.[94] Niels Christian Hansen[95] practised in Glasgow, Hamilton in Canada, Pittsburgh and California, and returned to become, over time, the manager of the iron foundry and machine-factory in Korsør.[96] Christian Adler-Nissen[97] had studied at the most common German school for Danish specialists in iron casting, the technical university in Aachen, but also worked at an ironworks in Berlin and as the casting engineer at Ludwig Loewe in the same city. Upon return, he became the manager of an iron foundry in Nykøbing Falster.[98] Returnee engineers also served at railways. Johannes Friis[99] had practised at the civil engineering office at the earlier-mentioned prominent construction company Holzmann in Frankfurt-am-Main and became the long-time manager of the railway between Næstved and Præstø.[100] Six of nine technicians noted for employment on the island of Bornholm had foreign experiences; they served at local railways, the town engineering office in the provincial capital of Rønne, a stonemasonry, and a sawmill.

Returnee engineers had leading positions at breweries and malt factories and sugar industries in Zealand, Lolland, and Funen. Gunner Falck Barfoed[101] had travelled in Germany and Austria and managed the sugar factory in St. Croix before he returned to take the position as manager in Sakskøbing.[102] Sugar industries in Assens on Funen and Holeby in Lolland were also managed by engineers with foreign experience. A chemical engineer had managed a brewery in the earlier-mentioned Swedish town Landskrona and returned to become the

89 Hannover, *Dansk Civilingeniørstat 1942*, 98.
90 PL, civil, 1896.
91 Hannover, *Dansk Civilingeniørstat 1942*, 77–78.
92 KM, mechanical, 1894.
93 *Danske teknika*, 179.
94 Hyldtoft, 'Perioden 1896–1930', 163.
95 KM, mechanical,1901.
96 *Danske teknika*, 125.
97 KM, mechanical, 1905.
98 Hyldtoft, 'Perioden 1896–1930', 166; Nørregaard, Ingeniørenes indsats', 32.
99 PL, civil, 1890.
100 Hannover, *Dansk Civilingeniørstat 1942*, 52.
101 PL, chemical, 1898.
102 Hannover, *Dansk Civilingeniørstat 1942*, 92.

long-time manager of a malt factory in Nakskov. Reginald Lyman[103] travelled in the German-speaking countries and returned to start a brewery in Svendborg.[104]

Employment of returned engineers in Odense followed three main lines: employment with the city's engineering office, self-employed as owner of an engineering firm, and employment at electrical manufacturer Thomas B. Thrige. Jens F. Engberg, who lectured about Taylorism and rational workshop organisation in the United States, served in Odense as works engineer for a car factory started by Thrige. Cay Allesen-Vernø[105] had been at General Electric in Schenectady and returned to become Thrige's head of transformers before he started his own business in Næstved in 1925.[106] Viggo Meyer[107] had worked for Brown Boveri in Mannheim and became one of Thrige's most prominent builders of electric motors and advanced to technical director in the late 1920s.[108]

In 1920, there were few returnee technicians, at least in this cohort, in the areas that had returned to Denmark from Germany, but this may be a source-related result. The region of Vejle hosted more, and a major concentration was Kolding. Returnee engineers worked in the town's engineering office and at a dairy while a third owned an engineering firm. The manager of a firm manufacturing apparatus for pasteurizing and, later, an American-style apparatus for churning, Karl Kristian Konstantin-Hansen,[109] had been in America in the late 1880s.[110] In the region of Ribe, to the west, there were a few returnee technicians working in the municipal harbour department in Esbjerg, while some worked for the road administration in Holstebro and for the national railway company in Struer, both in the region of Ringkøbing. Valdemar Birn[111] was on an educational journey to Germany before he took over as manager of the family business in Holstebro in 1903.[112] Silkeborg hosted, as mentioned, Jørgen Dreyer, whose experiences from Brandenburg and Austria were crucial when the paper factory started to produce high-class paper. Oscar Piper[113] in Horsens had studied combustion engines in France and Belgium and was the first to introduce a crude-oil two-stroke engine.[114]

---

103 PL, chemical, 1893.
104 Hannover, *Dansk Civilingeniørstat 1942*, 63.
105 KM, electrical, 1915.
106 *Danske Teknika*, 92.
107 PL, electrical, 1909.
108 Hyldtoft, 'Perioden 1896–1930', 46.
109 KM, mechanical, 1881.
110 *Danske Teknika*, 146; Hyldtoft, 'Perioden 1896–1930', 75.
111 KM, mechanical, 1902.
112 *Danske Teknika*, 100.
113 PL, mechanical, 1893.
114 Hyldtoft, 'Perioden 1896–1930', 140.

The largest concentration of returnee engineers in Jutland was in Aarhus, Denmark's second city. Herluf Forchhammer, mentioned in the beginning of this book, was the head of F. L. Smidth's local branch in 1910. In 1920, however, the number was still eighteen times lower than in Copenhagen. Theodora Hänschell[115] was one of few female engineers with foreign experience. She studied physiological chemistry at Uppsala University in Sweden before she became an assistant at Aarhus's municipal hospital. Many returnees in Aarhus worked for the national railways, but one was also manager for a private railway. Other engineers worked, for example, in municipal bodies, a local telephone company, and at a couple of foundries. Malthe Conrad Holst[116] travelled in Germany and northern France to examine different chemical methods.[117] He was employed by Aarhus oil factory and was a driving force in the usage of hardening methods to extract vegetable oils. Aarhus oil factory was the first in Scandinavia to introduce these methods.[118]

The county of Randers, north of Aarhus, hosted some returnee engineers. One of them worked in road and canal construction on the island of Anholt, between Jutland and the Swedish coast. Jacob Anker Bie[119] studied brewing in Germany and Austria and took over the management of the family brewery in Hobro.[120] Christian Boeck-Hansen,[121] one of the engineers who had worked for F. L. Smidth at the brickworks in southwest Sweden, became the leader of the cement factory in Mariager. Some returned engineers worked in the road and harbour administration in the provincial capital with the same name as the county, while others were builders and later managers of a railway car factory in the same town. Christian Rich-Christensen[122] had worked with water-power stations in Argentina and at a whaling station in South Georgia, before he returned to become the leading engineer at the power station in Gudenå in Viborg County.[123]

The northernmost part of Denmark, the counties of Aalborg, Thisted, and Hjørring, attracted few returnee engineers. We have mentioned the large rotary furnaces that engineer Poul Sehested Larsen noticed during a study trip in the United States, which were installed at a cement factory by Aalborg.

115 PL, chemical, 1914.
116 PL, chemical, 1896.
117 Hannover, *Dansk Civilingeniørstat 1942*, 77.
118 Hyldtoft, 'Perioden 1896–1930', 108.
119 PL, chemical, 1883.
120 Hannover, *Dansk Civilingeniørstat 1942*, 35.
121 PL, civil, 1907.
122 PL, construction, 1913.
123 Hannover, *Dansk Civilingeniørstat 1942*, 261.

They were the first of their kind in Europe. Around this city, some returned engineers worked as managers and in other engineering positions at cement factories and a few private railways as well as, for example, at a brewery, a sulphuric acid and superphosphate factory, and for the city in its harbour and the engineering office. We also find one our female graduates in Aalborg. Johanne Wille Jørgensen[124] studied in Germany in 1911 and became head of a laboratory examining products for the household.[125] Further north, the manager of the machine factory and iron foundry in Frederikshavn, one of Denmark's major manufacturers of boat motors before World War I, made many study and business trips abroad.[126] One of his employees had worked four years for a Danish company in Molde.[127] He had been to Norway, and that is also where we are going.

3.3 *Norwegian Concentrations in Bergen and Industrial Areas in the Southeast*

Towards the end of our period, ferries started to connect Frederikshavn with Gothenburg, but also with Kristiania[128] or Oslo, as the Norwegian capital has been called since 1925. We are arriving in a region, which was the centre of Norway's nineteenth- and early twentieth-century industrialisation. Table 15 shows that most returnee technicians worked in the capital; between 37 per cent in 1910 and 46 per cent in 1920.

Major concentrations of returned technicians, in real numbers, were the second city, Bergen, and Sør-Trøndelag, where Trondheim is located. The shares were generally higher in the counties following the coast from Kristiania to Bergen, except for Vest-Agder. Oppland also hosted relatively high shares. The shares were lower in the parts of Norway located north of Trondheim, with Troms County as a minor exception. Over time, Bergen hosted the highest shares of returnee technicians in her corps, about 70 per cent in 1910 and close to 60 per cent in 1920.

Among the foreign-experienced technicians working in the capital, we have mentioned Taylor advocate Lehmkuhl, furniture designer and architect Michalsen[129], and the Charlottenburg graduate, Sam Eyde, who started an

124 PL, chemical, 1907.
125 http://www.kvinfo.dk/side/597/bio/1514/origin/170/ (17 November 2017).
126 Hannover, *Dansk Civilingeniørstat 1942*, 188.
127 Hannover, *Dansk Civilingeniørstat 1942*, 196.
128 http://danmarkshistorien.dk/leksikon-og-kilder/vis/materiale/frederikshavn/?no_cache=1, (18 November 2017).
129 On returnee architects from the United States in Kristiania (Oslo), see: Per David Martinsen, Erfaringen fra Amerika. Norske arkitekter i USA 1880–1930 – et glemt kapitell?

TABLE 15 Foreign experience per region among graduates from technical schools in Norway 1880–1919, working in Norway in the years 1900, 1910, and 1920

County	1900	1910	1920
Kristiania (Oslo)	98 (49%)	189 (51%)	304 (43%)
Akershus	7 (58%)	10 (45%)	11 (34%)
Østfold	10 (59%)	18 (41%)	33 (42%)
Hedmark	3 (38%)	11 (48%)	18 (41%)
Oppland	2 (50%)	8 (38%)	13 (48%)
Buskerud	10 (50%)	21 (42%)	36 (39%)
Vestfold	3 (50%)	7 (54%)	20 (48%)
Telemark	7 (64%)	26 (53%)	36 (57%)
Aust-Agder	4 (80%)	5 (56%)	12 (46%)
Vest-Agder	2 (33%)	7 (58%)	9 (23%)
Rogaland	9 (56%)	13 (52%)	41 (47%)
Hordaland	2 (50%)	8 (47%)	21 (47%)
Bergen	23 (52%)	60 (69%)	107 (58%)
Sogn og Fjordane	2 (33%)	2 (25%)	9 (47%)
Møre og Romsdal	0 (0%)	10 (56%)	22 (48%)
Sør-Trøndelag	22 (46%)	52 (54%)	90 (40%)
Nord-Trøndelag	2 (29%)	5 (36%)	13 (59%)
Nordland	5 (28%)	10 (36%)	18 (37%)
Troms	2 (50%)	10 (63%)	9 (35%)
Finnmark*	1 (33%)	5 (50%)	8 (32%)
Unknown place	22 (35%)	31 (42%)	41 (32%)
NORWAY	**236 (46%)**	**508 (50%)**	**871 (43%)**

*Including two technicians in Svalbard in 1920.

SOURCES: SEE NORWEGIAN SOURCES, FIGURE 1.

engineering office. The most common employer of returnee engineers in Kristiania was, however, the Norwegian State Railways followed by the directorate for canal construction and electrical supply. Working for state authorities seems to have been more common here compared to in Stockholm and Copenhagen.

Eksempler fra remigrantenes arbeider i Oslo (Oslo 2008). Master Thesis University of Oslo.

The Norwegian Public Roads Administration and the Norwegian Industrial Property Office were other frequent workplaces. Some returnee technicians also served in the municipal bodies; at the gas and electricity works, tramways, street and sewer and public cleansing departments, and at the city engineering office. Most engineers in these departments had experience from Germany. Long-time administrative manager of the electricity works, Thomas Norberg-Schulz[130] studied in Charlottenburg and became a Norwegian pioneer for practical use of electricity. Many returned technicians also owned their own architect offices. One of the most common private employers was *Norsk Hydro* where the earlier-mentioned Christoffer Kahrs Kielland was technical manager. Kværner Bruk was also among the major private employers of returned engineers in the city. A few executive engineers and department heads at Kvæner had been in, primarily, Germany. Other major mechanical workshops in the city, such as Thune and Myren, also were among the employers who were interested in experience from other countries. Both places hosted leading engineers with experiences from other countries, of which Germany was the most common. Elektrisk Bureau, manufacturing telephones and telecommunications equipment, was another major employer; its administrative manager and executive engineer had studied in Hannover.

There were local areas in Norway hosting substantial numbers of engineers who had relatively high shares of returnees among them over time. Drammen, Kristiania's neighbouring town to the west, was a centre for wood processing and pulp and paper where many returnee technicians were employed. However, most of them worked for the national railway company or had other public employment. One town engineer, one harbour engineer, and one town architect had all spent time in Germany. Private companies that employed returnee technicians included one cellulose factory in which the technician had spent three years in America. Søren Johan Falch[131] and Knut Strand[132] had both been in Berlin and rose to responsible positions at Drammen's armature factory. This factory delivered to Sweden, Denmark, the Netherlands, Russia, Britain, and the United States, but—according to Falch's account—had its most important export market in Finland.[133] Norway's chemical industry employed a few returnee engineers. *Norsk Hydro*'s production of fertilizers in Notodden and Rjukan, west of Drammen, involved at least fifteen engineers

130 TTL, construction, 1884.

131 KTS, mechanical, 1894.

132 KTS, mechanical, 1912.

133 Bassøe, *Ingeniørmatrikkelen*, 488; Kristiania tekniske skole, *Skrift ved 50 års jubileet for ingeniørene fra K.T.S. 1894* (Oslo 1944).

with foreign experience in the early decades of the twentieth century. Another nearby concentration were the dual towns of Skien and Porsgrunn. These towns belonged to Norway's major industrial areas, but public employment was the most common among returnee technicians here too, especially in road administration. Skien was one of Norway's most important timber towns, and the wood-processing company Union was the most common private employer of returnee engineers.

Returnee engineers also served at several other mechanical workshops and shipyards, like Oscar Nørgaard[134] who returned from the United States and bought a mechanical workshop in Kristiansand. This town, on the southern tip of Norway, was also—as mentioned previously—the home of Anton Martin Grønningsæter who, based on his experience from Canada and the United States and an invention made by Swedish-American engineer Victor Hybinette, developed the nickel refining plant to one of the largest such facilities. Grønningsæter was not alone there; the nickel industry in Kristiansand employed many engineers with foreign experience.

Thorsen at the Stavanger Steamship Company, who introduced car ferries in western Norway, and Zimmer, who rationalised Rosenberg Mechanical Workshop, both upon returns from the United States, were touched upon earlier. To work for the city was, however, also the most common pattern among returnee engineers in Stavanger. Nils Haavardsholm,[135] the long-time executive engineer at the municipal power station and the electricity works, had studied in Aachen and worked more than ten years at dam construction in Germany. Rosenberg and a factory producing metal tins, cans, tubes, and boxes were among the private employers. Henrik Finne[136] spent a few years in Chicago before he returned to found and manage the just mentioned metal products factory for almost 30 years.[137] Not very far from Stavanger, returnees were involved with the smelting plants in Sauda, mainly producing manganese alloys, and in Odda, producing calcium carbide. Other returnees served in leading positions at the pyrite mines in Stord.

Bergen was one of the local communities in the Nordic area that had the highest shares of returned technicians. We may call into question how much emphasis we should give to the city's history as a Hanseatic centre, but one reasonable explanation lies in the significance of Bergen's shipyards. This possibly led to an inclination to go and work at, and thereby, study the technologically

---

134 KTS, construction, 1881.

135 KTS, construction, 1901.

136 BTS, mechanical, 1892.

137 Eskedal, *BTS-matrikkelen*, 11.

advanced shipyards abroad, as the earlier-mentioned Christian Stoltz Dekke did in Glasgow. We have also mentioned car pioneer Wilson, and dye and clean works owner Wilhelm Kahrs. Tramway manager Andreas Falkenberg[138] and long-time engineer at the electricity works, Justus L. Reimer[139] were also representative of the common pattern of being employed in municipal bodies. Both had studied electro-technology in Liege and practised in Berlin. Returnee engineers in Bergen basically worked in the same departments as in Kristiania; power and electricity, cleansing, sewer and street administration. Like in the capital, employment with the state railways was also common. The fact that a mechanical workshop was the most frequent private employer, and that it was common to own an architect office are two other traits that resemble Kristiania's. All in all, however, it seems to have been somewhat more common among returnee technicians in Bergen to start their own enterprise. Another difference was that returnees to Bergen often took employment at the technical school. The director and several head teachers and other teachers at the technical school had experienced Germany. This did not happen at the Kristiania equivalent to the same extent.

Many returnees also worked in municipal bodies or other public employments, like road administration and the national railways, also in smaller urban settings. The eastern towns of Lillehammer and Hamar were places where returnee engineers served the national and regional road administrations. Mining was also a sector in Norway heavily populated by returned technicians, and not far from the two neighbouring towns mentioned above lay the mines Tynset, Folldal, Orkdal, Løkken, Røros, and Meråker, where returnees served in leading positions.

In nearby Trondheim, it was even more common than in Bergen that returnee engineers taught at the technical school, and Heje and Heggstad became, as mentioned, professors at the Norwegian Institute of Technology after they had studied in Germany. There were at least seven returnee technicians who had reached such a position in 1920. One was Sverre Pedersen,[140] who had studied in Hannover and Charlottenburg and was city architect before he was appointed a professor in city planning. Pedersen became known as an innovator in city planning and was, for example, inspired by British garden cities.[141] Some

138 BTS, mechanical, 1893.

139 BTS, construction, 1903.

140 TTL, architecture, 1901.

141 Helga Stave Tvinnereim, 'Sverre Pedersen —Utdypning (Norsk Biografisk Leksikon)—Store Norske Leksikon', *Store Norske Leksikon*, http://snl.no/.nbl_biografi/Sverre_Pedersen/utdypning, (15 November 2017).

architects with foreign experience also started their own offices in Trondheim; we mentioned Guldahl, and Osness earlier, and there were at least five more returned architects opening offices in the city, and these offices also employed other returnees. Like in Kristiania and Bergen, it was also common among Trondheim's returned technicians to serve in municipal bodies as well as for the national railways. Holmsen returned, as mentioned, with experience from British shipbuilding to a successful management of a mechanical workshop. Leading engineers with the firm *Jernbeton* had, in 1920, experiences with reinforced concrete from Germany, America, and other countries. This firm was among the most frequent private employer of returnee technicians in the city.

North of Trondheim, the numbers and shares of returnee technicians were, as mentioned, lower, but some of them also found their way there. As elsewhere, many served in the municipal bodies. Steinkjer, Namsos, and Bodø employed town engineers and town architects with foreign experiences, and so did Narvik, Tromsø, and Harstad further north. All six observed technicians in Harstad in 1910 had been abroad and four of them worked in road administration, but Kaarbø returned, as mentioned, from a shipyard in Grimsby to take over the family workshop. Water-power stations constituted a frequent employer of technicians with foreign experience in Norway, often in the eastern and western parts of the country. Bernt Lund,[142] however, managed the power station in Glomfjord in Nordland and used his experiences from studies in Zurich and employment in the United States.[143] Close to Glomfjord lay the Sulitjelma and the Dunderland mines, where some returnee engineers worked. Dunderland employed, as mentioned, Bertel Skjerdal with experience from American mining and became known for American methods. Skjerdal was also employed by the King's Bay Kull Co. in Svalbard, managed the first mines at the same time that he was responsible for the build-up of the village Ny-Ålesund.[144] He was one of nine Norwegian engineers noted for employment on these latitudes near the North Pole; four had foreign experiences, and three had been to mines in America.

The Sydvaranger mining company also employed a returnee technician from the United States, but they also employed two brothers who had experience from Russia, or rather Estonia, but it still belonged to the Russian Empire when the Knudtzen brothers were there. Johan Jentoft[145] was employed with a waggon factory and as executive engineer at an electrotechnical company in

---

142 KTS, mechanical, 1899.
143 Bassøe, *Ingeniørmatrikkelen*, 319.
144 https://nbl.snl.no/Bertel_Sherdahl, (18 November 2017).
145 TTL, mechanical, 1897.

Reval before he came to Kirkenes in 1907. He became executive engineer and later manager. Harald[146] found employment at the same electrical factory in Estonia and participated later, according to a genealogical website about the family, in Russian investigations of iron ore on Novaya Zemlya.[147] Harald also became one of the mining company's leading engineers.[148]

According to the same genealogical website, Harald and his brother Ole met a Swede with knowledge of cinematography onboard *Hurtigruten*—the Norwegian Coastal Express—on their way from Kirkenes to the school in Trondheim. The three youngsters debarked the ship before it even had arrived in Trondheim and decided to go and get material in Stockholm to continue to Archangel to establish a cinema. The boys travelled through Finland.[149] We will now do the same.

3.4 *Finnish Returnee Technicians on the Industrial Island in Tampere*

Petsamo, east of Norway's Varanger Peninsula and today's Norwegian-Russian border, belonged to Finland from 1920 to 1947, but we have no information on technical employment there in 1920. A person crossing a border to Finland in the north during our studied period arrived in Oulu County, which encompassed around half the country. We will begin our survey through Finland in the north.

Table 16 shows that there were few returnee technicians in the northern county, and also in central inland counties such as Mikkeli and Kuopio. In these regions, technicians with foreign experience noted lower shares all through the period. However, we should keep in mind that almost all regional shares in Finland are high, compared to the Scandinavian countries.

Uusimaa, home to Helsinki, hosted the highest number of returnee technicians, followed by the counties hosting the other larger cities: Hämeen, home to Tampere, Viipuri, and Turku and Pori. The shares of technicians with international experience were, over time, higher in Turku and Pori and Hämeen, the latter an industrialised region. Tampere was described as Finland's only typical industrial city in the nineteenth century and the Grand Duchy's equivalent to Manchester. The city hosted a cotton factory, linen and iron manufacturer, and a paperworks, and these industries were some of the largest in the Nordic

146 TTL, mechanical, 1910.

147 http://s221243733.onlinehome.us/index.php/handelsmann-oscar-knudtzen-i-finnbukta-og-hans-familie, (18 November 2017).

148 Bassøe, *Ingeniørmatrikkelen*, 280.

149 http://s221243733.onlinehome.us/index.php/handelsmann-oscar-knudtzen-i-finnbukta-og-hans-familie, (18 November 2017).

TABLE 16 Foreign experience per region among graduates from technical schools in Finland 1880–1919 working in Finland in the years 1900, 1910, and 1920

County	1900	1910	1920
Uusimaa	76 (66%)	188 (68%)	324 (73%)
Viipuri	20 (71%)	31 (58%)	63 (73%)
Mikkeli	1 (25%)	1 (17%)	4 (57%)
Kuopio	3 (33%)	16 (76%)	18 (67%)
Hämeen	21 (78%)	46 (69%)	62 (83%)
Turku and Pori	19 (76%)	26 (70%)	67 (74%)
Vaasa	10 (63%)	22 (56%)	32 (71%)
Oulu	6 (38%)	13 (62%)	14 (63%)
Unknown place	1 (33%)	6 (75%)	9 (64%)
FINLAND	**157 (64%)**	**349 (66%)**	**593 (73%)**

SOURCES: SEE FINNISH SOURCES, FIGURE 1.

countries in the nineteenth century. Pertti Haapala writes that 'Tampere was a highly industrialised island in an agrarian society'.[150]

In Oulu County, a few returnee engineers were involved in railway building, for example around Tornio by the Swedish border. Otto Fridolf Nyberg[151] served there by the turn of the century and brought experiences from studies in Berlin as well as employment in Chicago and the building of the Trans-Caspian Railway in Russia.[152] Uuno Aleksanteri Karén,[153] who participated in the building of the railway between Ylivieska and Iisalmi around 1920, had also experienced railway building in Russia. The provincial capital, the city of Oulu, hosted some returned technicians of whom a few also were railway engineers. One was dyeing manager Iivari Karhi whose experience from the German dyestuff industry was mentioned in a previous chapter. A returnee from the United States managed a soap factory in Oulu, and two city engineers had been in Germany. The county architect between 1918 and 1927 had studied in Germany and travelled in Europe. A few teachers and the principal of Oulu's industrial school were returned technicians, as was the city's factory inspector.

150 Haapala, *Tehtaan Valossa*, 393.
151 SPO, civil, 1882.
152 https://www.genealogia.fi/genos-old/38/38_64.htm, (19 November 2017).
153 SPO, civil, 1903.

In Kokkola, between Oulu and Vaasa, Soini Pohjanpalo[154] returned from studies at Charlottenburg to be technical director for the mechanical workshop he owned together with his brother. The coastal town of Vaasa was a local community with relatively high shares of returnee technicians over time. Manager Bernhard Hahl[155] of the Vaasa cotton factory had deepened his knowledge of the textile industry as a student in Mühlhausen, today the French city of Mulhause, and through practise in England. He managed the factory during what is described as a successful period.[156] Another employer was the Vaasa Electrical Company, whose founder, Johan Vilhelm Samberg,[157] had studied in Gothenburg and Stockholm and travelled in Sweden, England, and Germany. He also served as the principal for Vaasa's industrial school, which employed a couple of returned technicians as teachers. One of them was Thor Lagerroos who also served as town architect and was mentioned in a previous chapter as one of few pre-1920s Finnish architects who had been in the United States. Lagerroos's predecessor as town architect also had foreign experience, and the same was true for two of Vaasa's town engineers. In Vaasa County, the technical leaders and executive engineers at the paperworks in Äänekoski had stayed in both Germany and the United States, and the manager of the local telephone company in Jakobstad had also been in America. In Varkaus, in Kuopio County, a few managers and executive engineers at the forest company Ahlström brought experiences from both Germany and Russia as well as some European countries. Many returned engineers were in the timber and pulp and paper industries and not seldom in Viipuri County. Establishments in, for example, Harlu in today's Russia as well as in Lappeenranta and Joutseno in the southeast corner of today's Finland employed returnee engineers. A railway engineer in Sortavala by Lake Ladoga was also a returnee, so were some leading engineers at electrochemical and electrometallurgical industries in southeast Finland.

Members of Viipuri's community of returnee technicians often worked at the city's industrial school, a pattern that was common also in Jyväskylä and Kuopio. Two of the principals of Viipuri's industrial school in the early twentieth century had experience from Germany. The architect for city planning, as well as a couple of city engineers and the head of the gas and electricity works, all had foreign experiences, and it was also common for the returnees to work for the national railway company. Some former migrants and travellers also had

154 SPO, mechanical, 1909.
155 SPO, chemical, 1897.
156 'Bernhard Hahl', *Hufvudstadsbladet*, 28 September 1923.
157 SPO, mechanical, 1883.

architect offices, and one of them was Uno Ullberg[158] who had travelled in Germany, Austria, Italy, France, Britain, and the Scandinavian countries. Ullberg is described as the man who introduced functionalism in Viipuri. Allan Schulman,[159] a long-time county architect in Viipuri, belonged to the few Finnish technicians who had experience from South America where he had served two years as an architect in Buenos Aires. During his time in Viipuri, he also travelled to Germany, the Netherlands, Sweden, and Denmark.[160] The county architect in Mikkeli had also travelled to Germany and other European countries. Mikkeli County had few returnee technicians, but one owner of a glassworks in Savonlinna was another of the returnees from Russia who made a good career. A few glassworks in southern Finland were managed by returnees, including the Iittala establishment near Hämeenlinna, whose long-time manager, Claes Oskar Norstedt,[161] had studied the glass industry in Scandinavia and Germany, taken a course at the technical university in Braunschweig, and served as technical manager at glassworks near Copenhagen. The Tervakoski paperworks was also located close to Hämeenlinna, the home of the earlier-mentioned fine-paper technician Johannes Andersin who had developed his skills in New England. Georg Candelin[162] had also acquired solid knowledge in pulp and paper technology in America and Russia and was Tervakoski's manager.

Town engineers in Hämeenlinna, Lahti, and Tampere also had acquired experience abroad, and so had, as mentioned, Tampere's city architect Lambert Petersson who was inspired by the German Neo-Renaissance after his studies in Dresden; he was the man behind the beautiful Ruuskanen building. It was common to work in Tampere's municipal bodies; a couple of city engineers, as well as the city geodesist, had also been in Germany. The city's textile and mechanical industry, however, employed most foreign-experienced technicians. The Tampere Linen and Iron Industry was a merge between these two industrial branches and the city's largest employer of foreign-experienced technicians. Verner Ryselin[163] practised in Germany and Switzerland and was employed by the locomotive department, where he constructed the first locomotives on an industrial basis in Finland. Later, he founded and became a manager for Lokomo,[164] where he also employed some other foreign-experienced engineers. In

158 SPO, architecture, 1903.
159 SPO, architecture, 1889.
160 'Allan Schulman Död', *Viborgs Nyheter,* 7 December 1937.
161 SPO, chemical, 1892.
162 SPO, mechanical, 1883.
163 SPO, mechanical, 1894.
164 'Ingeniör Verner Ryselin Död', *Hufvudstadsbladet,* 2 January 1940.

the textile industry, we have mentioned cotton factory manager Magnus Lavonius and tricot factory manager Harald Jensen, who had studied the textile industry in Germany. Nikolai Uschanoff[165] continued his studies at Chalmers in Gothenburg before he specialised in the textile industry in America. Uschanoff also spent time in Germany before he became a technical manager at the tricot factory.[166]

Returnee technicians not only worked in the clothing and textile industry in Tampere but also elsewhere in southern Finland, for example, in Hyvinkää by Helsinki. A railway in Porvoo, east of Helsinki, was managed by a returnee from Germany. We have mentioned a few returned technicians who were active in the capital; telegraphy pioneer Mäklin; Strömberg, who designed some of Finland's first large-scale electrical machines; Heikinheimo, head of the municipal electricity works; the German-inspired architects Lindqvist and Jung; and the only Nordic technician in Costa Rica, who returned to become part of an asphalt and cement factory. Ahlstedt, who had been at German shipyards and became head of all state-owned shipyards, is another example. Not least, the professors Aschan, Hjelmman, Holmberg, Kyrklund, and Wuolle at the Finnish Institute of Technology were important for technical development in Finland. Many returned to teach at the newly established technical university and, to some extent, also at the University of Helsinki. At least fourteen returned technicians were appointed professors at the Finnish Institute of Technology or the University of Helsinki.

To work in state authorities was more common in Helsinki compared to other Nordic capitals. The national boards for railways, road and canal construction, and public housing belonged to the most common employers. Three head architects at the Board for Public Housing had all been travelling extensively in Europe, including Germany. Municipal bodies constituted a common employment for returnee technicians also in Helsinki: the gas, water, and electricity works, the housing office, the municipal laboratory, and the city planning department all employed several former migrants and study travellers. It was also more common to start an architect office in Helsinki than in the other capitals. Strömberg's electrical firm, the dealer in agricultural machines, OY Agros, the Hietalahti shipyard, and returnee Gustaf Zitting's engineering bureau were among Helsinki's most frequent private employers of foreign-experienced technicians. Three engineers at Hietalahti had, for example, worked at larger shipyards in the United States; one was manager Karl Åström.[167] The experience

165 SPO, mechanical, 1906.

166 '50 År Fyller Den 16 December Ingeniör Nikolai Uschanoff', *Hufvudstadsbladet*, 16 December 1931.

167 SPO, mechaniccal, 1902.

he gained in the United States facilitated his career at Hietalahti, and Åström made icebreakers a specialisation at Hietalahti; this was an interest he had developed in America.[168] The metal industry, *Maskin och Brobyggnads AB,* also was among the Helsinki companies that employed returnee technicians. Robert Lavonius[169] had been both in Germany and the United States. He became a manager in the late 1910s and was inspired by Taylorism and modern work methods.[170]

The city engineer, the harbour engineer, and the managers for a dynamite factory and a granite company in Hanko, Finland's southernmost town, also had experience from abroad. Other southern Finnish stoneworks followed similar patterns, for example, in Lojo, Särkisalo and Uusikapunkki. Emil Sarlin[171] graduated in chemical engineering in Helsinki but studied mining at the Royal Institute of Technology in Stockholm and practised in some of Sweden's major mining districts. Later, he travelled in the United States, Germany, and Austria-Hungary. Sarlin became the long-time manager of a lime industry in Pargas, south of Turku, and was a pioneer for this industry in Finland. The iron industry was another branch in which foreign-experienced technicians were present; they managed ironworks in southwestern Finland. Albert von Julin's modernisations of the Koskis ironworks were mentioned in a previous chapter.

The southwest town of Pori was a minor concentration of returnee technicians. In 1920, twelve of sixteen technicians active in the town had been abroad. They included the technical managers at one of the larger mechanical workshops in Finland, John M. Gylphe.[172] He had worked in Germany and travelled in the United States and contributed to rationalising and modernising the workshop. In Turku, the most common pattern was to work in municipal bodies; two city engineers had been in Germany, and Jung moved, as mentioned, to Turku and became the architect for city planning after his duty in Helsinki. The long-time head of the gas works, Karl Gustrèn,[173] had improved his knowledge through studies in Germany and employment with Nobel in Saint Petersburg. In a speech by Gustrèn's grave in 1934, Turku city's technical director claimed that he had brought knowledge that was very unusual in Finland at the time of his appointment in the mid-1890s.[174] Another common pattern was to serve as

168 '70-åringar', *Åbo Underrättelser,* 7 February 1950.
169 SPO, mechanical, 1905.
170 Lavonius, 'Pohjois-Amerikan Konepajojen', 131–135.
171 SPO, chemical, 1898.
172 SPO, mechanical, 1908.
173 SPO, mechanical, 1885.
174 'Vid Karl Gustréns Grav Den 17.2.34.', *Åbo Underrättelser*, 28 February 1934; Fellman, *Uppkomsten,* 217.

a teacher at Turku's industrial school and the Swedish language university *Åbo Akademi*. Turku's iron industry was one of the most common private employers of returned technicians; one vice president had been in the United States and a couple of technical managers in Germany. Crichton-Vulcan's shipyard also was among the Turku companies employing returnee technicians, for example, the head of docks and a few of their departmental heads. A bank architect in Mariehamn in the Åland Islands was a returnee and the only technician working in this Swedish speaking Finnish archipelago northeast of Stockholm. We are almost back where we started this Nordic survey.

4 Summary

A somewhat generalised interpretation of the geographic dispersion patterns is that the more industrialised areas had somewhat higher shares of foreign-experienced technicians. We can assume that a more heavily industrialised area was more interesting to an engineer, regardless of whether he had foreign experience or not, but we may also hypothesise with the idea that this was even more relevant for technicians with experience from leading industrial countries who possibly had observed and learned about more advanced technology and means of production.

Return rates of engineers and architects to the Nordic countries were high; more than two-thirds of the technicians from Denmark, Norway, and Sweden returned as did nine out of ten of their Finnish colleagues. These rates were considerably higher than for ordinary emigrants, and this is also valid if we look only at the technicians going to North America. Clearly, target migration of technicians suited well in a region on the edge of large-scale industrialism and where two of the countries also were in the process of gaining full independence. To acquire experience abroad in regions that, in different fields, were ahead of the Nordic countries was a means for technicians to contribute to development back home besides promoting their own careers. To leave the native country, therefore, may, in this context, be interpreted in terms of patriotism; the latter implies that a return happened.

The strong Finnish demand for technical expertise contributed to the country's high return rate. Finnish graduates study-travelled, as mentioned, much more than their Scandinavian colleagues, and even the Finnish migration had a kind of study-trip character. This is also reflected in shorter durations abroad; the average Finnish duration of stay lasted somewhat over three years, whereas the Norwegian, Danish, Swedish, and pan-Nordic durations abroad lasted in average between four and five years, and the Swedes generally stayed abroad longest. Finland's steps towards independence from the Russian Empire possibly implied a stronger will to contribute; engineers rose, as Myllyntaus states, to key positions in Finland after the 1917 independence. These conclusions raise, of course, the question: why were the patterns not the same in Norway, which also reached full independence during our studied period? One interpretation is that Norway's industrial breakthrough, after all, preceded Finland's. Another answer relates to the stronger Norwegian connection to transatlantic emigration in general. Ireland was the only European country that lost a higher share of her population to North America. Finnish technical migration—like mobility at large—was more short distance, and migration studies often conclude that return rates decrease with increasing migration distances.

Return was, however, dependent not only on country of education. An upper- or upper-middle-class family background generally implied higher return rates, even though majorities of the graduates with lower social origin also returned. Family ties mattered for careers back home. Seemingly, it was also positively correlated with a return to have a foothold in the capital cities. Capital-born graduates returned to somewhat greater extents than colleagues born in other cities, towns, and rural areas, but it was even more important that the graduate had studied in a capital city. The patterns were less marked in Denmark, where Copenhagen hosted both a higher and a lower technical institute, and in Norway, where the country's most prestigious school—informally before 1910 and formally thereafter—was located to Trondheim.[175] In Sweden, the Royal Institute of Technology was the country's most prestigious school, and since the capital did not host a technical upper-secondary school, it is also the only Stockholm-based school included.

Timing seems to have asserted influence on the return decision. We can observe that graduates departing between 1890 and 1919 returned more often than those travelling in the 1880s and 1920s, but Denmark shows a different pattern with decreasing return rates over time. The lower return rate for the 1920s is easily explained by the stoppage year 1930. Departing young and shortly after graduation also implied higher return rates, even if the rates for graduates travelling the second and the third year after graduation was higher than for those who departed the same year or the year after. It was less a target migration if a graduate travelled more than ten years after leaving school and after his or her 30th birthday, even if the age differences in Finland were small. All in all, younger graduates fresh from school not only had more time to go abroad but they also had more time to return for a career

175 All Finnish technicians in the cohort studied in Helsinki.

back home. That is also what the returnees did; being born in the same country in which the graduate studied implied return, whereas foreign-born graduates, in a way, also went back. However, as this study takes its point of departure in the country of education instead of the country of birth, these graduates are not defined as returnees, even if return migration was what they de facto did.

Generally, architects returned more than all engineering specialisations, which partly can be interpreted in the same way as the Finnish return; architects study-travelled more than engineers, and architect migration was also more study-trip-like. Mining engineers and metallurgists returned somewhat more than civil and construction engineers and mechanical, electrical, and naval engineers, but specialisation brought about more national differences than other characteristics. Mining engineers and metallurgists had, for example, the highest return rate in Sweden, even higher than architects, but the lowest in Norway. Chemical engineers from these two countries acted similarly and returned to a lesser extent than colleagues from Finland and Denmark in relation to the other specialisations.

The German-speaking countries stand out as the major target migration goal; graduates choosing Germany, Switzerland, or Austria as their first destination returned more compared to those who chose other destinations. In Finland, however, migrants to other European destinations were even more likely to return. There is a correlation with distance; migrants to Europe went back more than those who chose an overseas destination, but other Nordic countries show lower rates than more distant European destinations, which can be explained by more labour-market migration to these countries. The graduates going back from the schools in Sweden and Denmark contribute to the pattern.

To bring about change, the returnee technicians must have the possibilities to implement new knowledge and technology in their home societies. In all four countries, it was more common that graduates with foreign experience climbed from the group higher professionals, that is, the one they are in when they start work after graduation, to the group higher managers. Foreign experience was rewarding, but we should underline that (1) a more finely meshed system is needed to discuss this more deeply, and (2) foreign experience was not the only thing that mattered for one's career; age, specialisation and the place where a technician worked was also important. Interestingly, the differences between migrants and study travellers were smaller than one could have expected and—perhaps even more remarkable—the destination seemingly did not matter at all. It was the foreign experiences that were important, not necessarily how and where a technician had acquired them. Nevertheless, as a manager, executive engineer, or the like, a returnee technician had the power required to act in accordance with his or her ideas.

We are pursuing the idea that significant shares of technicians with foreign experience provide a base for change and impact. The shares of internationally experienced technicians in the early twentieth-century Nordic area point in this direction. About 30 per cent of the technicians had migration experience, and total mobility experience lay about 10 per cent higher. This must be considered significant, especially given the fact that about 1 per cent or even smaller shares of the entire Nordic population had resided in other countries. We may also underline that these percentages do not include technicians who had taken their entire education abroad, nor do they include foreign-born engineers and architects. Thus, the shares were, in reality, even higher. There are, however, no reasons to assume that this was a special Nordic characteristic. Studies of other countries have shown that many technicians worked and studied abroad, but as far as the author's knowledge reaches, this is the first attempt to estimate the presence of foreign experience among national corps of technicians.

As for the Nordic comparison, Denmark and Sweden lay below average and Norway above. Finland, however, had much higher shares than the other countries when it came to total foreign experience. We may connect this to what Myllyntaus has called an 'obsession' with foreign studies and regular travelling abroad among Finnish technicians; the differences between the countries indicate that this can be called a distinctive Finnish feature in a comparative Nordic context. There are also reasons to assume that an addition of technicians that had taken the entire education abroad would increase the differences between the less and the more industrialised Nordic countries. Compare, for example, Sam Eyde and the other Norwegian technicians taking all their studies abroad that Bassøe has registered, and the students at Chalmers who went back to Finland and Norway. The lower level of technical education drew several individuals to study abroad, but these countries could, to a certain extent, be compensated through higher shares of technicians with foreign experience.

Civil and construction engineers generally had acquired less experience abroad than other specialisations, a pattern that was relevant in the entire area except in Norway. Architects belonged to the groups that had acquired the most foreign experience, especially around 1920 when they were much ahead of the other specialisations. This applied more to Norway than to Sweden and Finland. Another architect trait was the means; it was the only specialisation where study-travelling accounted for more. This, in turn, applied more to artistic architects from Sweden and Finland than to their technical Norwegian colleagues. There were some national patterns. Chemical engineers were generally the most internationally experienced technicians in Denmark and Finland but were overtaken by architects in Norway around 1910. Mining engineers and metallurgists were the most experienced in Sweden but were almost at the other end in Norway. Civil and construction were overtaken by mechanical and electrical engineers as the least experienced specialisation in Denmark but was, over all time, the least experienced in Sweden and Finland.

Experience was often acquired in the leading industrial and educational countries, which indicates high-quality training and experience of technologically advanced environments. Over time, Nordic space, and specialisation, the experience patterns, were strongly dominated by the German-speaking countries; primarily Germany, accounting for about 90 per cent of the moves to these three countries. The technical universities made the German superiority over other experience regions extraordinarily strong in Norway and Finland. Germany's leadership in the chemical industry implied that the region also dominated this specialisation markedly. This combination led to a more significant German domination among Finnish and Norwegian chemical engineers compared to among their colleagues in Sweden and Denmark. However, the differences between German and other experiences were also large among architects and civil and construction engineers, the latter a group that often completed their studies and took an architectural degree in Germany.

Transatlantic experiences came second in all Nordic countries, except early on in Denmark and Finland. North America grew in relative importance over time, and Sweden was the country where these experiences were most frequent both in numbers and shares; the differences between American and German experiences were also smaller than in the other countries at all three observation points. North America had one of its specialisation strongholds among mechanical and electrical engineers together with naval architects. This applied to the entire Nordic area, but somewhat less to Denmark. This is not surprising. The mechanical engineers and their sub-categories were probably the ones that were most interested in rational workshop organisation in a more or less Taylorist or Fordist spirit.

Mining engineers and metallurgists constituted, however, the relatively most American group, perhaps a combination of the fact that this group only was present in the high emigration countries Sweden and Norway and that studies of mechanised mining in Michigan, Minnesota, Montana, and the Adirondacks attracted. German experiences still dominated over time in Norway, whereas this specialisation actually had an American majority in Sweden in 1910 and 1920, two of few country-time-specialisation cut-off points with non-German domination. Civil and construction engineers diverged by country when it came to American experiences. They constituted the main alternative to German experiences in Denmark and Norway but were less frequent in Sweden and Finland. Borgstedt's statement on American architecture, that it set aside the *eternal laws of beauty,* is reflected in the fact that architects had acquired experience in the United States less often. Nevertheless, Borgstedt's Swedish countrymen were actually somewhat more likely to bring this experience in their luggage, partly because some of them had visited the 1893 Chicago fair.

In general, Finnish technicians had not acquired less transatlantic experience than their colleagues from the neighbouring countries. Chemical engineers in Finland had, for example, often been in New England and Finnish mechanical and electrical engineers had more or less experienced rational American workshops and shipyards as often as their Swedish and Norwegian colleagues. What distinguished the Finnish technicians from their colleagues from Denmark, Norway, and Sweden was their extensive collection of experience in Europe. Russia, of course, made the most pronounced difference between the Grand Duchy and the other Nordic countries. Calculated as experience percentages, however, Finnish technicians had acquired more experience in more or less every European country. Not only had they more often been in German-speaking Europe, but also in neighbouring Nordic countries and France, Italy, and other Mediterranean and eastern and central European areas. The cosmopolitanism in general, as well as the high share of architects in Finland, contributed to this pattern. Architects had acquired more experience in European countries that were located outside German-speaking regions, Russia, the Nordic countries and Britain, but the latter two regions were also overrepresented in their experience pattern. The Swedish classicists constitute one example of attraction; the English garden cities constitute another.

Danish technicians and those with the mechanical, electrical, and naval specialisation also often had experiences from Britain. Denmark's pattern is partly due to F. L. Smidth's, Christiani & Nielsen's, and other's London offices, whereas the latter is, to some extent, explained by a continuously relatively strong British position in shipbuilding. Nevertheless, being the third experience around the turn of the century, Britain declined some in shares and relative importance to 1920. Other Nordic countries now constituted a more frequent experience, so did other European destinations. This was an echo of the relative decline in Britain's industrial world position, even if this primarily happened before the turn of the century. Chemical engineers had very rarely experienced Britain; Chandler's statement that the country 'remained out of the game almost entirely' is thus reflected in this context.

In numbers, of course, most returnee technicians served in and around capitals and the other larger cities, but this was mainly because most technicians, regardless of experience, followed such a pattern. The main pattern seems to be that returnee technicians aimed to work in more industrialised regions, be it larger cities, midsize towns, or smaller communities. Norway is possibly a minor exception; the region around Kristiania (Oslo) was the most industrialised, but the second city of Bergen hosted significantly higher shares of returnee technicians compared to the capital. However, if we look at individual industrial communities around the Oslofjord, like Drammen, we can see high shares of returnee technicians. Copenhagen is one example of the industrial pattern, another is the county of Västmanland in central Sweden, hosting the largest electrotechnical company in the country as well as some of Sweden's leading ironworks. Tampere, the heavily industrialised island in the agrarian Finnish society is also a clear example. Tampere's county of Hämeen had higher shares of returnee technicians, even if the shares generally were high in Finland, except for the inland regions Mikkeli and Kuopio and the northern region Oulu at the beginning of the twentieth century.

Oulu was not the only peripheral region with a lower share; most northern regions had more problems attracting returnee technicians compared to the core regions to the south. This was also true for northernmost Jutland in Denmark, and not only for the Arctic Circle regions in Norway, Sweden, and Finland. Norway's Troms constituted one exception from the pattern, as did Sweden's Västernorrland and Gävleborg, but they are more industrialised and clearly south of the Arctic Circle, even if they traditionally are counted to Norrland, the northern main part of the country.

This is not to say, however, that returnee technicians were absent from the north. In 1910, there were nearly twenty in Sweden's two northernmost regions, Norrbotten and Västerbotten, thirteen in Oulu, and twenty-five in Norway's Nordland, Troms, and Finnmark. Other regions hosted, as we have seen, significantly higher numbers, but these technicians were still present from Svalbard and the Arctic Circle counties to the island of Bornholm and the Danish border towns with Germany, and from today's ceded Finnish areas around the city of Viipuri and the shores of Lake Ladoga to the west Norwegian coastal regions around Bergen and Stavanger. In chapter six, we also learned about technicians' activities in Iceland. All of them had, of course, foreign experience; Iceland possessed no higher technical education. If they could act as living examples of innovation and force their neighbours, read colleagues, 'to evaluate the old and traditional against the new', they became a major force when the Nordic countries industrialised and, in the long run, developed into modern welfare societies.

CHAPTER 8

Summary and Concluding Discussion

This study aims to investigate the pre-1930 transnational mobility patterns of 12,376 graduates with at least three years of education from technical schools in Sweden (5,530), Norway (2,924), Denmark (2,821), and Finland (1,101) between 1880 and 1919 and the implications of these patterns for the economic and industrial development in the Nordic countries. This represents about 80 per cent of the Nordic graduation suiting these criteria; about 70 per cent in Sweden and more than 90 per cent in the neighbouring countries. We departed from Kristensen's taxonomy, in which transnational mobility is divided into two subcategories: 'labour-market-related mobility' and 'mobility for learning purposes'. In a second step, we divided the latter into two sub-subcategories: 'placements abroad' and 'study travel'. This division assumes that graduates stating that they had been on a study trip normally were not employed or enrolled in a university and did not involve themselves in the concrete work-processes that is one of the prerequisites for a placement abroad. The study visits were, hypothetically, short and we have also assumed that a placement abroad was of greater value for acquiring knowledge and experience.

There were Nordic citizens taking their entire technical education abroad. For source-related reasons as well as difficulties in reaching appropriate comparability, they are not included in the statistical base of this study, but we can safely state that the effect was stronger in Norway and Finland. Transnational mobility was a common feature; more than every second student went abroad, and roughly every sixth mobile graduate was only a study traveller. These patterns characterised technicians in many countries; it was not a Nordic 'speciality'. The 'bird-of-passage identity' that Dmitri Gouzévitch and other researchers have ascribed to technicians also characterised Nordic technicians in the decades around 1900. Technicians certainly constituted more of a *peregrine profession* compared to other professional groups, even if there is a need for deeper comparative historical studies. The male, upper- and upper-middle-class character of the profession implied, as Berner argues, better possibilities to move freely over borders and through different environments. Even educated women were often expected to return to housewifery. Children of wealthy parents could count on economic support for foreign intermissions and utilise a wider contact net. Stays abroad were often a means to create an identity as a mature, upper-class man.

 | DOI:10.1163/9789004385207_009

Transnational mobility of technicians was stronger from Finland and Norway, partly due to slower industrialisation implying fewer domestic employment opportunities. Next to Ireland, Norway was also the European country most affected by transatlantic emigration in general, whereas Russia proper and especially Saint Petersburg to some extent played America's role in Finland. Norway lacked higher technical education before 1910, that is, there was a need to go abroad to acquire a university degree in engineering or architecture. Helsinki's polytechnic school also had a limited curriculum which had an effect similar to the Norwegian: For a long time, it was impossible to earn a degree in electrical engineering in the two countries. This low-level education sometimes also made domestic employers prefer foreign technicians in leading positions. Early twentieth-century tsarist policies, like conscriptions into the Russian army, contributed to Finnish transnational mobility, but the decisive factor in giving the Grand Duchy the Nordic area's highest percentage was extensive study travel. This organised strategy to collect information and adopt foreign technology was a specific Finnish trait that was weaker in Scandinavia. Study travel was also an architect pattern, and this specialisation made up a much larger share in Finland. Regarding labour-market mobility and placements abroad, that is, migration, Norway notes the highest share. Transnational technical mobility from Sweden and Denmark was roughly on the same level. A particular Danish trait was the payroll migration, to go abroad to work for a domestic company, even if the trait also existed in Sweden, and to a very small extent, in Finland and Norway. The Danish world position in, for example, the cement business implied extensive travel as experts to almost every corner of the earth.

Younger freshmen had more time to acquire experience. Consequently, transnational mobility decreased with increasing age at graduation. Belonging to the mechanical, electrical, and naval or the mining engineer and metallurgist group implied much learning abroad when it came to rational workshop organisation and automation, while chemical engineers had world-leading Germany nearby. The international unrest of the 1910s contributed to a decline in technicians' transnational mobility, but Finland shows an opposite pattern as study travel aimed at picking up ideas increased as the country advanced step by step towards, the independence it finally achieved. However, labour-market mobility and placements abroad decreased from all Nordic countries among the 1910s graduates. The inauguration of technical universities and the improvement of domestic engineer and architect education added an extra factor in Finland and Norway.

We may speak of a mutual mobility system for Nordic technical school graduates, with some important national variances. Basically, they were a part of the Atlantic migration system that emerged after the Congress of Vienna in 1815, but the Finnish graduates showed more traits of the eastward Russo-Siberian system. Russia's attraction in Finland comes as no surprise: the country was a Grand Duchy under the Russian Tsar until 1917, and the border to Russia proper ran just outside the imperial capital of Saint Petersburg. The Nordic technicians' mobility system was one in which mobility for learning purposes was more frequent than labour market mobility. The study trips are always for learning, and the choices of destinations combined with the significant migrant return rates of around 70 per cent for the Scandinavian countries and 90 per cent for Finland, often after relatively short intermissions abroad, point to a system where target migration or target mobility was the dominating trait. We may use the return rates as indications that placements abroad clearly dominated over labour-market mobility among the 'migrants'. A somewhat generalised and heedless assumption is that about 15 to 20 per cent of the transnational mobility was study travelling; about twenty to 25 per cent were labour-market mobility and around 60 per cent placements abroad.

One consequence of the major learning pattern is the importance of the German-speaking countries and North America as destinations. Steen's letter to Edström shows how technicians evaluated experiences in Germany and the United States: he claimed that he was not 'mature for successful work' in Sweden before he had practised some years in both countries. There were still important differences, and transatlantic mobility was also influenced by 'non-learning' factors.

We will return to this in a while but begin by stating that the German-speaking countries constituted the largest overall Nordic destination and were number one also in Finland and Norway. The Nordic area had long-time cultural connections to Germany, and especially the elites saw the country as a model: this is reflected in the fact that the shares choosing the German-speaking countries increased with higher social origin. Germany, Austria and Switzerland were, however, places to go for most technicians that wanted to acquire knowledge and experience abroad. The learning character is reflected in the facts that these countries were more common destinations for younger graduates, freshmen travelling shortly upon graduation, and study travellers. Many travellers were architects interested in Art Nouveau and neo-renaissance as well as town-planning in the spirit of Austrian architects Otto Wagner and Camillo Sitte. Chemical engineers often went to Germany, due to her prominence in the field. Norwegian and Finnish graduates also choose these countries more often than Swedish and Danish; this is a combination of more deficiencies in the domestic technical education and the establishment of German and Swiss technical universities as world leading in the late nineteenth century. Zurich's extraordinary well-reputed *Eidgenössische Technische Hochschule* was known for its broad education and its focus on engineering practises. *Technische Hochschulen* in Berlin-Charlottenburg, Dresden, Darmstadt, Hannover and Munich were also forerunners in fields like electrical, chemical and hydraulic engineering as well as architecture. Brewing in Bavaria and spinning and weaving in Reutlingen were other fields of study. Some well-known Nordic technicians were enrolled at these schools, and the studies were often a base for good careers in all Nordic countries.

Technical school enrolment was the main pattern from Finland and Norway, whereas Swedish and Danish graduates preferred the other type of mobility that falls under our definition of a placement abroad (which is somewhat broader than Kristensen's learning through employment). German-speaking Europe was 'strong' in every field and bridges in Hamburg and prominent construction companies in Frankfurt-am-Main, as well as mining in Silesia and Austria, attracted many Nordic graduates even if this region was a less likely choice for civil and construction engineers as well as mining engineers and metallurgists. Berlin was the most frequented place. Germany's capital attracted in a variety of fields, but most important were the rational factories that early had taken influences from at least adjusted Taylorist or Fordist ideas of efficiency and often applied corporate social welfare. Nordic technicians were thus able to study this closer than in America and not only in Berlin, but also at workplaces such as Baltic Sea coast shipyards, Ruhr's steel and ironworks, and Swiss workshops. Some returned to rationalise Nordic industries, and there were also returnees from Germany's successful chemical industry located primarily in today's North Rhine-Westphalia and Hesse, which also were the most common targets besides Berlin. World-leading colour works such as the one in Höchst near Frankfurt-am-Main and successful producers of dyestuff like BASF in Ludwigshafen were 'educational' places and so were, for example, sugar plants and paperworks. This development, and the cultural, linguistic and geographic proximity, contributed to make the German-speaking countries the most important part of the Nordic technicians' transnational mobility system.

Labour-market mobility to the German-speaking countries was very minor, even if some Danes represented F. L. Smidth in Germany and Austria and were employed by Danish companies in harbour construction along the Baltic Sea coast. North America, the system's other main 'leg', was in part a different case. Travelling for domestic companies like Sweden's SKF and F. L. Smidth was a minor pattern also in a large American context, even if this migration was not necessarily weaker to the United States and Canada than to other countries. Learning mobility was still the main trait, but it was more one-sided compared to the stream to German-speaking Europe. Study travelling did, of course, exist and was facilitated by technological improvements in communications, but North America was generally too far away to be considered a place to go for a shorter visit without employment. Towards the 1920s, schools like the Massachusetts Institute of Technology grew in importance for Nordics who wanted to complete domestic technical education. This can be viewed in the light of Germany's loss of prestige after World War I, and France also grew as a 'student destination'.

For most of the time, however, North America diverged from German-speaking Europe as a destination only for employment. Placements abroad dominated and industrial America, primarily located in a rectangle with Boston, Washington, DC, St. Louis, and Chicago/Milwaukee as corners, offered many interesting study objects and was an area for learning. This region also stretched into southernmost Ontario. This was the area where the lion's share of Canadian industry was located and where most Nordic engineers in Canada worked. Canada as well as the United States belonged to the countries most frequented by Nordic technicians; especially from the Scandinavian countries. Niagara Falls offered, of course, interesting technical development on both sides of the border and the town of Shawinigan was a model for communities built up around electric power stations.

It was nevertheless almost always the case that Nordic technicians who were in Canada also were south of the border, and the United States hosted a significantly higher number. America's technological development, discovered at events like the 1876 and 1893 fairs in Philadelphia and Chicago and often described in admirable terms in technical journals and the like, attracted technicians from a multitude of European countries, including the Nordic ones. Like in the German case, America attracted in all fields. Even if architects were divided in their view of America and were unlikelier travellers than engineers, at least Swedish returnees took influence from Richardson's round-arch architecture. Developments in reinforced concrete inspired construction engineers to cross the Atlantic and return to start businesses based on the experience. Chemical engineers could deepen their knowledge in a field like electrochemistry and experience from pulp and paper industries in New England played a role in modernisations of Finnish counterparts. Anton Martin Grønningsæter had served in both the United States and Canada, and his knowledge of an electrolytic refining process helped develop a Norwegian nickel works into one of the world's largest.

The two specialisations that most often chose North America were, however, the mechanical, electrical, and, naval group and the mining-metallurgical one. They acquired a multitude of experiences and brought back ideas on the construction of turbines and generators, the use of car decks on ferries in western Norwegian fjord regions and environmental ideas on the dispersion of pollution to a northern Swedish smelting plant. They were expected to serve at industries that supposedly were more receptive to rationalisations echoing the ideas of Taylor and Ford. General Electric and Westinghouse, the United States' two electrotechnical giants, were targets for Nordic engineers of whom many returned to found electric businesses and to reorganise already existing ones; the Americanisation of ASEA, Sweden's largest electrical manufacturer, is the most obvious example. Shipyards around Boston, Philadelphia, and in Virginia as well as Carnegie Steel's plants near Pittsburgh are other examples of industries that attracted with far-reaching rationalisation. Engineers brought back experience from these places, acted in accordance with what they had learned, and introduced labour-saving actions in the Nordic area. The combination of rationalisation and corporate social welfare was another inspirational source. This included profit-sharing systems, workers parliaments with limited power at shareholder meetings, and the like. Danish engineer Engberg's description of it as a 'road to a promised land' can symbolise the Nordic admiration. It was an inspirational source for returnee engineers in responsible positions at steel and ironworks, mining companies, and shipyards.

The Swedish engineer's statement at the 1901 returnee technician meeting and Bjork's observation that engineers—with their return aims and in terms of orientation towards larger cities and industrial settings—diverged from emigrants in general were relevant. The typical technician going to America was not, to quote Lars O. Olsson's translation of Hugo Hammar's statement, 'an ordinary emigrant who went out at a venture'. Most of them had, as Bjork underlines, 'every intention of returning to the homeland'. Nevertheless, transatlantic crossings also involved significant labour-market mobility influenced by the contemporaneous mass emigration from Europe. This was a major difference compared to the streams to Germany and her linguistic neighbours. The combination of placements abroad and labour-market mobility is reflected in the fact that the United States and Canada were—like German-speaking Europe—early destinations, but remained equally popular also for those who departed between six and ten years after graduation. Another observation is that birth in a rural area was positively correlated with the choice of North America: Nordic emigration in general also had a rural character. Technician return to lifework back home was lower from North America than from Europe and especially from the German-speaking countries, but still much stronger among technicians than among Nordic emigrants in general. The flag arrangement over the entrance to the Swedish Engineers Society of Chicago thus symbolised a view of America differing from the target migrant's. The United States was the place for the lifework. A return was only to enjoy the autumn of life back home. Some of the settling technicians became well-known, like Swedish radio pioneer Ernst F. W. Alexanderson and Finnish city planning architect Eliel Saarinen.

The two main destinations dominated but were far from the only places where the Nordic technicians went. Nordic neighbouring countries were nearby and thereby easily accessible, not least for study travel, and they also constituted the 'third' mobility pattern. The reasons why they were an important part of the transnational technical mobility system were connected to this geographical proximity, but the learning aspect was still important. Other Nordic countries offered inspiring architects, hydroelectric plants, a large scale electrical industry like ASEA, and—as a Norwegian engineer indicated after a visit to a Swedish workshop in the 1920s—how ideas on rationality could be adjusted to Scandinavian conditions. Graduates who studied in another Nordic country went back to contribute to technical development in Finland, Norway, and—not least—in the build-up of Iceland's infrastructure. Expert migration, like the ones aimed at establishing telecommunications and cryolite mining in Greenland contributed, so did Nordic companies with activities in the neighbouring countries like the Danish involvement in the cement business and Swedish participation in Norwegian railway electrification.

Britain, despite the loss of some of her previous world position, remained important in this mobility system as some Nordic companies had major foreign branches in London and many also went to British subsidiaries of American electrotechnical companies. London-based Swedish gun factory owner Nordenfeldt also contributed, so did advanced shipyards like the ones around Glasgow and Newcastle-upon-Tyne and mining schools and industry near Sheffield. Architects interested in Art Nouveau found their way to Britain, and so did those with an interest looking closer into the garden city movement and were attracted to work with exciting colleagues in town planning.

Russia was also a major part of this mobility system. Finland's political status, geographic proximity, and the fact that Russia proper was a kind of Finnish 'substitute America' contributed significantly to the pan-Nordic pattern, while the empire was a minor Norwegian destination. Mining, radio engineering, and railway building are examples of fields employing, primarily, Finnish engineers. Domestic Nordic employers like ASEA and the Finnish railways also brought engineers to Russia since it was a major region of foreign activity. Last, but not least, another very important Nordic connection was the Swedish Nobel brothers employing countrymen and Finland-Swedes in the Saint Petersburg workshops and the oil ventures in Baku.

France, Italy, Belgium, and the Netherlands were the most frequented other European destinations, but there were also technicians going to see Gaudi's architecture and Gothic Cathedrals or work for domestic or other foreign companies in the Iberian Peninsula; to study Roman and Byzantine buildings, and construct railways in the Ottoman Empire, and to represent Nordic companies in newly independent countries in eastern and central Europe after World War I. As for Italy, it was first and foremost the classical study trip destination for architects. Parks, gardens, ornaments, mural paintings, early Renaissance and the anonymous so-called *architettura minore* were inspirations for Finland's Ekelund couple and many of their colleagues. Mobility to France, by far the most visited destination in this group, was more manifold, even if architects interested in steel constructions and practical and logical architecture constituted a major part. Nordenfeldt's gun factory moved from London to Paris, and Swedish engineers started to work for them in the French instead of the British capital. French iron-concrete technology was an important source of knowledge. Denmark's Christiani & Nielsen, founded on this experience, later sent engineers to Paris and several other Danish, Swedish, and even Norwegian companies were present in France. Studies, formal and informal, in fields such as aeronautics, marine engineering, and civil engineering contributed to make France a major part of the system, so did employment in various fields such as mechanical engineering. Belgium was a centre for Art Nouveau architecture, offered valuable studies and practise in mining, electrical engineering and the textile industry around Liege. Brussels hosted an engineering firm partly owned by a Swedish technician, renowned in the construction of pig iron and coke-fuelled blast furnaces. Antwerp was the home of major European subsidiaries of large American corporations. The Netherlands attracted with interesting study objects in water management, peat trade, and exciting architecture, like modernist Berlage's commodity exchange in Amsterdam. Some Nordic technicians also represented domestic and other foreign companies and were employed in Dutch mechanical industries.

Argentina was the non-North American overseas country that attracted most, but at least from Scandinavia, there were also technicians working in countries such as Brazil, Chile, Mexico, South Africa, China, and Australia. As for Argentina, the country experienced a nineteenth-century economic boom and the construction of railways as well as drainage and water supply systems in Buenos Aires involved engineers from all over the world. Infrastructure building was one trait attracting Nordic technicians to remote parts of the world; exploration of natural resources and mining was another.

Variations in this mobility system did, of course, exist. There were variances in the means of travelling, even if 'migration', that is, placements abroad and labour-market mobility dominated. Finnish graduates were much likelier to go on a study trip compared to their Scandinavian colleagues. This, together with the higher return rate, points to the Finnish system as even more directed towards learning compared to the Scandinavian counterparts. The architects were most characterised by study travelling. This trait had long been a tradition in their education. The significantly larger share of architects in the Finnish cohort contributes, but this does not tell the whole story. Study trips to acquire knowledge and technology were, as mentioned, a pronounced Finnish strategy and embraced all specialisations.

The choice of destinations diverged with different criteria. Country of education was one; German-speaking Europe was, as mentioned, a more important destination in Norway and Finland because of the deficiencies in domestic technical education. North America was more important in Sweden and Norway as these countries were more characterised by general transatlantic emigration. Also, Swedish industrialisation was, in a Nordic context, comparably large-scale and based on larger corporations and this could have made it more relevant to study rationalisation in, to quote ASEA manager Edström, 'the best place in the world to study [...] how to cheaply run and organise an industry'. Intra-Nordic mobility was most important in Finland because of the extensive study travelling; it was easy to pass by Sweden and/or Denmark on the outward and/or the homeward journey, and these two countries were interesting to architects. Also, there was a sense of affiliation as most Helsinki graduates had Swedish as their native tongue. The British Isles constituted a more important destination in Denmark as many of their early multinational companies had some of their largest foreign subsidiaries in London, and this type of migration was a 'Danish' trait in the Nordic context. Russia was a significantly more important destination in Finland because of the political ties, geographical proximity, and the empire's role as Finland's 'substitute America'. Latin America and the Caribbean constituted a Norwegian destination but was an even likelier choice in Denmark. The Danish West Indies contributed to the pattern. However, Africa, Asia, and Oceania had a special position in Denmark since the ties to Asia were stronger than in the rest of the Nordic area; the colonies in Siam and Shanghai are examples.

Early graduation made, on the one hand, Britain a likelier choice as more of the early and mid-nineteenth-century British industrial leadership remained. Late graduation increased, on the other hand, mobility to remote destinations in Africa, Asia, and Oceania, possibly because communication technology, as well as Danish colonialism, had developed. The two main destinations had their high points in the middle of the period: the German-speaking countries were important for graduates of the 1890s as the decade largely coincided with Germany's development as an industrial and educational 'superpower', but also with economic crisis on the other side of the Atlantic contributing to make the

Americas a less attractive alternative. The earliest decade of the twentieth century implied better times in the United States and Canada, and North America was a more common destination for graduates leaving school between 1900 and 1909 than for those who left earlier and later, even if German-speaking Europe also remained strong. World War I contributed to a loss of German prestige, and the latest graduates were the ones least likely to choose Germany, Austria, and Switzerland.

The German-speaking countries, as well as Europe at large, were also more upper-class and urban destinations, while graduates of rural and middle-class origin were prone to choose North America. Being born outside the country of education implied intra-Nordic mobility; these graduates were simply going back home. The mobility system was also influenced by the point in the career when the departure took place. Graduates who departed shortly upon graduation chose German-speaking countries, Britain, and North America because of the stronger learning character in this mobility. North America remained, however, popular for a longer period, reflecting the duality between target migration and traditional emigration in the transatlantic mobility pattern. Intra-Nordic as well as destinations in Africa, Asia, and Oceania and partly also other European countries show a different pattern and were more common choices for technicians departing later in their career. This implies, in part, labour-market mobility and migrations for permanent settlement, but also journeys as experts, like the Danish cement technicians. All in all, we may conclude that mobility implying transfer of ideas and technology back to the Nordic countries occurred early in the career, while journeys implying transfer from the Nordic area often were performed by experienced technicians and occurred later. The means of travelling also influenced the choice of destination. Nearby countries in the Nordic area and Europe were likely choices for study travellers, while remoter areas were almost only for migrants. Russia was the only exception to this rule.

Different specialisations had different 'model countries'. Large and rational electrical and mechanical workshops in Berlin and other places, mechanised shipyards along the Baltic Sea coast and opportunities to study a subject such as electrotechnology at renowned universities created interest for the German-speaking countries in the mechanical, electrical, and naval group. Germany's leadership in dyeing, pharmaceuticals, brewing, and other chemical sub-branches, as well as the possibilities to specialise in chemistry at technical universities and other educational institutes, made the German-language area a more common choice for chemical engineers. American rationality was most interesting and applicable for the mechanical, electrical, and naval group as well as for mining engineers and metallurgists. Upon return, these groups could utilise their experiences at shipyards, large-scale electrical workshops, and the like as well as at steel and ironworks and in mines. These groups went to North America as they had a lot of interesting things to see at shipyards, electrical industries and steel and iron districts like the one around Pittsburgh.

Architects were a different story; they rarely thought that America lived up to what Borgstedt called the 'eternal laws of beauty'. Instead, they flocked to the highly admired Italy, France's steel constructions and monumental buildings, Gaudi's Spain as well as Art Nouveau and modernist architectural centres such as Belgium and the Netherlands. Roman and Byzantine architecture also attracted and was one reason that architects also were likely to visit Africa. Visits to places like Algeria, Tunisia, Egypt, and Morocco were often parts of the Mediterranean study trips. We have already mentioned that Nordic neighbour countries, particularly Sweden and Denmark, also often were major parts of the architects' mobility system.

Civil and construction engineers and mining engineers and metallurgists had one thing in common; they looked more to overseas destinations outside North America than other specialisations. The former group was attracted by large infrastructure building projects in remote countries like the nineteenth-century Argentine railway constructions as well as water supply and drainage projects. Similar projects were going on in countries such as Brazil, Chile, South Africa, Siam, and China. The patterns for mining engineers and metallurgists were often connected to exploitation; they travelled for American and European mining companies to search for and develop deposits in far away regions like Congo, Madagascar, and Chile, where America's Guggenheim Exploration Company employed many Nordic technicians in the world's largest copper mines of Chuquicamata.

One striking trait in these technicians' transnational mobility system was the extent of return to the Nordic countries. If we add study travellers, the shares get even higher. This was clearly a specific type of movement. Even if the transatlantic return shares of technicians were lower than from the German-speaking area and other nearby European countries, they were still considerably higher than for ordinary emigrants to North America. Once again, this underlines that learning mobility and the placement abroad pattern clearly dominated over labour-market mobility. We may speak of a largely circular mobility system as a large majority certainly belonged to the category 'migrants with return intensions who did, in fact, return' as outlined by Bovenkerk. A minority were migrants with intensions to settle permanently who also settled for life. There were almost certainly migrants who changed their minds in both directions, but it is very difficult to decide how many they really were. This pattern, consciously going abroad to acquire information about cutting-edge technology and other ideas, seems to have been more organised in Finland than in Scandinavia, but was useful in a region on the edge of large-scale industrialisation wherein two countries were undergoing a process of becoming fully independent. For the technicians, this was a means of contributing to domestic technical development alongside the promotion of the individual career. Somewhat paradoxical, moving abroad in this context can be interpreted as a patriotic act. Thereby, the return rates reflect the engineer dualism Håkon With Andersen argues for: their legitimacy and credibility are based in international experience, but they are also a group participating extensively in the national race on technical, industrial and economic development. This can be understood in light of late nineteenth and early twentieth century development nationalism: focusing a nation's future prominent position in a modern industrialised world, rather than 'resting on memories from old and great days'.[1]

The extent of circularity in the mobility system was, of course, governed by different criteria. The overwhelmingly male character of the profession may have contributed to the pattern as return migration generally was more of a male than a female phenomenon, and temporary stays abroad were often a part of the maturity process for young industrialists and businessmen born into the upper class. It comes as no surprise that the return increased if the graduates came from the higher societal layers. Family ties must have mattered for domestic employment opportunities and careers in all the countries. Prestige and possibilities to utilise social networks also contribute to explaining why graduates from schools in capital cities returned more often: Stockholm and Copenhagen definitively hosted their countries' most prestigious schools, while the Norwegian schools were, at least formally, on 'equal terms' before the 1910 technical university inauguration in Trondheim.[2]

The fact that younger graduates with comparably longer careers ahead of them returned more often is obvious, and the same is true for the higher return rates among graduates who were born in the same country as they studied in. Timing asserted influence. Return rates were generally higher for graduates who departed shortly upon graduation. This is in line with the learning pattern, and the same is valid for the fact that a choice of the German-speaking countries significantly increased the return. Distance also mattered as graduates going to Europe came back more often than those to overseas destinations. As for year of departure, it was positively correlated with return on the

1 'Free' translation of the beginning of the second verse of the Swedish national anthem, 'Du tronar på minnen från fornstora dar'.

2 This is not applicable for Finland as all graduates studied in Helsinki.

pan-Nordic level to travel between 1890 and 1920, but Denmark diverged with decreasing return rates over time.

Transnational mobility of Nordic technicians was primarily for learning purposes, and Finland suited this pattern to an even greater extent. Finnish graduates often returned after shorter foreign intervals compared to their Scandinavian colleagues. Why did Finnish graduates return more than their Scandinavian colleagues? Myllyntaus has underlined how study travel became an institution in Finland in the nineteenth century and the return rates can be interpreted as parts of the same pattern. The Finnish mobility system was more short-distance and directed towards learning; this spurred return. North America played a smaller role than in Scandinavia, and this also contributed to a higher return rate. Russia, Finland's 'substitute America', was not a place for permanent settlement in the same way.

What were the impacts of these returning technicians on industrial development and technology transfer? Before continuing, we must underline that mobility and migration were far from the sole channels for technology transfer. Previous research has pointed to import of machinery and equipment from abroad, foreign investments, licenses and patents, scientific publications, and so forth. Seely, however, underlines that 'historical analysis emphasizes that successful transfers rest on the exchange of people', and this is our perspective. This study does not set out to weight different transfer channels against each other.

One main idea is that substantial shares of returnee technicians provided the preconditions for technology transfer and change, an idea 'fetched' from Bovenkerk's framework of factors that made it possible for return migrants to make an impact and connected to the observations of the 1980s Norwegian returned emigrant project and Hunger's studies of today's Indian software industry. We can also connect to Ghosh's statement that the returnees must be willing and given the opportunities to act as agents of transfer and change. These technicians can be viewed as next to elite groups in national and, not least, in local contexts. This implies that they had education and that they were expected to obtain more influential positions than 'ordinary' returnee emigrants. The return of this social group asserted, therefore, more influence than return migration in general. The fact that foreign experience facilitated the career paths also indicates that the returnee technicians were willing to use their experience in their post-return professional life and that they were given the chances to do it.

There are some criteria in Bovenkerk's model. The Nordic median duration of stay abroad was three years: Finland and Norway note two years. The average duration was four and a half years; this was also the average for Scandinavia, whereas the Finnish average was just above three years. We may interpret these intervals as 'somewhere in between' too short and too long sojourns in accordance with Bovenkerk's framework. Staying away between two and five years implies an appropriate time to learn about foreign technological development and undertake university studies, without being too alienated from the cultural and technical development back home. Here, we may also connect to the placements abroad pattern, which is supposed to constitute about 60 per cent of this transnational mobility. We have initially hypothesised that this pattern had stronger importance than study trips, as it implied involvement in real work processes and not 'only' observation. This pattern points to less 'efficiency' in Finland, considering the importance of study travelling. This was, however, not necessarily the case.

We will return to the fact that there were much higher shares of foreign-experienced technicians in Finland than in the other countries, but we may underline already at this point that the Finnish study trips had clear technology-adopting targets and Bovenkerk has emphasised that 'motives and how migration and return is organised' assert influence on the possibilities to make an impact. The organised study trips were clearly a part of a strategy connected to technological and economic development as, for example, Hietala and Myllyntaus have underlined. The more organised Finnish study travel pattern implies, on the one hand, that the differences between placements abroad and study trips may have been more relevant in Scandinavia. On the other hand, we have reached another conclusion indicating that we may have exaggerated these implication differences at the beginning of the book: the incidence of upward occupational mobility. Foreign experience was important, but migration experience was not necessarily more valuable than study trips. Of course, this does not tell the whole truth about impact implications; at best a small part of it. The career effects of placements abroad were, however, somewhat dubious, and there were also criteria not connected to foreign experience that mattered.

The 'difference between the home environment and the adopted environment' is another factor in Bovenkerk's framework; so is the 'quality of the training and skills acquired abroad'. Ghosh has pointed out that the training must be appropriate for the home country: it cannot be too advanced to be applicable upon return, but it is also useless with a foreign intermission if the training on the same level can be obtained at home. The Nordic countries were ready to receive and integrate new technology. Swedish industrial products had, as Ahlström has stated, been recognised at international fairs since the mid-nineteenth century and the professionalisation of Swedish engineering almost equalled the German pattern; Denmark is described by Nielsen and

Wagner as unusually skilled when it came to adoption and filtering of foreign technology, and the same qualities have been ascribed to Finland by Hjerppe, Jalava, and Kirby and to Norway by Nerheim. We may interpret the Nordic countries as receptive to new technology; there was, of course, a gap to leading industrial countries like Germany, Britain, and the United States, but it was not insurmountable. The quality of training and skills were appropriate; this is also indicated by the fact that foreign experience was important for occupational careers.

Also important, of course, are the views of the countries these technicians returned from. The views of the 'models' were often very positive: Germany was referred to as a 'high-tech' country at the 1909 annual meeting of the Swedish Technical Association, and this admiration was, as Björck has stated, connected to the technical education system and an industrially oriented research policy. Switzerland was also viewed as a country of advanced technology, especially in power transmission. It possessed one of Europe's most prestigious technical universities and was also admired for its democracy and independence. Austria and Germany were homes of admired and exciting architects, and Austrian mining education was viewed as highly qualitative. It must have been prestigious for a mechanical or electrical engineer to be able to certify employment with Siemens, AEG, Borsig, or Loewe or studies in Zurich, Berlin-Charlottenburg, or Darmstadt and for a naval architect to state that he had been employed at one of the heavily mechanised shipyards in Hamburg, Kiel, Bremerhaven, or Stettin. A chemical engineer had much prestige if he brought experience from brewing in Bavaria or from the dyeing and colouring works near Frankfurt-am-Main. A construction engineer who had been with Holzmann in the same city must have been well-off as well as an architect who had studied Art Nouveau in Munich or was well-acquainted with Wagner's and Sitte's Vienna-based architecture and city planning.

The view of the United States was generally also positive, and this was, of course, connected to the contemporaneous mass emigration. Rumours were spread in Nordic districts about people who had crossed the Atlantic and realised a kind of 'American Dream': Such success stories were probably more frequently told than stories about failures. The use of American examples, including technology, and returnees were sometimes held up as positive examples in response to public concern about the drainage of the domestic population. The United States was thus also looked upon as a country of excellent technology, and this view was facilitated by reports from international fairs like the ones in Philadelphia in 1876 and Chicago in 1893. Electricity had done 'marvels', and the tools were very modern. It was such reports that led to statements about America's technological superiority like the one at the

1902 lecture in Kristiania. For a mechanical or electrical engineer, employment at General Electric, Westinghouse or Allis-Chalmers was prestigious; a naval architect brought a lot of 'symbolic capital' if he could prove that he worked at one of the modern shipyards around Boston or Philadelphia and the same was valid for a mining engineer, metallurgist, or chemical engineer who could prove that he been employed in the outmost automated steel and ironworks in the Pittsburgh area.

Britain, too, kept a good reputation in shipbuilding and a work certificate from a shipyard around Newcastle-upon-Tyne or Glasgow as well as degrees in naval architecture from the university in the latter city were valuable assets, so was employment with Hennibique in Paris for someone who wanted to work in iron and concrete construction. An engineer who had studied aeronautics in France brought valuable experience, so did an architect who had seen steel constructions in the same country, ornaments, mural paintings and classical environments in Italy, Byzantine and Roman architecture in the Mediterranean, Art Nouveau buildings in Belgium and, not to forget, Danish brick architecture and the inspiring Nordic classicists in Sweden. Germany and the United States were admired in almost every field, with a minor exception for the divided view on American architecture, but other countries were also admired in certain fields of technology and architecture.

This is important to remember. We saw that experience of foreign travelling was important for career mobility, but we could not find any differences between migrants and study travellers nor between experience of different countries. We must, of course, also add important 'non-mobility' variables: age, experience, urban or rural origin, specialisation, and whether the technician worked in a rural or smaller urban area contra a big city. Nevertheless, graduates able to prove foreign experience more often climbed up the ranks.

Roughly 40 per cent of the technicians in this cohort who worked in the Nordic countries in the early decades of the twentieth century had foreign experience, and most of them had lived abroad for some years. This is substantial, especially considering that neither technicians who took their entire education abroad nor foreign-born engineers and architects are included. Thus, the share was, in reality, higher. If these technicians acted in accordance with what they learned, they could have been 'living examples of innovation'. Where, when, and in what fields were there more such examples? An employee working in Norway and especially Finland was more likely to meet a colleague or superior with foreign experience than someone working in Sweden and Denmark. In Scandinavia, this chance did not change much over time, whereas the earliest years of independent Finland offered somewhat better possibilities for such interaction compared to the final decades of the Grand Duchy. The chemical field offered the best possibilities in Denmark and Finland and better chances in Norway than in Sweden. Over time, however, the shares of Danish and Norwegian chemical engineers with foreign experience diminished. Civil and construction engineering offered the least possibilities, but it was different in Norway and became 'better' in the other countries over time. In Sweden, the mechanical and electrical field, as well as mining and metallurgy, offered the best chances to meet internationally experienced technicians; the chances in mining also increased over time. The mechanical group offered least chances in Denmark and mining-metallurgy was the least experienced group in Norway. Architects had good chances to meet colleagues that had been in other countries, more in Norway and Finland—where the shares also increased over time—but less in Sweden.

Experience was, as the mobility statistics already have shown, often collected in the leading industrial and educational countries and experience from the German-speaking countries dominated. This was valid for all countries and over time, but stronger in Norway and Finland. In 1920, more than 40 per cent of the Finnish engineers and architects had experience from German-speaking Europe, and Finland had at all times the highest German share. The differences compared to Norway increased in the 1910s; both countries had improved their domestic technical education, but Finnish study travelling was nevertheless intensified in the years around the independence.

German experiences in the Nordic countries also decreased some from 1900 and 1920, while American experiences increased. The gap became smaller, but there were still almost twice as many 'German' technicians in the Nordic countries in 1920. The differences between the two main experience areas were largest in Finland, where experiences from other Nordic and other European countries, Russia and, early in the period, also Britain generally were more common than American experiences. British and other European experiences were also more common than American in Denmark around 1900. Technicians in Denmark and Sweden had less experience from German-speaking Europe; the shares were generally somewhat higher in Sweden, but the challenge from other experience areas, read North America, was also stronger. Sweden notes the smallest German-American difference. From a general perspective and in line with previous research, this still points to Germany as the main source of technical and industrial knowledge in the Nordic countries and that this was more pronounced in Norway and Finland. America's impact seems to have been stronger in Sweden than in the neighbouring countries, which is in line with previous assumptions of more large-scale industrialisation. Main 'alternative' sources of knowledge were other Nordic countries, other European countries, and Britain. These experiences were most common in Finland.

Russia played a relatively minor role on the pan-Nordic level but was a common experience in Finland. Here, of course, we may underline the difficulties in pasting national origin marks on technology, architecture, and organisation. Technology does not develop in a vacuum but often in interaction between different environments based in different countries. Many technicians served Nordic subsidiaries and companies from other countries, and an American or German company in, say, Spain most likely applied similar technology as they did at their plants back home.

The German-speaking countries dominated in all specialisations except among mining engineers and metallurgists, where it became more common with experience from North America on to the twentieth century. This is, however, a Swedish pattern that caught the pan-Nordic. Norwegian mining engineers and metallurgists still had more experience from German-speaking Europe. The only other exceptions are Swedish architects who had collected somewhat more experience in other European countries and civil and construction engineers working in Finland in 1920; they acquired about the same amount of experience in Scandinavia. In terms of shares, architecture and chemical engineering were the most German. This is hardly surprising considering German chemical leadership and the opportunities to study Art Nouveau, neo-Renaissance, and city planning. The shares were, on the one hand, somewhat higher among architects, but German-speaking Europe was, on the other hand, more superior among chemical engineers. The mechanical, electrical, and naval group, as well as civil and construction engineers, lay in between, whereas mining engineers and metallurgists had collected the least experience in the German-speaking parts of Europe. Norway diverged as civil and construction engineers had much experience from this region and Swedes and Finns in the mechanical, electrical, and naval group had comparably more experiences from Germany, Austria, and Switzerland.

North America's strongholds were the mechanical, electrical, and naval group and mining engineers and metallurgists, a pattern we have connected to their interest in studying rationalisation and a stronger potential to introduce such organisations upon return. America was, as stated, a less common experience in architecture, but seems to have been comparably more important for Swedish architects than for their Nordic colleagues, which some of the Richardson-inspired buildings may indicate. We can also note relatively strong American experiences among the civil, construction, and chemical engineers in Denmark and Norway, and the textile and pulp and paper technicians in New England also made the United States an important experience among chemical engineers in Finland.

As for experiences from other countries and regions, we can mention a few things. Mining engineers and metallurgists, as well as architects, had somewhat more experience from overseas countries outside North America, but Latin America was most common among Danish chemical engineers and Norwegian civil and construction engineers. Architects had acquired a lot of experiences in the other Nordic countries, Britain and other European countries. Thus, many architects had experienced British Art Nouveau and garden cities as well as Nordic classicism and Danish brick architecture. Experience from other European countries was, however, the pronounced architect pattern, and many had, thus, been in Italy, France, and other architectural centres. This region was also a relatively common experience for mining engineers and metallurgists, while experience of British shipbuilding and from the American electrical subsidiaries in Britain was dispersed in the mechanical, electrical, and naval group. Few chemical engineers had been there, but the neighbouring Nordic countries were a common experience, and the group was also the one that had acquired the most experience in Russia.

Is it then possible to say that the impact of these technicians stretched beyond local scenes and limited time periods? Yes, is the most reasonable answer. The Norwegian project concluded that it was a necessity for returnee emigrants to act after similar patterns in different places and over time to reach this extended impact.

In the earliest decades of the twentieth century, there were foreign-experienced technicians serving all over the Nordic countries: from the low south Scandinavian plains to the open spaces of the Arctic north, from the—until 1917—intra-empire borderlands between Russia proper and the Grand Duchy of Finland to the west Norwegian fjords, and even in Iceland. Some of them became teachers and professors at the educational institutes in the focus of this study and became important for technical development in their respective countries. They were present in the capitals: in Stockholm, Copenhagen, Kristiania, Helsinki, and Reykjavik, in other larger urban areas, but also in smaller local communities. Some served in municipal bodies. Many worked with gas and electricity and had experience from the German-speaking countries and sometimes from North America and other countries. Among them were city engineers who contributed to modernisation with drainage, water, and gas supply as well as improved building methods to make residential houses safer. Among them were electrical pioneers like Denmark's Angelo. Among them were also architects inspired by Art Nouveau, neo-Renaissance, and other styles from the European continent, like Tampere's city architect Petersson. Returnee engineers managed harbours in Rønne on the Danish island of Bornholm, and the electricity works in most towns along the north Swedish coast. Foreign-experienced engineers built up Icelandic infrastructure from a minimum at the turn of the century and established telecommunications in Greenland. Engineers with foreign experience built railways by the Finnish-Swedish border and served as town engineers in Steinkjer in central Norway.

Some returnees started businesses like Christiani's and Nielsen's Copenhagen-based iron and concrete construction company, partly founded on experience from Hennibique in Paris, and Kreuger's and Toll's similar company in Stockholm, based on international experience of reinforced concrete. Selmer had similar experiences and started a business in Kristiania, and many of the leading engineers of the Trondheim-based firm Jernbeton had experience in reinforced concrete in Germany and the United States. Many architects also had their own offices. Selim A. Lindqvist and the Ekelund couple in Helsinki, Brag and Östberg in Stockholm, Ewe, and Melin in Malmö as well as Osness in Trondheim are some examples, and they were all inspired by Art Nouveau and architecture in the German-speaking countries, Western Europe as well as France, Italy, and other Mediterranean countries. Östberg's Stockholm City Hall took inspiration from Venice's Doge Palace while Boberg's main post office in the Swedish capital had American models, and so had Michalsen's Norwegian furniture, designed for serial production. Ullberg's office in Viipuri is described essential for the introduction of functionalism in the city; the owner had travelled in Scandinavia, German-speaking Europe, France, Italy, and Britain.

There were, of course, other founders, for example of electrotechnical businesses. Strömberg studied in Germany and founded what has been described as Finland's most superior electrical company, and he brought German dynamos and lamps to the country. He also taught electrotechnology at the Helsinki Polytechnic. In Vaasa, an electrical business was started by a returnee from university studies in Sweden and travel in Germany and Britain. Luth returned from France and Britain and founded a business in Stockholm, whereas returnees from the United States started businesses in places like Næstved in Denmark as well as in southern Norway. Some engineering firms, construction companies in Gävle and Sundsvall in northern central Sweden, dye and clean works in Bergen, Gothenburg, and Oulu and a locomotive workshop in Tampere are further examples of businesses started by returnee technicians.

Power stations in the Nordic countries sometimes employed foreign-experienced technicians in responsible positions like the one at Trollhättan in southwest Sweden, described by Fridlund as a way to create 'Sweden's Niagara'. The power station Porjus, in the far north of Sweden, had returnees in its management and took influences from the Canadian environment in Shawinigan. Engineers in responsible positions at power stations like the ones at Rjukan and Svelgforss in southeast Norway were also returnees; the same was true for the plant in Glomfjord in the north. Hlìðdal utilised his experience from studies and practise in Germany to construct a water power station upon return to Iceland, and an Icelandic engineering student from Denmark was responsible when Reykjavik erected a water-power-based electricity works.

Danish technicians were involved in the cement business in other Nordic countries, but there were also returnees managing some factories. One returnee introduced revolving furnaces from Germany at a cement factory in Malmö. A Finnish colleague became part-owner of a Helsinki-based asphalt and cement company after he had participated in railway building in Costa Rica. Superphosphate innovations based on experience from Germany, France, and Britain were the foundation of an industry in Helsingborg in southern Sweden. A returnee from Germany improved carbide and chlorate production and was essential to the development of Sweden's electrochemical industry as manager of a plant in Avesta in the central part of the country; a returned student from Dresden had a vital role starting a carbide-factory in Medelpad in northern central Sweden after he had studied how calcium cyanamide could be produced cheaply.

Sawmills and pulp and paperworks also employed technicians who had been abroad. This happened in Drammen west of Kristiania, Roskilde, and Næstved in Zealand, and in Silkeborg in central Jutland where a returnee from Germany and Austria was essential to the re-build of the factory and reorientation towards high-class paper. The pattern was reflected at similar industries near Östersund in northwest Sweden, along the coast in lower northern Sweden, at establishments in Lappeenranta and other places in southeast Finland and Tervakoski north of Helsinki where Andersin utilised his experience from the United States and his skills in fine paper technology. The textile industry employed returnee technicians in responsible positions in 'Sweden's Manchester', Norrköping in the southeast, in its Finnish 'centre' in Tampere, but also in places such as Hyvinkää by Helsinki. Sugar refining in Sweden's southernmost province Scania and in Lolland and Funen; breweries in Stockholm, Gothenburg, Kristiania, and Copenhagen are other examples. In Hobro in central Jutland, a returned student from Germany and Austria took over the family brewery.

Another chemical industry employing many returnee technicians was *Norsk Hydro,* whose production of fertilisers in places like Notodden and Rjukan in southeast Norway utilised foreign experience. Kielland had studied in Berlin-Charlottenburg and became technical manager for Hydro and another aluminium producing company. In Kristiansand, on the southern tip of Norway, the nickel works was developed into one of the world's largest under

Grønningsæter's management. Mining and quarrying also made use of technicians with experience from abroad. The quarry in Ivigtut, the only significant source of the mineral cryolite in the world, was a main employer of Danish technicians in Greenland. The mining community of Grängesberg in the important mining district of Bergslagen in central Sweden was described as 'miniature America' and was managed by a returnee. In northern Sweden, mining around Skellefteå, Kiruna, and Malmberget utilised returnees and American experience, for example, when the height of a smokestack at a smelting-plant was decided to try to send the environmental problems to 'Russia'. The mines in Dunderland in northern Norway became known for American methods and were managed by a returnee from the United States.

Steel and iron making was a connected field that took a lot of influence. Leading engineers at Bergslagen's ironworks had experienced an environment like Carnegie Steel's around Pittsburgh and rationalised the production like Esselius and Magnusson did in Sandviken. Such principles—in a Taylorist or Fordist spirit, if not in a pure sense and sometimes even introduced before Taylor and Ford made their appearances–were also applied by many returnees in mechanical workshops such as Bolinders in Stockholm. It was an even more pronounced pattern in the specialised workshops. There are many indications of inspirations from Taylor, Ford, or their predecessors. Mechanical workshops were rationalised by returnees from the United States, Germany, and sometimes other countries such as Belgium. Shipyards and other mechanical industries in cities like Sweden's Gothenburg, Malmö, and southeast Karlskrona; Norway's Bergen, Trondheim, Stavanger, Horten, Moss and Fredrikstad; in Helsinki, Tampere and Pori, in southwest Finland, and Køge near Copenhagen as well as south-Jutland's Horsens and Vejle and Frederikshavn in the north of the peninsula are indications. Workshops in Copenhagen were organised and reorganised by returnee technicians who were inspired by Taylor and Ford in a broad sense and the Ford Company's own establishment in Copenhagen was, of course, one of them. Odense hosted a car factory where Engberg, the lecturer describing American work organisation as a 'promised land', was works engineer. Joakim Lehmkuhl, Norway's most outspoken advocate of Taylorism, had studied in Cambridge, Massachusetts, and started a workshop in Oslo in the mid-1920s, but understood the difficulties for 'scientific management' in Scandinavia. A few Norwegians worked close to Taylor, and some of them returned. Following the mid-1920s article in the Norwegian technical journal, Sweden had managed to create at some places the 'ideal' application of Taylor's system to Scandinavian conditions. This was a report from another workshop, but it could, perhaps, have been written also about ASEA. Under the management of returnee J. Sigfrid Edström, the company was re-organised along American principles, and this included many returnees in responsible positions in the different workshops and other departments. Most of them had been at General Electric and Westinghouse, but many also at the large electrotechnical companies in Berlin and around Germany. Stockholm-based Separator was another company known for Taylorism as they employed Erik August Forsberg, described as Sweden's leading Taylorist. Forsberg shows that a person did not necessarily need to go to America to be inspired by Taylor. His foreign experience was mainly as head of the draftsman's office at a Saint Petersburg workshop.

There is a multitude of examples of Nordic returnee technicians. One returnee from the United States manufactured American-style apparatus for churning in Kolding in southern Jutland; one Norwegian who returned from studies at Chalmers in Gothenburg started a motor factory, became the country's first authorised car inspector and a pioneer for Norwegian motorisation, and one Finn who also studied at Chalmers started a shipyard in Vaasa. A female Danish technician returned from Germany to establish a chemical laboratory in Nykøbing Falster; a Norwegian woman technician had a position in Switzerland before she returned to a chocolate factory in Kristiania, while a female Finnish architect went to London and later received awards for her furniture. One technician worked at a chemical laboratory in London and returned to manage margarine factories in southern Sweden. An 'adventurous' Finn projected and constructed sea canals and worked in the harbour in Vladivostok and projected a narrow-gauge railway as the only free man on the island of Sakhalin before he escaped the Russian-Japanese war and returned to a responsible position with the Finnish national railways. The first mathematics professor at the Norwegian Institute of Technology in Trondheim was an educated engineer and had studied with some of the most famous French mathematicians in Paris. *Norsk Hydro*'s mercantile manager had studied electrotechnology in Liege, and a Norwegian colleague had been at electro-steelworks in France and around Torino in Italy and returned to modernise an electro-steelworks in Stavanger. This list could be much longer. We are not arguing that their employment patterns necessarily diverged significantly from technicians without foreign experience. However, they were everywhere in the technician corps and can, therefore, be viewed as 'living examples of innovation' for their surroundings.

Nevertheless, some points must be made. The impact, at least as measured in representation, seems to have been stronger in industrialised regions such as Copenhagen, a capital whose industrial dominance over the rest of country hardly had any European counterpart and definitively no Nordic one. This concentration of returnee technicians was, thus, dependent on Copenhagen's position as an industrial centre rather than the city's status as Denmark's capital. The shares in Stockholm, Helsinki, and Kristiania lay, roughly, on the national average. As for Norway, this may sound surprising and somewhat contradictory as the Oslofjord dominated domestic industrialisation, but we can note that there were 'high' shares of returnee technicians in nearby heavily industrialised communities like Drammen. Finland's 'industrial island' Tampere and ASEA's Swedish hometown Västerås are two more examples of this 'industrial' concentration of returnee technicians.

A second point to be made is that peripheral regions—the northernmost, Arctic, parts of Finland, Norway, and Sweden and also northernmost Jutland in Denmark—had difficulties attracting these experienced technicians. Peripheries generally had slower industrialisation. However, there are single examples of influence and importance of returnee technicians from the northern parts of the countries like the build-up of the mining industry around Skellefteå and Kiruna in northernmost Sweden.

A third point is that examples of individual influential returnee technicians in a sense prove that a quantitative approach is problematic. Falkman, Palén, and their colleagues in Skellefteå's mining were too few to raise the percentage of returnee technicians in upper-northern Sweden, but the company has been ascribed immense importance for industrial development in the area. However, let us still use the quantities of returnee technicians as an indication of influence. The strongest impact was in Iceland, where all educated technicians had foreign experience. These technicians were primarily essential to infrastructure building. This conclusion is, however, self-evident. Let us also underline that this is—of course—not an argument for a lower level or absence of domestic technical education. The presence of such education is a sign of a country's technical and economic development.

Nevertheless, countries with deficiencies in their domestic technical education could compensate with foreign education, which created experiences that could not be acquired at home, especially in countries with a lower level of industrialisation. Let us use the quantities and ascribe the greatest impact of foreign-experienced technicians to Finland and conclude that it also was stronger in Norway than in Denmark and Sweden. Myllyntaus has described an 'obsession' among Finnish technicians to travel regularly to foreign countries and to study abroad, and this must have been favourable for the country's transition to industrialism. The differences in shares of foreign-experienced technicians between the two less and the two more industrialised Nordic countries were even larger, since there were more technicians taking their entire education abroad; the important Norwegian industrialist Sam Eyde is one example. Norway and Finland may have had more deficiencies in their domestic technical education, but they were compensated by larger shares of returnees from acknowledged high-quality foreign technical educational institutes.

In comparing Finland and Norway, we bump into problems as the Finnish duration of absence often was shorter and many were study travellers. It is difficult to say, but the significantly higher foreign-experienced shares in Finland must have made a difference. Short-term sojourns often implied university studies, and the organised Finnish study trips often had very specific targets when it came to technology. The incidence of upward occupational mobility for foreign-experienced technicians was also somewhat stronger in Finland than in the other countries. Thus, we end up sticking to the idea that the impact was greatest in Finland. It seems clear that the two less industrialised countries gained more from this mobility, which indicates that the rule of impact in industrialised areas largely is irrelevant when comparing countries. National sentiments play an important role. If appropriate employment can be found back home; the industrialisation level of the native country is hardly very important in itself. 'Industrial islands' in less industrialised countries, like Tampere, were however important targets for returnee technicians.

This brings us to the end of our journey. Denmark's Herluf Forchhammer, Finland's Fjalar Witting, Norway's Haakon Hauan, and Sweden's Julius Körner took part in an activity that embraced more than every second Nordic technician during the industrial breakthrough in the decades around 1900 and was especially dispersed in Norway and Finland. All were in German-speaking Europe, and Fjalar was the only one of them who did not cross the Atlantic; he was, instead, in Saint Petersburg. Herluf also found his way to Britain and Turkey. This reflects the major destination patterns, North America's smaller and Russia's larger role in Finland, and that Danes more often went to other destinations and also served domestic companies. All returned and gained some importance, which also echoes some of the main lines in this study.

We have given some examples of transfer of technology and other ideas through these foreign-experienced technicians. However, there are still innumerable stories to be told about returnee and immigrant technicians at breweries, mechanical workshops, mines, paperworks, textile industries, railways, architectural offices, universities, and municipal engineering all around the Nordic area. There are archives to go through and qualitative studies to be made. It is enough for hundreds of researchers in the ages to come. This book ends here, however, with the hope that it has cast a little light on what truly was the late nineteenth and early twentieth century's 'peregrine profession'.

Appendices

Appendix 1: International destinations of Nordic technical school graduates, 1880–1930 (graduation 1880–1919)

DESTINATION	TOTAL	%	Sweden	%	Denmark	%	Norway	%	Finland	%
German-speaking countries	**3015**	**45**	**1153**	**42**	**433**	**33**	**910**	**50**	**519**	**63**
Germany	2797	42	1067	39	406	31	822	45	502	61
Switzerland	477	7	158	6	53	4	143	8	123	15
Austria	346	5	129	5	66	5	46	3	105	13
Liechtenstein	1	0	0	0	0	0	1	0	0	0
North America	**2628**	**39**	**1230**	**45**	**450**	**34**	**763**	**42**	**185**	**22**
United States	2561	38	1210	44	431	33	737	40	183	22
Canada	264	4	94	3	51	4	98	5	21	3
Other Nordic countries	**1349**	**20**	**470**	**17**	**232**	**18**	**215**	**12**	**432**	**52**
Sweden	638	16			144	11	142	8	352	43
Norway	393	8	221	8	66	5			106	13
Denmark	378	7	122	4			69	4	187	23
Finland	220	4	184	7	14	1	22	1		
Iceland	32	0	1	0	23	2	7	0	1	0
Greenland	23	0	2	0	21	2	0	0	0	0
Nordic countries, undefined	*72*	*1*	2	0	2	0	*3*	0	*65*	8
Europe*	**1199**	**18**	**462**	**17**	**241**	**18**	**210**	**12**	**286**	**35**
France	622	9	249	9	139	11	110	6	124	15
Italy	267	4	110	4	20	2	47	3	90	11
Belgium	261	4	108	4	48	4	43	2	62	8
The Netherlands	182	3	56	2	41	3	30	2	55	7
Spain	90	1	53	2	7	1	17	1	13	2

(Continued)

(Continued)

DESTINATION	TOTAL	%	Sweden	%	Denmark	%	Norway	%	Finland	%
Hungary	67	1	24	1	12	1	11	1	20	2
Poland	63	1	22	1	19	1	3	0	19	2
Czechoslovakia	52	1	21	1	4	0	8	0	19	2
Turkey	35	1	9	0	15	1	8	0	3	0
Estonia	24	0	1	0	1	0	0	0	22	3
Greece	18	0	11	0	2	0	2	0	3	0
Yugoslavia	13	0	6	0	2	0	4	0	1	0
Latvia	10	0	1	0	3	0	0	0	6	1
Luxembourg	9	0	2	0	1	0	3	0	3	0
Romania	9	0	3	0	3	0	0	0	3	0
Portugal	7	0	2	0	1	0	2	0	2	0
Bulgaria	5	0	2	0	1	0	1	0	1	0
Lithuania	3	0	0	0	0	0	0	0	3	0
Monaco	1	0	0	0	0	0	0	0	1	0
Europe, undefined	*116*	2	*30*	1	*10*	1	*10*	1	*66*	8
Britain	**994**	**15**	**409**	**15**	**242**	**18**	**176**	**10**	**167**	**20**
Russia	**521**	**8**	**187**	**7**	**94**	**7**	**28**	**2**	**212**	**26**
Latin America & the Caribbean	**411**	**6**	**124**	**5**	**142**	**11**	**137**	**8**	**8**	**1**
Argentina	189	3	52	2	63	5	70	4	4	0
Brazil	67	1	29	1	22	2	15	1	1	0
Chile	55	1	11	0	14	1	30	2	0	0
Mexico	47	1	21	1	8	1	17	1	1	0
Danish West Indies	28	0	0	0	28	2	0	0	0	0
Uruguay	17	0	3	0	5	0	8	0	1	0
Cuba	15	0	3	0	5	0	6	0	1	0
Bolivia	14	0	6	0	1	0	7	0	0	0
Colombia	10	0	4	0	2	0	4	0	0	0
Panama	10	0	5	0	4	0	1	0	0	0
Peru	10	0	5	0	1	0	4	0	0	0
Paraguay	7	0	3	0	0	0	4	0	0	0

(Continued)

(Continued)

DESTINATION	TOTAL	%	Sweden	%	Denmark	%	Norway	%	Finland	%
Dominican Republic	5	0	2	0	2	0	1	0	0	0
South Georgia	3	0	0	0	1	0	2	0	0	0
Ecuador	2	0	1	0	0	0	1	0	0	0
Haiti	2	0	0	0	1	0	1	0	0	0
Puerto Rico	2	0	1	0	1	0	0	0	0	0
Trinidad & Tobago	2	0	0	0	1	0	1	0	0	0
Costa Rica	1	0	0	0	0	0	0	0	1	0
Honduras	1	0	0	0	0	0	1	0	0	0
Venezuela	1	0	0	0	0	0	1	0	0	0
South America, undefined	*29*	0	*19*	1	*7*	1	*1*	0	2	0
Africa, Asia, & Oceania	**401**	**6**	**127**	**5**	**171**	**13**	**89**	**5**	**14**	**2**
China	80	1	28	1	41	3	10	1	1	0
South Africa	63	1	20	1	8	1	35	2	0	0
Dutch East India	58	1	3	0	47	4	8	0	0	0
Australia	47	1	15	1	20	2	11	1	1	0
Japan	33	0	23	1	6	0	3	0	1	0
Siam	32	0	1	0	31	2	0	0	0	0
British India	29	0	11	0	10	1	6	0	2	0
Algeria	15	0	6	0	2	0	5	0	2	0
Iran	12	0	4	0	6	0	1	0	1	0
Malaysia	10	0	4	0	5	0	1	0	0	0
Morocco	9	0	3	0	1	0	4	0	1	0
Singapore	9	0	1	0	8	1	0	0	0	0
Egypt	8	0	2	0	3	0	1	0	2	0
Philippines	8	0	2	0	3	0	2	0	1	0
New Zealand	7	0	1	0	4	0	1	0	1	0
Congo	4	0	0	0	0	0	4	0	0	0
Ghana	4	0	1	0	2	0	1	0	0	0

(Continued)

(Continued)

DESTINATION	TOTAL	%	Sweden	%	Denmark	%	Norway	%	Finland	%
Ethiopia	3	0	0	0	0	0	3	0	0	0
Iraq	3	0	1	0	1	0	1	0	0	0
Madagascar	3	0	2	0	0	0	1	0	0	0
Namibia	3	0	2	0	1	0	0	0	0	0
Rhodesia	3	0	0	0	0	0	3	0	0	0
Burma	2	0	2	0	0	0	0	0	0	0
Fiji	2	0	0	0	2	0	0	0	0	0
Kenya	2	0	1	0	1	0	0	0	0	0
Tunisia	2	0	2	0	0	0	0	0	0	0
Angola	1	0	0	0	0	0	1	0	0	0
Guam	1	0	1	0	0	0	0	0	0	0
Korea	1	0	0	0	1	0	0	0	0	0
Mongolia	1	0	1	0	0	0	0	0	0	0
Mozambique	1	0	0	0	0	0	1	0	0	0
New Caledonia	1	0	0	0	0	0	1	0	0	0
Niger	1	0	0	0	0	0	1	0	0	0
Nigeria	1	0	0	0	1	0	0	0	0	0
Senegal	1	0	0	0	1	0	0	0	0	0
Solomon Island	1	0	0	0	0	0	1	0	0	0
Somalia	1	0	0	0	0	0	1	0	0	0
Sudan	1	0	0	0	0	0	1	0	0	0
Tanzania	1	0	0	0	0	0	1	0	0	0
Unknown destination	*139*	*2*	*43*	*2*	*39*	*3*	*37*	*2*	*20*	*2*
TOTAL	**6711**	**100**	**2744**	**100**	**1323**	**100**	**1820**	**100**	**824**	**100**

*Except the Nordic and German-speaking countries, Britain and Russia.

SOURCES: see figure 1

Appendix 2: Highest share per country of mobile engineers and architects from Sweden, Denmark, Norway and Finland, 1880–1930

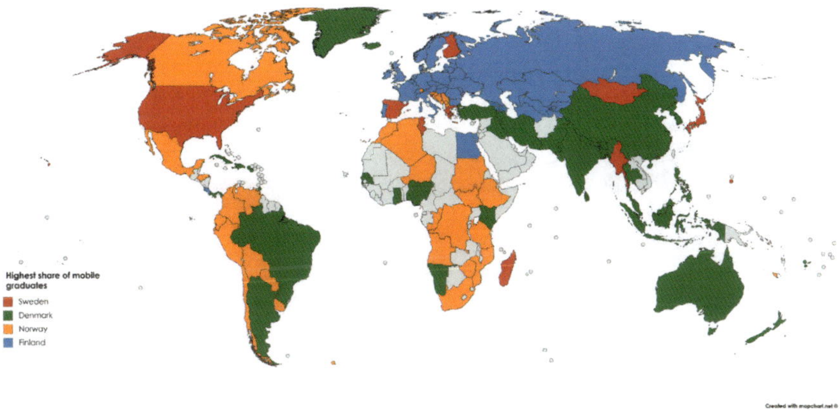

SOURCES: SEE FIGURE 1

Appendix 3: Destinations of Nordic mechanical and electrical engineers and naval architects, 1880–1930

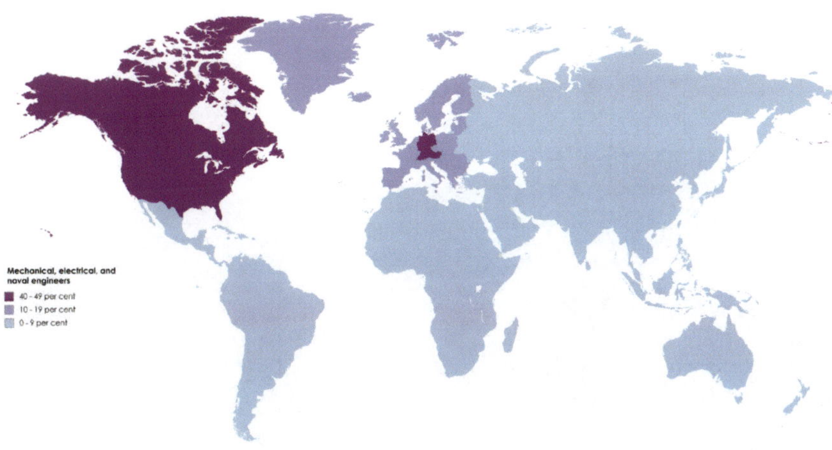

SOURCES: SEE FIGURE 1

Appendix 4: Destinations of Nordic civil and construction engineers, 1880–1930

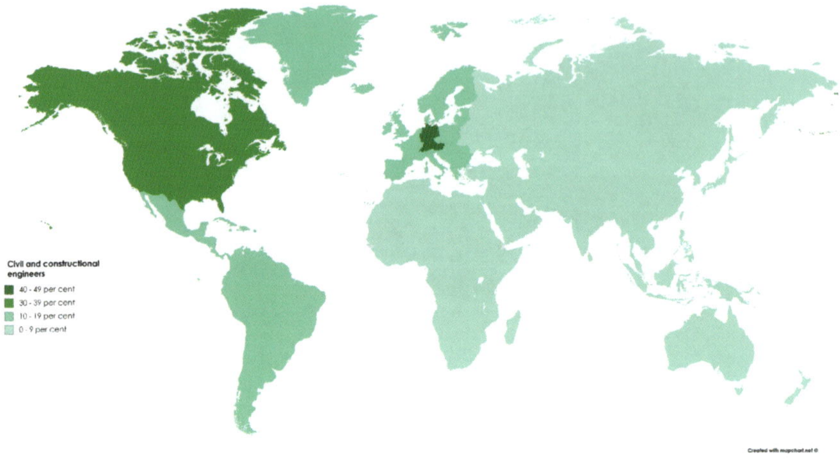

SOURCES: SEE FIGURE 1

Appendix 5: Destinations of Nordic chemical engineers, 1880–1930

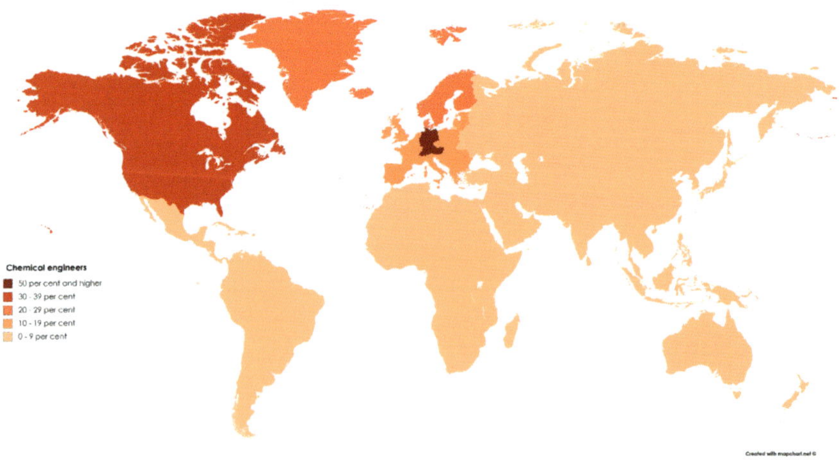

SOURCES: SEE FIGURE 1

Appendix 6: Destinations of Nordic mining engineers and metallurgists, 1880–1930

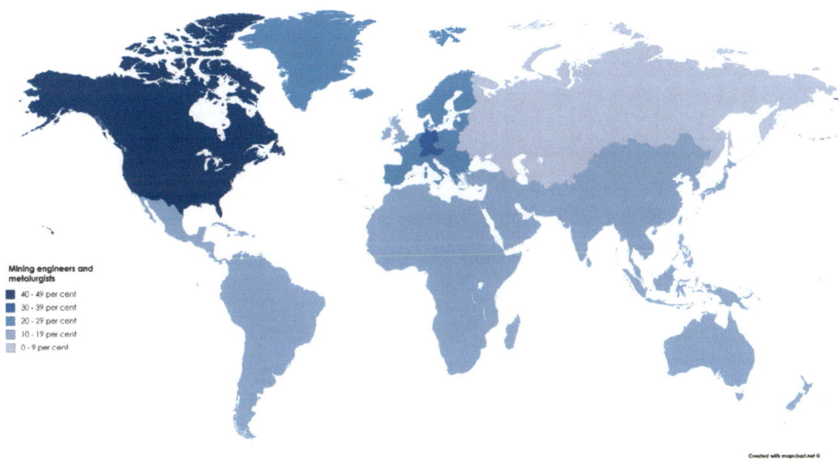

SOURCES: SEE FIGURE 1

Appendix 7: Destinations of Nordic architects (educated at technical schools), 1880–1930

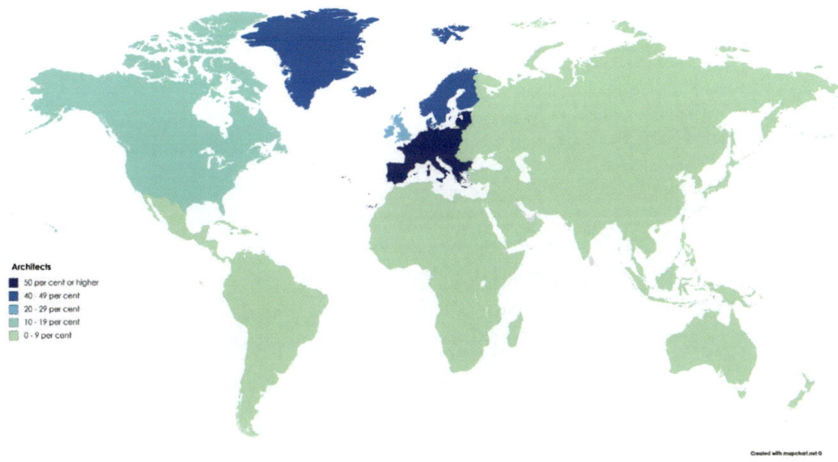

SOURCES: SEE FIGURE 1

Sources and References

Printed Sources

Denmark

Danske Ingeniører fra Teknika (København 1945).

Danske Teknika og deres Dimittender gennem 50 Aar, Ingeniør- og Konstruktørsammenslutningen gennem 35 Aar (København 1931).

Hannover, Aage, (ed.), *Dansk Civilingeniørstat 1942, biografiske Oplysninger om Polytekniske Kandidater 1829–1941* (København 1942).

Hannover, Aage, (ed.), *Dansk civilingeniørstat 1955: biografiske Oplysninger om polytekniske Kandidater 1829–1955* (København 1956).

Jespersen, R. (ed.), *Biografiske oplysninger angaaende Den Polytekniske Læreanstalts kandidater 1829–1929* (København 1930).

Finland

Matrikel öfver Tekniska realskolans och Polytekniska skolans i Helsingfors samt Polytekniska institutets i Finland lärare och elever 1849–1897 (Helsingfors 1899).

Heiniö, Sulo (ed.), *Matrikel öfver Polytekniska institutets i Finland lärare och elever 1898–1908* (Hämeenlinna 1918).

Talvitie, Y[rjö Juho] (ed.), *Suomalaisten teknikkojen seuran nimikirja 1896–1936* (Helsinki 1936).

Suomen Insinöörejä ja arkkitehtejä: Ingenjörer och arkitekter i Finland: 1948 (Vaasa 1948).

Laati, Iisakki, and Blomstedt, Yrjö (eds.), *Kuka kukin oli: henkilötietoja 1900-luvulla kuolleista julkisuuden suomalaisista* (Helsinki 1961).

Profiles—pioneering women architects from Finland (Helsinki 1983).

Norway

Alstad, O. (ed.), *Tillegg til Trondhjemsteknikernes Matrikkel* (Trondhjem 1932).

Alstad, O. (ed.), *Trondhjemsteknikernes matrikel: biografiske meddelelser om samtlige faste og hospiterendeelever av Trondhjems tekniske læreanstalt 1870–1915: med ca. 1300 ungdomsportrætter* (Trondhjem 1916).

Bassøe, Bjarne (ed.), *Ingeniørmatrikkelen: norske sivilingeniører 1901–55 med tillegg* (Oslo 1961).

Brinchmann, Chr., Anders Daae, and K.V. Hammer. *Hvem er hvem?: haandbok over samtidige norske mændt og kvinder* (Kristiania 1912).

Brochmann, Georg (ed.), *Vi fra NTH: de første ti kull: 1910–1919* (Stavanger 1934).

Eskedal, Leif (ed.), *BTS-matrikkelen: ingeniører uteksaminert ved Bergen tekniske skole 1875–1975* (Bergen 1975).

Fougner, Eiliv (ed.), *Norske ingeniører og arkitekter: kort oversigt over Den norske ingeniørforenings og Norskearkitekters landsforbunds historie, samt biografiske oplysninger om de to organisationers nulevende medlemmer med portrætter* ([Kristiania 1916).

Ingeniørene fra K.T.S 1897–1947 ([Oslo] 1947).

Ingeniører fra 1896 fra Kristiania Tekniske Skole: biografiske Oplysninger samlet til 25 Aars Jubileet 1921 (Kristiania 1921).

Jansen, Einar and Paulus Svendsen (eds.), *Norsk biografisk leksikon* (Oslo 1954).

Kristiania tekniske skole, *Festskrift i anledning af Kristiania tekniske skoles 25-aars jubilæum i juni 1898* (Kristiania 1898).

KTS: 50 årsberetning om Ingeniørkullet fra Kristiania tekniskeskole 1896 ([Oslo] 1946).

Steenstrup, Hj. *Hvem er hvem?* (Oslo 1930).

Steenstrup, Hj. *Hvem er hvem?* (Oslo 1937).

Tidsskrift for kjemi, bergvesen og metallurgi (1950).

Sweden

Bodman, Gösta. *Chalmers tekniska institut: matrikel 1829–1929* (Göteborg 1929).

Civilingenjörer och arkitekter utexaminerade från Chalmers tekniska högskola: jämte redogörelse för Chalmersska ingenjörsföreningens och Chalmersska forskningsfondens verksamhet (Göteborg 1904–1930).

Forsberg, Emil and Emil Adlers (eds.), *Tekniska föreningen i Örebro 1875–1925: minnesskrift utgiven med anledning av föreningens femtioåriga verksamhet* (Stockholm 1925).

Govert Indebetou and Erik Hylander (eds.), *Svenska teknologföreningen 1861–1936: biografier* (Stockholm 1937).

Malmö teknologförbund: Minnesalbum utg. i anledning av Malmö tekniska läroverks 75-åriga verksamhet: 1853–1928 (Malmö 1928).

Tekniska föreningen i Örebro. Medlemsförteckning, m.m. Örebro (1926–1932).

Archival resources

Brages Pressarkiv, Helsinki, Finland.
Biografiska avdelningen [Biographical department].

Oslo City Archives, Norway.
Library: Beretning om Kristiania Tekniske Skole virksomhet, 1891, 1898, 1902, 1908

Oslo Tekniske Skoles arkiv:
volume G0001: Beretning om Kristiania Tekniske Skole virksomhet, 1889–1890, 1893–1897, 1899–1901, 1903–1907, 1909–1910
volume G0002, Beretning om Kristiania Tekniske Skole virksomhet, 1911–1915.

References

http://www.byhistoriskforening.org/AArringer-i-byhistorien/1900/1921, 19 November 2017.
http://www.colorantshistory.org/GermanDyeIndustry.html, 14 June 2018.
http://danmarkshistorien.dk/leksikon-og-kilder/vis/materiale/frederikshavn/?no_cache=1, 18 November 2017.
http://denstoredanske.dk/Dansk_Biografisk_Leksikon/Naturvidenskab_og_teknik/Ingeni%C3%B8r/Poul_Larsen, 8 October 2018.
https://www.genealogia.fi/genos-old/38/38_64.htm, 19 November 2017.
http://www.hestories.info/company-history-and-historical-facts-concerning-makers-of-elec.html?page=2, 17 November 2017.
http://www.kvinfo.dk/side/597/bio/1514/origin/170/, 17 November 2017.
https://nbl.snl.no/Bertel_Sherdahl, 18 November 2017.
http://www.rosekamp.dk/DBL_All/DBL_15_text.pdf, 16 November 2017.
http://s221243733.onlinehome.us/index.php/handelsmann-oscar-knudtzen-i-finnbukta-og-hans-familie, 18 November 2017.
'100 år sedan Gottfr. Strömbergs födelse', *Björneborgs Tidning*, 30 December 1963.
'50 År fyller den 16 December ingeniör Nikolai Uschanoff', *Hufvudstadsbladet*, 16 Februari 1931.
'50-åring', *Åbo Underrättelser*, 8 September 1934.
'60 år', *Hufvudstadsbladet*, 16 December 1923.
'60 år. Arkitekt E. G. Hedman', *Hufvudstadsbladet*, 19 December 1927.
'60-åringar', *Åbo underrättelser*, 22 February 1937.
'75-åring', *Åbo Underrättelser*, 11 October 1954.
Aarsbog, Sindre W. Med Mars og Merkur: en analyse av norsk deltakelse i VOC basert på skipssoldbøker 1633–1794 (Trondheim 2003). Master Thesis, Norwegian University of Science and Technology.
Ahlström, Göran. *Technological development and industrial exhibitions 1850–1914: Sweden in an international perspective* (Lund 1995).
Ahlström, Göran. 'Världsutställningar och teknikspridning 1850–1914', in: Bengt Berglund, Per-Olof Grönberg, and Tomas Nilson (eds.), *Historiska perspektiv på tekniköverföring 1800–2000* (Göteborg 2006) 69–76.
Ahonen, Kirsi, Marjaana Niemi, and Jaakko Pöyhönen. *Tietoa, taitoa, asiantuntemusta: Helsinki eurooppalaisessa kehityksessä 1875–1917. 3, Henkistä kasvua, teknistä taitoa* (Helsinki 1992).
Akenson, Donald Harman. *Ireland, Sweden and the great European migration, 1815–1914* (Liverpool 2012).
Åkerman, Sune. 'Theories and methods of migration research', in: Harald Runblom and Hans Norman (eds.), *From Sweden to America: A History of the Migration* (Uppsala 1976) 21–76.

"Albert von Julin,' *Åbo Underrättelser,* 14 April 1918.
'Allan Schulman Död', *Viborgs Nyheter.* 7 December 1937.
Allardt, Erik. 'Finlandssvenskarna', in: Ulf Lindström and Lauri Karvonen (eds.), *Finland: en politisk loggbok* (Stockholm 1987) 93–115.
Almquist, Johan Axel, Bertil Boëthius, and Bengt Hildebrand. *Svenskt Biografiskt Lexikon. Bd 9, Cornell-Dal* (Stockholm 1931).
Alvstad, Cecilia. 'Migrants on skis: Norwegian-Latin American return migration in the 1890s', in: Steinar A. Sæther (ed.), *Expectations unfulfilled: Norwegian migrants in Latin America, 1820–1940* (Leiden 2015) 77–101.
Amdam, Rolv Petter. 'Industrial espionage and the transfer of technology to the early Norwegian glass industry', in: Kristine Bruland (ed.), *Technology Transfer and Scandinavian Industrialisation* (New York, NY 1991) 73–93.
Amundsen, Otto Delphin. *Den kongelige norske Sankt Olavs orden* (Oslo 1947).
Andersen, Håkon With. *Fabrikken* (Oslo 2004).
Andersen, Håkon With. *Fra det britiske til det amerikanske produksjonsideal: forandringer i teknologi og arbeid ved Aker mek. verksted og i norsk skipsbyggingsindustri 1935–1970* (Trondheim 1986).
Andersen, Håkon With. 'Den nye vitenskapsprofesjonen: teknologene', in: Anton Fredrik Andersen and Guri Hjeltnes (eds.), *Universitet, samfunn og politikk: 18 innlegg om universitets- og vitenskapshistorie* (Oslo 1997) 151–176.
Andersen, Håkon With. 'Germany and the Education of Norwegian Engineers. With some Reflections on the Role of Engineers as a Social Group', in: *Bürgertum und Bürokratie im 19. Jahrhundert: Technologie, Innovation, Technologietransfer: Bericht über das 2. deutsch-norwegische Historikertreffen in Bonn, Mai 1987* (Oslo and Berlin 1987) 100–114.
Andersen, Håkon With and John Peter Collett. *Anchor and Balance: Det norske Veritas 1864–1989* (Oslo 1989).
Andersen, Ketil Gjølme. *Flaggskip i fremmed eie: Hydro 1905–1945* (Oslo 2005).
Andersen, Willy. 'Knud Fæster 24. Juli 1887—30. Januar 1966', *Grønland,* 6 (1966) 223.
Anderson, Sven A. and Augustus Jones, 'Iron in the Adirondacks', *Economic Geography* 21:4 (1945) 278–279.
Andersson, Ronnie, and Harald Theorin. 'Civilingenjörerna stannar i Sverige', *Välfärdsbulletinen* 2 (1999) 6–7.
Andersson, Ronnie, and Harald Theorin. 'Kompetensflykt - myt eller verklighet?' *Välfärdsbulletinen* 4 (1998) 6–7.
Anduaga, Aitor. 'The engineer as a "linking agent" in international technology transfer: The case of Basque engineers trained in Liège.' *Engineering Studies* 3:1 (2011) 45–70.
Antoniou, Yannis, Fotini Assimacopoulou, and Konstantinos Chatzis. 'Greek Engineers: Two Centuries of History (19th-20th Centuries)', in: Ana Cardoso de Matos,

et. al. (eds.), *The Quest for a Professional Identity: Engineers between Training and Action* (Lisboa 2009) 513–523.
Appelberg, Axel. 'Elektriska ugnar vid General Electric Co:s försökslaboratorium, Schenectady. Ur reseberättelse af fil. dr A. Appelberg,' *Teknisk tidskrift. Afdelningen för kemi och bergsvetenskap* 1 (1911) 7–9.
'arkitekt | Nationalencyklopedin', n.d. http://www.ne.se/lang/arkitekt. 14 June 2018.
'Arkitekt i tredje generasjon. Arkitektprofesjonen har gått i arv i familien Guldahl i Trondheim. Lars Axel Guldahl utøver faget i tredje generasjon, godt over 100 år etter at bestefaren Axel Guldahl senior tegnet sine første hus i Trondheim.' *Adresseavisa*, 17 December 2002.
'Aschan, Ossian—Uppslagsverket Finland', https://www.uppslagsverket.fi/sv/sok/view-103684-AschanOssian, 8 October 2018.
Åsgård, Lasse, and Christer Ellgren. *Ericsson: historien om ett svenskt företag* (Stockholm 2000).
Assimacopoulou, Fotini, Konstantinos Chatzis, and Anna Mahera. 'Elève en France, enseignant en Grèce. Les ensignants de l'Ecole polytechnique d'Athènes (1837–1912) formès dans des ècoles d'ingènieurs en France', in: Ana Cardoso de Matos, et. al. (eds.), *The Quest for a Professional Identity: Engineers between Training and Action* (Lisboa 2009) 25–41.
Baudet, Jean C. 'The training of engineers in Belgium, 1830–1940', in: Robert Fox and Anna Guagnini (eds.), *Education, Technology and Industrial Performance in Europe, 1850–1939* (Cambridge 1993) 93–114.
Beijbom, Ulf. *Amerikaminnen: berättelser i utvandrarbygd* (Stockholm 1996).
Berg, Bjørn Ivar. 'Bergseminaret på Kongsberg (1757–1814) og annen tidlig kunnskapsformidling i bergfag', in: Anne Kristine Børresen and Astrid Wale (eds.), *Vitenskap og teknologi for samfunnet?: bergfagene som kunnskapsfelt* (Trondheim 2005) 101–111.
Berger, Stefan. 'Comparative History', in: Stefan Berger, Heiko Feldner, and Kevin Passmore (eds.), *Writing History: Theory & Practice* (London 2003) 161–179.
Berglund, Bengt. 'Från förlag till automat. Textilindustrins utveckling och organisation i Sjuhäradsbygden 1750–2000', in: Bengt Berglund, Per-Olof Grönberg, and Tomas Nilson (eds.), *Historiska perspektiv på tekniköverföring 1800–2000* (Göteborg 2006) 189–195.
Bergström, Anders. *Arkitekten Ivar Tengbom: byggnadskonst på klassisk grund* (Stockholm 2001).
Bendtsen, P. H., and Povl Vinding. 'Per Kampmann', http://denstoredanske.dk/Dansk_Biografisk_Leksikon/Naturvidenskab_og_teknik/Ingeni%C3%B8r/Per_Kampmann, 8 October 2018.
Berner, Boel. *Sakernas tillstånd: kön, klass, teknisk expertis* (Stockholm 1996).
'Bernhard Hahl'. *Hufvudstadsbladet*, 28 September 1925.
'Bertel Jung 50 år'. *Hufvudstadsbladet*, 11 July 1922.

Bing, Morten and Espen Johnsen, *Nye hjem: bomiljøer i mellomkrigstiden* (Oslo 1998).

Birkeland, Bent. 'Richard Birkeland—utdypning (Norsk biografisk leksikon)—Store norske leksikon', *Store norske leksikon*, n.d. https://nbl.snl.no/Richard_Birkeland, 8 October 2018.

Bjerg, Hans Christian. 'Aage Westenholz - Dansk Biografisk Leksikon', http://denstoredanske.dk/Dansk_Biografisk_Leksikon/Naturvidenskab_og_teknik/Ingeni%C3%B8r/Aage_Westenholz, 8 October 2018.

Björck, Henrik. *Staten, Chalmers och vetenskapen: forskningspolitisk formering och sociala ingenjörer under Sveriges politiska industrialisering 1890–1945* (Nora 2004).

Björck, Henrik. 'Bilder av maskiner och ingenjörskårens bildande: tekniska tidskrifter och introduktion av ny teknik i Sverige, 1800–1870', *Polhem* 5 (1987) 267–310.

Bjork, Kenneth O. *Saga in Steel and Concrete: Norwegian Engineers in America* (Northfield, MN 1947).

Björklund, Kim. 'En sentida Grand Tour. Resan 1921–1922 och bildningsresornas tradition', in: Hilding Ekelund, Eva Ekelund, and Kim Björklund, *Italia la bella: arkitekterna Hilding och Eva Ekelunds resedagbok 1921–1922* (Helsingfors 2004) 9–27.

Björkman, Axel. 'Skurubron', *Teknisk tidskrift. Afdelningen för väg-och vattenbyggnadskonst* 11 (1915) 120–133.

Björnsson, Sveinbjörn. 'Menntun íslenskra verkfræðinga. Sveinbjörn Björnsson tók saman úr frumdrögum Jóns E. Vestdal og ýmsum öðrum heimildrum', in: Jón E. Vestdal, Stefán Bjarnason, and Steingrím Jónsson (eds.), *Verkfræðingatal: Æviágrip íslenzkra verkfræðinga og annarra félagsmanna verkfræðingafélags Íslands* (Reykjavík 1956).

De Blij, Harm J. and Peter O. Muller, *Geography: Regions and concepts* (New York, NY 1991).

Bodman, Gösta. *Chalmers tekniska institut: matrikel 1829–1929* (Göteborg 1929).

Bodman, Gösta. 'Ett och annat ur Chalmers-matrikeln', in: Gösta Bodman (ed.), *Chalmers tekniska institut 1829–1929: Minnesskrift utgiven med anledning av institutets 100-åriga tillvaro* (Göteborg 1929) 189–214.

Boje, Per. *Ledere, ledelse og organisation 1870–1972*. Dansk industri efter 1870 5 (Odense 1997).

Boje, Per. 'Teknikumingeniører og dansk økonomisk vækst', in: Henrik Harnow, et. al. (eds.), *Industriens Pionerer: Teknikumingeniørernes uddannelseskamp og betydning* (København 2011) 279–321.

'Bør danske Ingeniører søge til hollandsk Indien?' *Ingeniøren* 50 (1911) 506–508.

Borgstedt, Martin. 'Reseberättelse af Arkitekten Martin Borgstedt öfver nyare byggnadskonstruktions-sätt, byggnadsmaterialer, m. m. i Nordamerikas Förenta Stater (1892)', 1892. Reseberättelser 269. Royal Institute of Technology Library, Stockholm.

Bowallius, Marie-Louise. *Den förändrade synen på amerikansk teknologi: rapportering och värdering av amerikansk teknologi i Teknisk Tidskrift 1870–1893* (Stockholm 1980).

Bovenkerk, Frank. *The Sociology of Return Migration: A Bibliographic Essay* (The Hague 1974).

Bramsen, Christopher Bo. *Open doors: Vilhelm Meyer and the establishment of General Electric in China* (Richmond, VA 2000).

Brandes, Stuart D. *American welfare capitalism, 1880–1940* (Chicago, IL 1976).

Brändström, Anders, Jan Sundin, and Lars-Göran Tedebrand. 'Two Cities: Urban Migration and Settlement in Nineteenth-century Sweden', *The History of the Family* 5:4 (2000) 415–429.

Braun, Hans-Joachim. 'A Technological Community in the United States: The National Association of German-American Technologists, 1884–1941', *Amerikastudien: American Studies* (1985) 447–463.

Braun, Hans-Joachim. 'Franz Reuleaux und der Technologietransfer zwischen Deutschland und Nordamerika am Ausgang des 19. Jahrhunderts', *Technikgeschichte* 48:2 (1981) 112–130.

Braun, Hans-Joachim. 'Technologietransfer im Maschinenbau von Deutschland in die USA 1870 bis 1939', *Technikgeschichte* 50:3 (1983) 238–252.

Brissach, Ingrid J. 'NTHs første veiproffesor. Kolbjørn Heje var som sin forgjenger Sem Sæland opptatt av «kunnskapens praktiske anvendelse» og hadde selv mye praktisk erfaring.' *Adresseavisa*, 13 November 2007.

Brittain, James E. *Alexanderson: Pioneer in American Electrical Engineering* (Baltimore, MD 1992).

Brochmann, Georg. 'A/S Myrens Verksted', *Teknisk ukeblad* 13 (1924) 114.

Brochmann, Georg. 'Kværner brug', *Teknisk ugeblad* 46 (1923) 303–305.

Brochmann, Georg. 'Verkstaden i Kristinehamn', *Teknisk ugeblad* 36 (1923) 296–297.

Bruland, Kristine. *British Technology and European Industrialization: the Norwegian Textile Industry in the Mid Nineteenth Century* (Cambridge 1989).

Bruland, Kristine, 'Skills, Learning and the International Diffusion of Technology: A Perspective on Scandinavian Industrialization', in: Maxine Berg and Kristine Bruland (eds.), *Technological Revolutions in Europe* (Cheltenham 1998) 161–187.

Bruland, Kristine. 'The Norwegian Mechanical Engineering Industry and the Transfer of Technology, 1800–1900', in: Kristine Bruland (ed.), *Technology Transfer and Scandinavian Industrialisation* (New York, NY 1991) 229–295.

Brunnström, Lisa. *Den rationella fabriken: om funktionalismens rötter* (Umeå 1990).

Burton-Skårdal, Dorothy. 'Rapport om prosjektet: Hjemvendte emigranter i Agder', in: Hans Try and Berit Eide Johnsen (eds.), *Arbeidsvandringar på Agder* (Kristiansand 1990) 103–122.

C. K. 'Gode utsikter for bergingeniører i Chile', *Teknisk ugeblad* 9 (1924) 92.

Cardoso de Matos, Ana. 'Asserting the Portuguese Civil Engineering Identity: The Role Played by the École des Ponts et Chaussées, 1825–1866', in: Ana Cardoso de Matos, et. al. (eds.), *The Quest for a Professional Identity: Engineers between Training and Action* (Lisboa 2009) 177–208.

Cardoso de Matos, Ana, and Maria Paula Diogo. 'Bringing It All Back Home: Portuguese Engineers and Their Travels of Learning', *Journal of History of Science and Technology* 1 (2007) 155–181.

Carlestam, Gösta. *Samhällsbyggarna vid Storsjön: en plats i utkanten blir en stad i världen: en diskussion om stålstaden Sandviken - och - om mål och medel i samhällsplaneringens Sverige 1862–1984* (Gävle 1986).

Carlson, Oscar. 'Om kalkkväfvetillverkningen. Föredrag vid Ljungaverkens invigning den 24 september 1912', *Svensk Kemisk Tidskrift* 24 (1912) 134–139.

Carlsson, Sten. *Swedes in North America: 1638–1988: Technical, Cultural and Political Achievements* (Stockholm 1988).

Carlsson, Sten. 'Chronology and Composition of Swedish Emigration to America', in: Harald Runblom and Hans Norman (eds.), *From Sweden to America: A History of the Migration* (Uppsala 1976) 114–148.

Carlsson, Sten. 'Swedish Engineers in Chicago', in: Philip J. Anderson and Dag Blanck (eds.), *Swedish-American life in Chicago: Cultural and Urban Aspects of an Immigrant People, 1850–1930* (Uppsala 1991) 181–192.

Carr, William. *A History of Germany 1815–1990. Fourth Edition.* (London 1991).

Chandler, Alfred D. 'The United States: Engines of Economic Growth in the Capital-Intensive and Knowlegde-Intensive Industries', in: Franco Amatori et al. (eds.), *Big Business and the Wealth of Nations* (Cambridge 1997) 63–101.

Chandler, Alfred D., and Takashi Hikino. *Scale and Scope: the Dynamics of Industrial Capitalism.* (Cambridge, MA 1994).

Chrimes, Michael M. 'British and Irish Civil Engineers in the Development of Argentina in the Nineteenth Century', in: Malcolm Dunkeld, et. al. (eds.), *Proceedings of the Second International Congress on Construction History* (Cambridge 2006) 675–694.

Christiernin, Georg. 'Elektroteknikens pionjärer i Finland', *Tekniska föreningens i Finland förhandlingar* 4 (1930) 83.

Christensen, Sverre A, and Harald Rinde. *Nasjonale utlendinger: ABB i Norge 1880–2010* (Oslo 2009).

Christiani, Rudolf. 'Elbtunnelen', *Ingeniøren* 62 (1912) 508.

Cinel, Dino. *The National Integration of Italian Return Migration, 1870–1929* (Cambridge, MA 1991).

Collins, George R., and Christiane Craseman Collins, *Camillo Sitte: the birth of modern city planning* (Mineola, NY 2006).

Dalsgaard Larsen, Keld. *Dansk papirindustri: mennesker, teknologi og produktion 1829–1999* (Silkeborg 2000).

Dansk Arkitektstat 1941 (København 1941).

'Danske ingeniører til nederlandsk Indien', *Ingeniøren* 3 (1912) 40.

Danske Teknika og deres Dimittender gennem 50 Aar, Ingeniør- og Konstruktørsammenslutningen gennem 35 Aar (København 1931).

Day, C. R. 'The Making of Mechanical Engineers in France: The Ecoles d'Arts et Métiers, 1803–1914', *French Historical Studies* 10:3 (1978) 439–460.

Devik, Olaf. *N.T.H. femti år: Norges tekniske høgskoles virksomhet 1910–1960* (Oslo 1960).

'Dir. John M. Gylphe 50 år', *Hangö*, 21 August 1930.

'Direktör Carl Jahn', *Hufvudstadsbladet*, 31 March 1937.

Dixi X. 'Patriotisme', *Teknisk ugeblad* 17 (1885) 85.

Djupedal, Knut. 'Some Thoughts on the Influence of Returned Migrants in Norway', in: *Göteborgs-emigranten* 6 (Göteborg 1997) 204–210.

'Dödsfall', *Tammerfors aftonblad*, 1 February 1928.

Dribe, Martin, and Maria Stanfors. *Demografins grunder* (Stockholm 2005).

'Edvard Hubendick, Http://www.ne.se/edvard-hubendick, Nationalencyklopedin, 21 November 2017.

Èfmertovà, Marcela. 'Les professeurs èlectrotechniques tchèques dans le monde: formation et impact des travaux scientifiques dans les annèes 1918–1938', in: Ana Cardoso de Matos, et. al. (eds.), *The Quest for a Professional Identity: Engineers between Training and Action* (Lisboa 2009) 513–523.

Eilertsen, Thorvald. 'Moderne amerikansk Værkstedsledelse (Scientific Management)', *Ingeniøren* 50 (1913) 329–336.

Eiserman, Bengt. *Minnen och episoder* (Stockholm 1937).

Ekelund, Hilding. 'Extracts from Travel Diaries', in: Timo Tuomi, Kristina Paatero, and Eija Rauske (eds.), *Hilding Ekelund: (1893–1984): arkkitehi=arkitekt=architect* (Helsinki 1997) 91–93.

Ekelund, Hilding, Eva Ekelund, and Kim Björklund. *Italia la bella: arkitekterna Hilding och Eva Ekelunds resedagbok 1921–1922* (Helsingfors 2004).

'Elektroingeniører til nederlandsk Indien,' *Ingeniøren* 8 (1914) 39.

'Elektroståltillverkning i Amerika', *Teknisk tidskrift. Afdelningen för kemi och bergsvetenskap* 10 (1912) 136–137.

Eliassen, Edward Martin. 'Modern Driftsledelse ved amerikanske Maskinverksteder', *Teknisk ugeblad*, 15, 16 (1913) 178–180, 182–183.

Eliasson, Pär. 'Svenska studenter i Tyskland 1372–1800', in: Kurt Genrup (ed.), *'Förtyskningen' av Sverige* (Umeå 1994) 43–64.

Ellert, Søren Jensen. 'Overproduktion af Ingeniører', *Ingeniøren* 22 (1905) 144–145.

'En teknikens pionjär i Finland', *Hufvudstadsbladet*, 22 March 1931.

Engberg, Jens F. 'Danske Ingeniører i Amerika', *Ingeniøren* 23 (1920) 188–189.

Engberg, Jens F. 'Social Udviking (sic!) i amerikansk Industri', *Ingeniøren* 6 (1920) 32–34.

Engblom, Alex. 'Rationell arbetsledning och dess praktiska tillämpning. Föredrag vid Svenska Ingenjörskongressen i Chicago 1915 af civilingenjör Alex. Engblom', *Teknisk tidskrift. Veckoupplagan* 39, 40 (1916) 352–357, 362–366.

Engelstoft, Povl, and Svend Dahl (eds.), *Dansk biografisk leksikon. XIX Quaade—Rongsted* (København 1940).

Engman, Max. *Lejonet och dubbelörnen: Finlands imperiella decennier 1830–1890* (Stockholm 2000).

Engman, Max. 'The Finns in St Petersburg', in: Max Engman (ed.), *Ethnic Identity in Urban Europe* (Aldershot 1992) 99–130.

Engström, Eric. *Bokbindargesällen Karl Stellan Söderströms gesällvandring 1843–1858: lärande i skråväsendet speglat i personliga document* (Stockholm 1995).

Epstein, Stephan R. 'Craft guilds, apprenticeship, and technological change in preindustrial Europe.' *Journal of Economic History* 58:3 (1998) 684–713.

Ericsson, Tom. *Mellan kapital och arbete: småborgerligheten i Sverige 1850–1914* (Umeå 1988).

Eriksson, Patrick. 'Petrelius, Albert - Uppslagsverket Finland', https://www.uppslagsverket.fi/sv/sok/view-103684-PetreliusAlbert, 8 October 2018.

Eriksson, Ulf. *Gruva och arbete: Kiirunavaara 1890–1990. Avsnitt 1, 1890–1920* (Uppsala 1991).

Erlendsdottir, Guðbjörg Lilja. *Trafiksäkerhet och tätortsplanering: en analys av Reykjavíks lokalgatunät med GIS* (Lund 2003).

Espelund, Arne. 'Elias Anton Cappelen Smith—utdypning (Norsk biografisk leksikon), in: Store norske leksikon.', https://nbl.snl.no/Elias_Anton_Cappelen_Smith, 8 October 2018.

Esser, Raingard. 'Südniederländische Textilarbeiter im England des 16. und 17. Jahrhunderts', in: Klaus J. Bade, et. al. (eds), *Enzyklopädie. Migration in Europa. Vom 17. Jahrhundert bis zur Gegenwert* (Paderborn 2007) 1029–1031.

Estey, James Arthur. *Business Cycles: Their Nature, Cause and Control* (New York, NY 1941).

Eurenius, A. G. J. *Matrikel öfver eleverna i Tekniska elementarskolan i Norrköping från skolans början år 1857 till och m. v. till 1912* (Norrköping 1912).

Evans, Chris, and Göran Rydén. 'Kinship and the Transmission of Skills: Bar Iron Production in Britain and Sweden, 1500–1860', in: Maxine Berg and Kristine Bruland (eds.), *Technological Revolutions in Europe* (Cheltenham 1998) 188–206.

'F. L. Smidth & Co. 1882–1922', *Ingeniøren* 1 (1922) 10–13.

Fagerberg, Mats. 'Det amerikanska systemets införande i den svenska järnmanufaturindustrin. Diskussionen i svenska tidskrifter under 1880-talet', in: Bengt Berglund, Per-Olof Grönberg, and Tomas Nilson (eds.), *Historiska perspektiv på tekniköverföring 1800–2000* (Göteborg 2006) 161–188.

Fellman, Susanna. *Uppkomsten av en direktörsprofession: industriledarnas utbildning och karriär i Finland 1900–1975* (Helsingfors 2000).

'Femtio år', *Tammerfors Aftonblad*, 7 January 1928.

'Femtioåring. Ingeniör K. E. Ranko', *Hufvudstadsbladet*, 6 February 1928.

'Femtioåringar', *Åbo underrättelser*, 22 August 1930.

'Fjalar Witting', *Tammerfors aftonblad*, 30 April 1935.

Fløgstad, Kjartan. *Eld og vatn: nordmenn i Sør-Amerika* (Oslo 1999).

Flora, Nicola, Paolo Giardiello, and Gennaro Postigilone. 'Journey to Italy', in: Nicola Flora, Paolo Giardiello, and Gennaro Postiglione (eds.), *Sigurd Lewerentz: 1885–1975* (Milano 2002) 35–43.

Fogelberg, Torbjörn S. and Friedrich D. Holl, *Wanderungen Deutscher Glashüttenleute und Schwedens Glasindustrie In den letzten fünf Jahrhunderten* (Växjö 1988).

'Förteckning öfver i Amerika vistande finska tekniker inom textilbranschen', *Tekniska föreningens i Finland förhandlingar* (1914) 212.

Fridlund, Mats. *Den gemensamma utvecklingen: staten, storföretaget och samarbetet kring den svenska elkrafttekniken* (Eslöv 1999).

Fritz, Martin. 'British Influence on Developments in the Swedish Foundry Industry around the Turn of the Eighteenth Century', in: Kristine Bruland (ed.), *Technology Transfer and Scandinavian Industrialisation* (New York, NY 1991) 59–72.

Fritz, Martin, and Birgit Karlsson. *SKF—a global story: 1907–2007* (Stockholm 2006).

G. Sn., 'Moderna arbetsmetoder och verktygsmaskiner i Nordamerikas Förenta Stater. Ur en reseberättelse af ingenjör W. Fagerström,' *Teknisk tidskrift. Afdelningen för mekanik*, 1 (1915) 3–4.

G. W. 'Teknikerna och utvandringen.' *Teknisk tidskrift. Veckoupplagan*, 51 (1910) 415.

Gårdlund, Torsten. *Industrialismens samhälle* (Stockholm 1942).

De Geer, Hans. *Rationaliseringsrörelsen i Sverige: effektivitetsidéer och socialt ansvar under mellankrigstiden* (Stockholm 1978).

Gerlow, Poul. 'Berliner Elektricitäts-Werke. Af en Rejseberetning', *Ingeniøren* 40 (1907) 279–287.

Ghosh, Bimal. 'Return migration. reshaping policy approaches', in; Bimal Ghosh (ed.), *Return Migration: Journey of Hope or Despair?* (Geneva 2000) 181–226.

Glete, Jan. *Ägande och industriell omvandling: ägargrupper, skogsindustri, verkstadsindustri 1850–1950* (Stockholm 1987).

Glete, Jan. *ASEA under hundra år: 1883–1983: en studie i ett storföretags organisatoriska, tekniska och ekonomiska utveckling* (Västerås 1983).

Goodwin, Jason. *Otis: Giving Rise to The Modern City* (Chicago, IL 2001).

Gouzévitch, Dmitri. 'The rise of state engineering administrations in eighteenth century Russia', *Engineering Studies* 3:3 (2011) 195–213.

Gouzévitch, Irina, and Dmitri Gouzévitch. 'Travelling Interchanges between the Russian Empire and Western Europe: The Travels of Engineers During the First Half of the Nineteenth Century', in: Ana Simões, Ana Carneiro, and Maria Paula Diogo (eds.), *Travels of Learning: A Geography of Science in Europe* (Dordrecht 2003) 213–231.

Grelon, Andrè. 'The Training and Career Structures of Engineers in France, 1880–1939', in: Robert Fox and Anna Guagning (eds.), *Education, Technology and Industrial Performance in Europe, 1850–1939* (Cambridge 1993) 42–64.

von Greyerz, Walo. 'F. W. Taylor—revolutionären', *Teknisk tidskrift. Veckoupplagan*, 50 (1915) 510–513.

Grimnes, Ole Kristian. *Sam Eyde: den grenseløse gründer* (Oslo 2001).

Grönberg, Per-Olof. *Learning and Returning: Return Migration of Swedish Engineers from the United States, 1880–1940* (Umeå 2003).

Grönberg, Per-Olof. 'Den vandrande professionen. Nordiska ingenjörers internationella mobilitet runt förra sekelskiftet', in: Jan Frode Hatlen and Pål Thonstad Sandvik (eds.), *En sann historiker. Festskrift til Svein Henrik Pedersen* (Trondheim 2007) 89–96.

Grönberg, Per-Olof. 'Die internationale Migration und die Rückwanderung skandinavischer Ingenieure zwischen 1880 und 1930', in: Dittmar Dahlmann and Reinhold Reith (eds.), *Elitenwanderung und Wissenstransfer im 19. und 20. Jahrhundert* (Essen 2008) 121–157.

Grönberg, Per-Olof. 'International Migration and Return Migration of Swedish Engineers before 1930 and in the 1990s', in: Jørgen Carling (ed.), *Nordic Demography: Trends and Differentials* (Oslo 2002) 267–285.

Grönberg, Per-Olof. 'Internationale Migration und Rückwanderung von nordischen Ingenieuren, 1880–1930', *Technikgeschichte* 73:3/4 (2006) 169–206.

Grönberg, Per-Olof. 'Journeymen or Traditional Emigrants? Norwegian and Swedish Engineers and Architects in North America, 1880–1930', in: Philip J. Anderson and Dag Blanck (eds.), *Norwegians and Swedes in the United States: Friends and Neighbors* (St. Paul, MN 2011) 197–218.

Grönberg, Per-Olof. 'My kind of town?' Ethnicity and Class as Determining Factors for Return Migration or Permanent Settlement among Swedish Engineers in Chicago 1910–1930', in: Daniel Lindmark (ed.), *Swedishness reconsidered: Three centuries of Swedish-American identities* (Umeå 1999) 121–142.

Grönberg, Per-Olof. "Skorsten kallar vi det rör som mot Ryssland röken för': om konstruktionen av Europas högsta skorsten vid Rönnskärsverken', *Polhem* 2 (2005) 25–42.

Grönberg, Per-Olof. 'The Engineer Paradox. International Migration as a Patriotic Act', in: K. G. Hammarlund (ed.), *Borders as Experience* (Halmstad 2009) 96–105.

Grönberg, Per-Olof. 'Tillbaka till framtidslandet: ingenjörsmigrationen mellan Nordamerika och Norrland före 1940', *Oknytt* 21, no. 3–4 (2000) 42–82.

Grönberg, Per-Olof. 'To study or to work? A Comparative Perspective on Nordic Engineer Migration to German-speaking Europe, 1880–1930', in: K. G. Hammarlund and Tomas Nilson (eds.), *Technology in Time, Space, and Mind: Aspects ON Technology Transfer and Diffusion* (Halmstad 2008) 71–96.

H. I. H. 'Taylor's System indført i Tyskland', *Ingeniøren* 40 (1914) 356.

Haapala, Pertti. *Tehtaan valossa: teollistuminen ja työväestön muodostuminen Tampereella 1820–1920* (Tampere 1986).

De Haas, Hein. *Migration and development in southern Morocco. The disparate socio-economic impacts of out-migration on the Todgha Oasis Valley* (Nijmegen 2003).

De Haas, Hein. 'Migration and Development: A Theoretical Perspective', *International Migration Review* 44:1 (2010) 227–264.

Haglund, Wilhelm. *Levebröd: strövtåg i minnet och brukshistorien* (Stockholm 1978).

Hammar, Hugo. *Minnen. 1, Från Ölands Alvar till livets* (Stockholm 1937).

Hammar, Hugo. *Minnen. 2, Som emigrant i U.S.A.* (Stockholm 1938).

Hanisch, Tore Jørgen, and Even Lange. *Vitenskap for industrien: NTH - en høyskole i utvikling gjennom 75 år* (Oslo 1985).

Hansen, Wilh. 'Indledning', in: *Danske Teknika og deres Dimittender gennem 50 Aar, Ingeniør- og Konstruktørsammenslutningen gennem 35 Aar* (Kobenhavn 1931) 7–11.

Hansson, Joakim. 'The Architectural-Historical Journeys of Hilding Ekelund and Eva Kuhlefelt and Their Early Links with Sweden', in: Timo Tuomi, Kristina Paatelo, and Eija Rauske (eds.), *Hilding Ekelund: (1893–1984) arkkitehti = arkitekt = architect* (Helsinki 1997) 61–63.

Hansson, Staffan. *Porjus: en vision för industriell utveckling i övre Norrland* (Luleå 1994).

'Harald Jensen 60 år', *Tammerfors aftonblad*, 1 July 1924.

Hård, Mikael, and Andrew Jamison. *Hubris and hybrids: A Cultural History of Technology and Science* (New York, NY 2005).

Harnow, Henrik. *Den danske ingeniørs historie 1850–1920: danske ingeniørers uddannelse, professionalisering og betydning for den danske moderniseringsproces* (Århus 1998).

Harnow, Henrik. *Viden om - den danske tekniske rådgivnings historie 1850 til i dag* (Kobenhavn 2004).

Hayami, Yujiro, and Vernon W. Ruttan. *Agricultural Development: An International Perspective* (Baltimore, MD 1971).

Helk, Vello. *Dansk-norske studierejser fra reformationen til enevælden 1536–1660: med en matrikel over studerende i udlandet* (Odense 1987).

Helk, Vello. *Dansk-norske studierejser: 1661–1813* (Odense 1991).

Henningsen, Bernd, and Birgitta Högvall (eds.), *Skandinavien och Tyskland: 1800–1914: möten och vänskapsband* (Stockholm 1997).

Hermanrud, Torbj. 'Newport News Shipbuilding and Dry Dock Co', *Teknisk ukeblad* 47 (1915) 584.

Hietala, Marjatta. 'Finnische Wissenschaftler in Deutschland 1860–1950. Allgemeine bemerkungen mit besonderer Berücksichtigung medizinischer Kontakte', in: Edgar Hösch (ed.), *Deutschland und Finnland im 20. Jahrhundert* (Wiesbaden 1999) 373–394.

Hietala, Marjatta. 'The Diffusion of Infrastructural Innovations in Finland', in: Pekka Ylä-Anttila and Synnöve Vuori (eds.), *Mastering Technology Diffusion: The Finnish Experience* (Helsinki 1992) 253–281.

Higginson, John. 'Privileging the machines: American engineers, indentured Chinese and white workers in South Africa's deep-level gold mines, 1902–1907', *International Review of Social History* 52:1 (2007) 1–34.

Hjelm, Jonny. *Begåvningsreserven inom industrin: förslagsverksamhet i Sverige under 1900-talet* (Lund 1999).

'Hjelmman, Leonard - Uppslagsverket Finland', https://www.uppslagsverket.fi/sv/sok/view-103684-HjelmmanLeonard, 8 October 2018.

Hjerppe, Riitta, and Jukka Jalava. 'Economic Growth and Structural Change—A Century and a Half of Catching-up', in: Jari Ojala, Jari Eloranta, and Jukka Jalava (eds.), *The Road to Prosperity: An Economic History of Finland* (Helsinki 2006) 33–64.

Hochstadt, Steve. 'Migration in Preindustrial Germany', *Central European History* 16:3 (1983) 195–224.

Hodne, Fritz. *Norges økonomiske historie 1815–1970* (Oslo 1981).

Hoerder, Dirk. 'Migration in the Atlantic economies: Regional European origins and worldwide expansion', in: Dirk Hoerder and Leslie Page Moch (eds.), *European migrants: Global and local perspectives* (Boston, MA 1996) 21–51.

Hoerder, Dirk, Jan Lucassen, and Leo Lucassen. 'Terminologien und Konzepte in der Migrationsforschung', in: Klaus J. Bade, et. al. (eds.), *Enzyklopädie Migration in Europa: vom 17. Jahrhundert bis zur Gegenwart* (Paderborn 2007) 28–53.

Hoffstad, Einar. *Merkantilt biografisk leksikon. Hvem er hvem i næringslivet?* (Oslo 1935).

Hoffstedt, W. 'Anordning och drift af en ur teknisk, hygiensk och social synpunkt modern mekanisk verkstad', *Teknisk tidskrift. Afdelningen för mekanik och elektroteknik* 5 (1902) 83–87.

Hollberg, Malin. 'Storängen - samhällsplanering under sekelskiftet', in: *Nackaboken* 36 (Nacka 2000) 69–96.

Holmberg, C. E. 'Om tyska och amerikanska brosystem', *Tekniska föreningens i Finland förhandlingar* (1894) 77–98.

Hornung, Erik. 'Immigration and the Diffusion of Technology: The Huguenot Diaspora in Prussia', *American Economic Review,* 104:1 (2014) 84–122.

Hösch, Edgar (ed.), *Deutschland und Finnland im 20. Jahrhundert* (Wiesbaden 1999).

Hounshell, David A. *From the American System to Mass Production, 1800–1932: The Development of Manufacturing Technology in the United States* (Baltimore, MD 1984).

Hughes, Thomas Parke. *American Genesis: a Century of Invention and Technological Enthusiasm, 1870–1970.* (New York, NY 1989).

Hughes, Thomas P. *Networks of Power. Electrification in Western society, 1880–1930* (Baltimore, MD 1983).

Hunger, Uwe. 'The 'Brain Gain' Hypothesis: Third World Elites in Industrialized Countries and Socioeconomic Development in Their Home Country'. Working Papers. San Diego, CA: Center for Comparative Immigration Studies, UC San Diego, 2002.

Hunger, Uwe. 'Vom 'Brain-Drain' zum 'Brain-Gain'. Migration, Netzwerkbildung und sozio-ökonomische Entwicklung: das Beispiel der indischen 'Software-Migranten', in *IMIS-Beiträge* (Osnabrück 2000) 7–23.

Hvidt, Kristian. *Flugten til Amerika eller drivkræfter i masseudvandringen fra Danmark 1868–1914* (Aarhus 1971).

Hyldtoft, Ole. *Københavns industrialisering 1840–1914* (Herning 1984).

Hyldtoft, Ole. *Teknologiske forandringer i dansk industri 1870–1896. Dansk industri efter 1870* (Odense 1996).

Hyldtoft, Ole. 'Foreign Technology and the Danish Brick and Tile Industry, 1830–1870', in: Kristine Bruland (ed.), *Technology Transfer and Scandinavian Industrialisation* (New York, NY 1991) 201–227.

Hyldtoft, Ole. 'Perioden 1896–1930', in: Ole Hyldtoft and Hans Chr. Johansen, *Teknologiske forandringer i dansk industri 1896–1972* (Odense 2005) 15–241.

'Ingeniør E. Rua', *Ringerikets blad*, 22 April 1965.

'Ingeniör Karl Lilius död', *Hufvudstadsbladet*, 23 December 1935.

'Ingeniör Verner Ryselin Död', *Hufvudstadsbladet*, 2 January 1940.

'Ingeniører til Nederlandsk Indien', *Ingeniøren* 66 (1920) 493–494.

'Ingenjör Gustaf Zitting 60 år', *Allas krönika: illustrerad veckoskrift*, 6 February 1920.

Ingenjörer: Insinöörit: Tampereen teknillinen opisto, Tekniska läroverket i Helsingfors, Helsingin teknillinen opisto, Turun teknillinen opisto (Tampere 1957).

Jangfeldt, Bengt. *Svenska vägar till S:t Petersburg: kapitel ur historien om svenskarna vid Nevans stränder* (Stockholm 1998).

Jansson, Ingrid. *Svensk rapportering av amerikansk teknologi på världsutställningen i Philadelphia 1876* (Stockholm 1980).

Jensen, Laur. 'Patriotisme', *Teknisk ugeblad* 14 (1885) 67.

Jeppesen, Torben Grøngaard. *Danske i USA 1850–2000: en demografisk, social og kulturgeografisk undersøgelse af de danske immigranter og deras efterkommere* (Odense 2005).

Johansson, Alf O. *Arbetarrörelsen och taylorismen: Olofström 1895–1925: en studie av verkstadsindustrin och arbetets organisering* (Lund 1990).

'Johannes I. Åström,' *Hufvudstadsbladet*, 20 August 1933.

Jörberg, Lennart. *The Industrial Revolution in Scandinavia 1850–1914* (London 1970).

Jørgensen, Poul J. 'Forhold og Muligheder for Teknikere i U. S. A', *Ingeniøren* 8 (1927) 100.

Jung, Bertel. 'Om nya rörelser på arkitekturens område hemma och i utlandet', *Tekniska föreningens i Finland förhandlingar* (1901) 65–66.

K., 'Nordmand hædret i Amerika', *Teknisk ugeblad* 36 (1920) 480.

Kaarsted, Tage. *Admiralen: Andreas de Richelieu: forretningsmand og politiker i Siam og Danmark* (Odense 1990).

Kampmann, Per. 'Svensk-dansk jernbanebygning i Tyrkiet', *Teknisk Tidskrift. Allmänna avdelninge* 18, 19 (1932) 177–179, 189–192.

Kanigel, Robert. *The One Best Way: Frederick Winslow Taylor and the Enigma of Efficiency* (New York, NY 1997).

Kastrup, Allan. *The Swedish Heritage in America: The Swedish Element in America and American-Swedish Relations in Their Historical Perspective* (Minneapolis, MN 1975).

Kero, Reino. *Migration from Finland to North America in the Years between the United States Civil War and the First World War* (Turku 1974).

Kero, Reino. *Suureen länteen: siirtolaisuus Suomesta Yhdysvaltoihin ja Kanadaan* (Turku 1996).

Kildal, Arne. 'KORT BERLE. En av de mest fremtredende nordmenn i Amerika er død', *Nationen*, 7 May 1934.

Kim, Terri. 'Shifting patterns of transnational academic mobility: A comparative and historical approach', *Comparative Education* 45:3 (2009) 387–403.

King, Russell. 'Generalizations from the History of Return Migration', in: Bimal Ghosh (ed.), *Return Migration: Journey of Hope or Despair?* (Geneva 2000) 7–55.

Kirby, David. *Östersjöländernas historia, 1772–1993* (Stockholm 1996).

Kline, Ronald R. *Steinmetz: engineer and socialist* (Baltimore, MD 1992).

Kock, Karin. *Skånska Cement Aktiebolaget 1871–1931: Minnesskrift* (Uppsala 1932).

Kocka, Jürgen. 'The middle classes in Europe', *The Journal of Modern History* 67:4 (1995) 783–806.

Koht, Halvdan. *Amerika i Europa: impulser från väster i teknik, politik, kultur* (Stockholm 1950).

'Kolbjørn Heje - Store norske leksikon.', https://snl.no/Kolbj%C3%B8rn_Heje, 8 October 2018.

König, Wolfgang. 'Technical Education and Industrial Performance in Germany: A Triumph of Heterogeneity', in: Robert Fox and Anna Guagnini (eds.), *Education, Technology and Industrial Performance in Europe, 1850–1939* (Cambridge 1993) 65–87.

Korkiasaari, Jouni. *Suomalaiset maailmalla: Suomen siirtolaisuus ja ulkosuomalaiset entisajoista tähän päivään* (Turku 1989).

Krabbe, Thorvald. *Island og dets tekniske udvikling gennem tiderne* (København 1946).

Krantz, Olle. 'Industrialisation in Three Nordic Countries: A Long-Term Quantitative View', in: Hans Kryger Larsen (ed.), *Convergence? Aspects on the Industrialisation of Denmark, Finland and Sweden 1870–1940* (Helsinki 2001) 23–65.

Kristensen, Søren. *Learning by leaving: Placements abroad as a didactic tool in the context of vocational education and training in Europe* (Luxembourg 2004).

Kristensen, Søren. 'Support for Transnational Mobility for Young People', in: *Mobility and Migration of Labour in The European Union and Their Specific Implications for Young People* (Thessaloniki 1998) 99–125.

Kronborg, Bo, and Thomas Nilsson. *Stadsflyttare: industrialisering, migration och social mobilitet med utgångspunkt från Halmstad 1870–1910* (Uppsala 1975).

Kuusanmäki, Jussi. *Tietoa, taitoa, asiantuntemusta: Helsinki eurooppalaisessa kehityksessä 1875–1917. 2, Sosiaalipolitiikkaa ja kaupunkisuunnittelua* (Helsinki 1992).

Kuuse, Jan. 'Gottlieb M T Piltz', in: *Svenskt Biografiskt Lexikon. Bd 29, Pegelow-Rettig* (Stockholm 1997) 313.

'Kvindelige arkitekter og ingeniører i Amerika.', *Teknisk ugeblad* (1906) 304.

Kvist, Kjetil. 'Arbeidslyst - eventyrlyst: Trondhjems Tekniske Læreanstalts-elevers utferdstrang i 1890-årene' (Trondheim 2005) Masters Thesis, Norwegian University of Science and Technology.

'Kyrklund, Harald - Uppslagsverket Finland', https://www.uppslagsverket.fi/sv/sok/view-103684-KyrklundHarald, 8 October 2018.

Lagus, Hugo. *W. Rosenlew & Co. Aktiebolag: 1853–1928* (Pori 1928).

Laine, Silja. 'Americanism and Architecture in 1920s Helsinki, in: Christina Folke Ax (ed.), *Encountering Foreign Worlds: Experiences at Home and Abroad: Proceedings from the 26th Nordic Congress of Historians, Reykjavik 8–12 August 2007* (Reykjavik 2007) 25–30.

Lange, Even. *Norske ingeniører i Amerika 1900–1950: en moderne svennevandring* (Bekkestua 1988).

Lavonius, Robert. 'Pohjois-Amerikan konepajojen sisäisestä työnjärjestämisestä', *Teknillinen aikakauslehti* 6–7 (1913) 131–135.

Leeuwen, Marco H. D. van, Ineke Maas, and Andrew Miles. *HISCO: historical international standard classification of occupations* (Leuven 2002).

Lieberman, Sima. *The Industrialization of Norway 1800–1920* (Oslo 1970).

Lindblad, Hans. 'Emigrationen—folkets projekt', in: Olov Isaksson (ed.), *Utvandrare och invandrare i Sveriges historia 1846–1996* (Stockholm 1997) 26–38.

Lindblad, Hans. 'Impulser som förändrade Sverige', in: Ingvar Henricsson and Hans Lindblad, *Tur och retur Amerika: utvandrare som förändrade Sverige* (Stockholm 1995) 101–272.

Lindqvist, Mats. *Herrar i näringslivet: om kapitalistisk kultur och mentalitet* (Stockholm 1996).

Lindqvist, Svante. 'Social and Cultural Factors in Technology Transfer', in: Kristine Bruland (ed.), *Technology Transfer and Scandinavian Industrialisation* (New York, NY 1991) 15–36.

Lindqvist, Svante. 'Vad är teknik?', in: Bosse Sundin and Boel Berner (eds.), *I teknikens backspegel: antologi i teknikhistoria* (Stockholm 1987) 11–33.

Lintsen, Harry. 'Two centuries of central water management in the Netherlands', *Technology and Culture: The International Quarterly of the Society for the History of Technology* 43:3 (2002) 549–568.

Ljungberg, Jonas and Schön, Lennart 'Domestic markets and international integration: paths to industrialisation in the Nordic countries', *Scandinavian Economic History Review* 61:2 (2013) 101–121.

Lodenius, Elias. *Tammerfors linne och jern-manufaktur aktie-bolag 1856–1906: historik öfver Tammerfors linnefabrik och bolagets öfriga värk* (S.l. 1908).

Lomas, Robert. *The Man Who Invented the Twentieth Century: Nikola Tesla, Forgotten Genius of Electricity* (London 2000).

Lovoll, Odd S. *A century of urban life: The Norwegians in Chicago before 1930* (Northfield, MN 1988).

Lundgren, Anders. 'Kunskap och resande. Några aspekter på den kemiska industrins tidiga utveckling i Sverige', in: Bengt Berglund, Per-Olof Grönberg, and Tomas Nilson (eds.), *Historiska perspektiv på teknikövérföring 1800–2000* (Göteborg 2006) 79–92.

Lundh Nilsson, Fay. *Lönande lärande: teknologisk förändring, yrkesskicklighet och lön i svensk verkstadsindustri omkring 1900* (Stockholm 2007).

Maddison, Angus. *The World Economy: Historical Statistics* (Paris 2003).

'Magnus Lavonius 60 år', *Svenska pressen*, 4 November 1930.

Magnússon, Magnús S. *Iceland in transition: Labour and socio-economic change before 1940* (Lund 1985).

Mäklin, Hugo. 'Om de nyaste försöken i Tyskland röramde elektrisk drift å banor med stor hastighet', *Tekniska föreningens i Finland förhandlingar* (1904) 132–137.

Malm, Gösta. *I min krafts dagar* (Stockholm 1963).

Manz, Stefan. 'Technologietransfer und spezialistenwanderung: eine Augsburger lagerbrauerei in Glasgow, 1889–1959', *Zeitschrift für Unternehmensgeschichte* 45:2 (2000) 225–247.

Marstrand, Vilhelm. 'Island. I anledning af Dansk Ingeniørforenings Islandstur Juni 1927', *Ingeniøren* 24 (1927) 300–301.

Martinsen, Per David. Erfaringen fra Amerika. Norske arkitekter i USA 1880–1930 – et glemt kapitell? Eksempler fra remigrantenes arbeider i Oslo (Oslo 2008). Master Thesis University of Oslo.

Matrikel över ingenjörer, utexaminerade från Högre tekniska läroverket i Norrköping, förutvarande Tekniska gymnasiet resp. Tekniska elementarskolan, från skolans början år 1857 t.o.m. vårterminen 1961 (Norrköping 1962).

'Mattsson, (Guss) - Uppslagsverket Finland', https://www.uppslagsverket.fi/sv/sok/view-103684-MattssonGuss, 8 October 2018.

McCreary, Eugene C. 'Social welfare and business: The Krupp welfare program, 1860–1914', *The Business History Review* 42:1 (1968) 24–49.

Michelsen, Karl-Erik. *Viides sääty: insinöörit suomalaisessa yhteiskunnassa* (Helsinki 1999).

Misa, Thomas J. *A Nation of Steel: The Making of Modern America 1865–1925* (Baltimore, MD 1995).

Mitchener, Kris James, and Marc D. Weidenmier. 'The Baring Crisis and the great Latin American meltdown of the 1890s', *The Journal of Economic History* 68:2 (2008) 462–500.

Moch, Leslie Page. *Moving Europeans: Migration in Western Europe Since 1650* (Bloomington, IN 2003).

Moen, Eli. 'Norway's Entry into the Age of Papers: The Development of the Pulp and Paper Industry in the Drammen District', in: Kristine Bruland (ed.), *Technology Transfer and Scandinavian Industrialisation* (New York, NY 1991) 363–386.

Mohn, H. E. 'Notiser fra tyske verfter', *Teknisk ugeblad* 39 (1905) 400–401.

Mohn, H. E. 'Notiser fra tyske verfter. Bremer Vulkan, skibsbygning og maskinfabrik, Vegesack ved Bremen, *Teknisk ugeblad* 51 (1905) 529–530.

Mølmen, Aslaug. Kvinnelige pionerer: en studie av 21 kvinnelige teknikere ved Trondhjem Tekniske Læreanstalt 1883–1915'(Trondheim 2008) Masters Thesis, Norwegian University of Science and Technology.

Myllyntaus, Timo. *Electrifying Finland: The Transfer of a New Technology into a Late Industrialising Economy* (London 1991).

Myllyntaus, Timo. *The gatecrashing apprentice: industrialising Finland as an adopter of new technology* (Helsinki 1990).

Myllyntaus, Timo. 'Discovering Switzerland. Internationalisation among Nordic Students of Technology prior to World War II', in: Ana Simões, Ana Carneiro, and Maria Paula Diogo (eds.), *Travels of Learning: A Geography of Science in Europe* (Dordrecht 2003) 299–328.

Myllyntaus, Timo. 'Foreign Models and National Styles in Teaching Technology in the Nordic Countries', in: Irina Gouzévitch, André Grelon, and Anousheh Karvar (eds.), *La Formation des ingénieurs en perspective. Modèles de référence et réseaux de médiation - XVIIIe-XXe siècles* (Rennes 2004) 141–152.

Myllyntaus, Timo. 'Technological Change in Finland', in: Jan Hult and Bengt Nyström (eds.), *Technology & Industry. A Nordc heritage* (Canton, MA 1992) 29–52.

Myllyntaus, Timo. ' "The best way to pick up a trade". Journeys abroad by Finnish technical students 1860–1940', *Icon (London)* 2 (1996) 138–163.

'N. I. A. F. i New York', *Teknisk ugeblad* 20 (1903) 208–209.

Nerheim, Gunnar. 'G Jebsen—utdypning (Norsk biografisk leksikon).', https://nbl.snl.no/G_Jebsen, 8 October 2018.

Nerheim, Gunnar. 'Patterns of Technological Development in Norway', in: Jan Hult and Bengt Nyström (eds.), *Technology & Industry. A Nordc heritage* (Canton, MA 1992) 53–72.

Nerheim, Gunnar. 'Tysklands rolle i norsk industrialisering frem til andre verdenskrig', in: Jarle Simensen, et. al. (eds.), *Tyskland - Norge: den lange historien* (Oslo 1999) 114–122.

Neyens, Mieke. 'The good, the bad and the rational: desirable and undesirable migration to Cuba and Mexico', in: Steinar A. Sæther (ed.), *Expectations unfulfilled: Norwegian migrants in Latin America, 1820–1940* (Leiden 2015) 102–126.

Nielsen, Henry, and Michael F. Wagner. 'Technology in Denmark', in: Jan Hult and Bengt Nyström (eds.), *Technology & Industry. A Nordc heritage* (Canton, MA 1992) 1–28.

Nielsen, Lauritz. 'Om den underjordiske Bybane i Buenos Aires samt om Forholdene for Ingeniører i Argentina', *Ingeniøren* 104 (1912) 807.

Nielsen, Valdemar. 'Irrigationsanlæget paa Plantagen La Grange. Brev fra St. Croix', *Ingeniøren* 32 (1909) 299–301.

Nikula, Riitta. 'Women in the history of Finnish art', in: Marja Manninen and Päivi Setälä (eds.), *The Lady With the Bow: The Story of Finnish Women* (Helsinki 1990) 80–90.

Nilson, Tomas. 'Vacker, föredömlig, rationell'—bilder av tysk teknik i Teknisk Tidskrift 1890–1914', in: Bengt Berglund, Per-Olof Grönberg, and Tomas Nilson (eds.), *Historiska perspektiv på tekniköverföring 1800–2000* (Göteborg 2006) 43–67.

Nilsson, Staffan. 'Brevborgen vid Vasagatan', *SFV kulturvärden* 3 (2005) 23–31.

Nilzén, Göran. 'Ragnar Sohlman', *Svenskt biografiskt lexikon. Bd 32, Sehlstedt-Sparre* (Stockholm 2003) 632.

Nipen, Kjersti. 'Kan få norsk «brain drain»', *Aftenposten,* 4 October 2012.

Nipperdey, Thomas. *Deutsche Geschichte: 1866–1918. Bd 1, Arbeitswelt und Bürgergeist* (München 1990).

Nolan, Mary. *Visions of Modernity: American Business and the Modernization of Germany* (New York, NY 1994).

Nordahl, Per. 'Lost and Found - A Place to Be: The Organization of Provincial Societies in Chicago from the 1890s to 1933', in: Daniel Lindmark (ed.), *Swedishness reconsidered: Three centuries of Swedish-American identities* (Umeå 1999) 65–89.

Nordlund Edvinsson, Therese. *Att leda storföretag: en studie av social kompetens och entreprenörskap i näringslivet med fokus på Axel Ax:son Johnson och J. Sigfrid Edström, 1900–1950* (Stockholm 2005).

Nordstrom, Byron J. 'Trasdockan: The Yearbook of the Swedish Engineers' Society of Chicago', in: Philip J. Anderson and Dag Blanck (eds.), *Swedish-American life in Chicago: Cultural and Urban Aspects of an Immigrant People, 1850–1930* (Uppsala 1991) 193–212.

Nørregaard, Georg. 'Ingeniørernes indsats. Danske ingeniører fra tekniske skoler og teknika', in: *Danske Ingeniører fra Teknika, Udgivet af Ingeniør-Sammenslutningen i Anledning af Foreningens 50 Aars Jubilæum* (København 1945) 13–223.

Nuorteva, Jussi. *Suomalaisten ulkomainen opinkäynti ennen Turun akatemian perustamista 1640* (Helsinki 1997).

Nykänen, Panu. *Kortteli sataman laidalla. Suomen Teknillinen Korkeakoulu 1908 – 1941* (Porvoo 2007).

Nykänen, Panu. *Käytännön ja teorian välissä: teknillisen opetuksen alku Suomessa* (Espoo 1998).

Nykänen, Panu. *Turning the wheel. The History of Helsinki University of Technology TKK* (Porvoo 2008).

Odlaug, Arne, 'Ingeniør Bernt Berger. In Memoriam', *Teknisk ugeblad* 5 (1919) 62.

Ohrt, Heinrich. 'Den store Jærnbane- og Gade-Drejebro over Oberhafen i Hamborg', *Ingeniøren* 48 (1904) 320–321.

Olsson, Lars O. *Engineers as System Builders: The Rise of Engineers to Executive Positions in Swedish Shipbuilding and The Industry's Emergence as a Large Technological System, 1890–1940* (Göteborg 1995).

Olsson, Lars O. *Technology Carriers: The Role of Engineers in The Expanding Swedish Shipbuilding System* (Göteborg 2000).

Olsson, Lars O. 'Amerikaemigrationen och återvändande svenska ingenjörer, 1890–1930', in: *Göteborgs-emigranten* 6 (Göteborg 1997) 229–243.

Olsson, Lars O. 'To See How Things Were Done in a Big Way: Swedish Naval Archtects in the United States, 1890–1915', *Technology and Culture* 39:3 (1998) 434–456.

Østrem, Nils Olav. *Norsk utvandringshistorie* (Oslo 2014).

'Överste Lauri Stark död', *Svenska pressen*, 24 October 1940.

Palèn, Paul. 'Skärstensmältning i flamugn'm *Teknisk tidskrift. Afdelningen för kemi och bergsvetenskap* 4 (1908) 70–72.

Parmer, Trine. 'How Industrial Technology First Came to Norway', in: Kristine Bruland (ed.), *Technology Transfer and Scandinavian Industrialisation* (New York, NY 1991) 37–57.

Pauer, Erich. 'Technologietransfer und industrielle Revolution in Japan, 1850–1920', *Technikgeschichte* 51:1 (1984) 34–54.

Pedersen, Ingrid. *Litt om tegne- og arkitektundervisning i Norge før høiskolens tid: bidrag til en historisk oversikt* (Trondheim 1935).

Persson, Curt. *Hjalmar Lundbohm – En studie av ledarskap inom LKAB 1898–1921* (Luleå 2015).

Persson, Magnus. *Coming Full Circle? Return Migration and the Dynamics of Social Mobility on the Bjäre Peninsula 1860–1930* (Lund 2007).

Pollard, Sidney, and Paul Robertson. *The British Shipbuilding Industry, 1870–1914* (Cambridge, MA 1979).

Pollock, Walter. *The Bolinder Book* (Stockholm 1930).

Pommerin, Reiner. *Geschichte der TU Dresden* (Köln 2003).

'Professor C. E. Holmberg 75 år', *Västra Nyland*, 8 June 1935.

Profiles - Pioneering Women Architects from Finland (Helsinki 1983).

R. F. R, 'Planmæssig Arbeidsordning. Frederik Winslow Taylor. Kortfattet referat av advokat Ludvig Meyers foredrag i Polyteknisk forening den 14de november 1916', *Teknisk ugeblad* 47 (1916) 527.

Rahbek, K. L. 'Queensland som Arbejdsfelt for unge danske Ingeniører', *Ingeniøren* 66 (1918) 459–460.

Reich, Leonard S. *The making of American industrial research: science and business at GE and Bell, 1876–1926* (Cambridge, MA 1985).

Reith, Reinhold 'Circulation of skilled labour in late Medieval and early Modern Europe', in: Stephen R. Epstein and Maarten Prak (eds), *Guilds, innovation and the European economy, 1400–1800* (Cambridge 2009) 114–142.

Reith, Reinhold. 'Einleitung: Elitenwanderung und Wissentransfer', in: Dittmar Dahlmann and Reinhold Reith (eds.), *Elitenwanderung und Wissenstransfer im 19. und 20. Jahrhundert* (Essen 2008) 7–14.

Reuss, Martin. 'Learning from the Dutch. Technology, management, and water resources development', *Technology and Culture: The International Quarterly of the Society for the History of Technology* 43:3 (2002) 465–472.

Reuter, Jonatan. 'Ernst Gustaf Hedman', in: Jonatan Reuter (ed.), *Finlandssvenska tekniker: biografiska anteckningar under medverkan av flere författare* (Helsingfors 1925) 31–52.

Reynarsson, Bjarni. 'The Planning of Reykjavik, Iceland: three ideological waves—a historical overview', *Planning Perspetives* 14:1 (1999) 49–67.

Ringrose, Daniel M. 'Nomadic Technicians and Migration in the Francophone World', *Proceedings of the Western Society for French History* 32 (2004) 308–327.

Ringström, Jan. *Wäija sulfatfabrik och pappersbruk: en tillbakablick på 80 års pappersstillverkning i Dynäs: AssiDomän* (Bjästa 1998).

Riskerar Sverige en kompetensdränering?: om utlandsarbete och rörlighet bland unga akademiker: en rapport från Pensionsforum, Industriförbundet och SACO (Stockholm 2000).

Rogers, Everett M. *Diffusion of Innovations* (New York, NY 1962).

Rosales, Synnøve Ones. 'Opportunities for the few and select: Norwegians in Guatemala (1900–1940)', in: Steinar A. Sæther (ed.), *Expectations unfulfilled: Norwegian migrants in Latin America, 1820–1940* (Leiden 2015) 127–161.

Rosenberg, Nathan. *Inside the black box: technology and economics* (Seoul 1982).

Rosenberg, Nathan. *Perspectives on technology* (New York, NY 1976).

Rosenberg, Nathan. 'Economic development and the transfer of technology: Some historical perspectives', *Technology and Culture* 11:4 (1970) 550–575.

Rudi, Inga Berntsen. 'Trondhjems Mekaniske Verksted', http://www.kildenett.no/artikler/2008/120360496407, 15 November 2017.

Runeby, Nils. 'Americanism, Taylorism and social integration: Action programmes for Swedish industry at the beginning of the twentieth century', *Scandinavian Journal of History* 3:1-4 (1978) 21–46.

Runeby, Nils. 'Tyskland som teknisk förebild', in: Bernd Henningsen and Birgitta Högvall (eds.), *Skandinavien och Tyskland: 1800–1914: möten och vänskapsband* (Stockholm 1997) 389–396.

Rydberg, Sven. *Svenska studieresor till England under frihetstiden* (Uppsala 1951).

Rydén, Göran. 'Bergsbruket, moderniteten och den första svenska industrispionen', in: Bengt Berglund, Per-Olof Grönberg, and Tomas Nilson (eds.), *Historiska perspektiv på tekniköverföring 1800–2000* (Göteborg 2006) 139–159.

Rygert, Göran T. *Svenska arkitekter i USA 1846–1930: forskningsprojekt 1993–1996* (Stockholm 1996).

Sabadasz, Joel. 'The Mon Valley—Discovering the genesis of the modern American steel industry', *Cultural Resource Management (CRM)* 16:3 (1992) 27–30.

Sæther, Steinar A. 'Introduction', in: Steinar A. Sæther (ed.), *Expectations unfulfilled: Norwegian migrants in Latin America, 1820–1940* (Leiden 2015) 1–16.

Sæther, Steinar A. 'Making sense of a minor migrant stream', in: Steinar A. Sæther (ed.), *Expectations unfulfilled: Norwegian migrants in Latin America, 1820–1940* (Leiden 2015) 17–56.

Saloutos, Theodore. *They Remember America. The Story of the Repatriated Greek-Americans.* (Berkeley, CA 1956).

Samuelsson, Kurt. *From great power to welfare state: 300 years of Swedish social development.* (London 1968).

Sandvik, Pål Thonstad. *Falconbridge nikkelverk 1910–1929–2004: et internasjonalt selskap i Norge* (Kristiansand 2004).

Sandvik, Pål Thonstad. *Mekanisk industri i en europeisk periferi: Fabrikken ved Nidelven 1843–76* (Oslo 1994).

Schrier, Arnold. *Ireland and the American Emigration 1850–1900* (Minneapolis, MN 1958).

Schybergson, Per. 'Large Enterprises in Small Countries: Aspects on the Industrialisation of Denmark, Finland and Sweden', in: Hans Kryger Larsen (ed.), *Convergence? Aspects on the Industrialisation of Denmark, Finland and Sweden 1870–1940* (Helsinki 2001) 97–155.

Screen, J. E. O. 'The Finnish Army 1881–1901: A national force in a Russian context', *The Slavic and East European Review* 70:3 (1992) 453–476.

Screen, J. E. O. 'The Finnish Cadet Corps, 1819–1903: A reflection of Finno-Russian relations and the language conflict in Finland', *The Slavic and East European Review* 81:2 (2003) 217–235.

Seely, Bruce Edsall. 'Historical patterns in the scholarship of technology transfer', *Comparative Technology Transfer and Society* 1:1 (2003) 7–48.

Sejersted, Francis. *Demokratisk kapitalisme* (Oslo 1993).

Sejersted, Francis. *Socialdemokratins tidsålder: Sverige och Norge under 1900-talet* (Nora 2005).

Sekne, Ivar. *Hundre års oppfinnsomhet. Historien om Multiconsult 1908–2008* (Oslo 2008).

'Selim A. Lindqvist, en bortglömd arkitekt', *Hufvudstadsbladet*, 20 May 1967.

Sellergren, G. 'Nyare verktygsmaskiner på Chicago-utställningen', *Teknisk tidskrift. Afdelningen för mekanik och elektroteknik* 9 (1893) 16–18.

Sellergren, G. 'Nyare verktygsmaskiner på Chicago-utställningen', *Teknisk tidskrift. Afdelningen för mekanik och elektroteknik* (1894) 29–32, 53–57.

Semmingsen, Ingrid. *Veien mot vest. 2, Utvandringen fra Norge 1865–1915* (Oslo 1950).

'Sextioåringar', *Hufvudstadsbladet*, 24 November 1955.

'Sjuttioåringar', *Hufvudstadsbladet*, 16 June 1956.

Simensen, Jarle, and Ole Kristian Grimnes (eds.), *Tyskland - Norge: den lange historien* (Oslo 1999).

Simões, Ana, Ana Carneiro, and Maria Paula Diogo. 'Travels of Learning. Introductory Remarks', in: Ana Simões, Ana Carneiro, and Maria Paula Diogo (eds.), *Travels of Learning: A Geography of Science in Europe* (Dordrecht 2003) 1–18.

Sisson, William. 'A revolution in steel: mass production in Pennsylvania, 1867–1901', *The Journal of the Society for Industrial Archeology* 18:1-2 (1992) 79–93.

'Sjøbussens skaper er død', *Stavanger Aftenblad,* 22 January 1980.

'Sjuttioåring', *Hufvudstadsbladet*, 19 May 1937.

Skard, Sigmund. *USA i norsk historie: 1000–1776–1976* (Oslo 1976).

Skeie, Jon. 'Fredrik Selmer—utdypning (Norsk biografisk leksikon).', https://nbl.snl.no/Fredrik_Selmer, 8 October 2018.

Skrift ved 50 års jubileet for ingeniørene fra K.T.S. 1894 (Oslo 1944).

Smith-Meyer, Trond. 'Per Kure—utdypning (Norsk biografisk leksikon)', http://snl.no/.nbl_biografi/Per_Kure/utdypning, 15 November 2017.

'Some Important Events in the History of the General Electric Company', *Schenectady Works News*, 6 April 1923.

Sonnemann, Rolf. *Geschichte der Technischen Universität Dresden: 1828–1978* (Berlin 1978).

Sörlin, Sverker. *De lärdas republik: om vetenskapens internationella tendenser* (Malmö 1994).

'Sporvejsingeniør til Sumatra', *Ingeniøren* 8 (1913) 60.

'Standertskjöld, Hugo - Uppslagsverket Finland', https://www.uppslagsverket.fi/sv/sok/view-103684-StandertskjoeldHugo, 8 October 2018.

Stang, Gudmund. 'A measure of relative development?: A note on the education and dispersion of Scandinavian engineers 1870–1930', in: Bjørn L. Basberg, Helge W. Nordvik, and Gudmund Stang (eds.), *I det lange løp. Essays i økonomisk historie tilegnet Fritz Hodne* (Bergen-Sandviken 1997) 87–100.

Stang, Gudmund. 'Ble det for mange ingeniører?', in: *Trondheim ingeniørhøgskole 1912–1987: festskrift til jubileumsfeiringen 31. oktober 1987* (Trondheim 1987)33–42.

Staudenmaier, John M. *Technology's Storytellers: Reweaving the Human Fabric.* (Cambridge, MA 1985).

Steinmetz, Charles Proteus. Excerpt from a letter to Eskil Berg, 1895, General Electric Historical Achives, Schenectady, NY, USA.

Stén, Johan. 'Elektroteknikern som blev ett varumärke. Om Axel Gottfrid Strömberg (1863–1938)', in: Henry M. Ericsson (ed.), *Finlandssvenska tekniker. Bd 4* (Helsingfors 2003) 44–61.

Storhaug, Hans. 'The 1910 Norway Census and return migration: The case of Nedstrand', *AEMI—Journal*, 2 (2004) 1–8.

Stråth, Bo. *Union och demokrati: de förenade rikena Sverige och Norge 1814–1905* (Nora 2005).

Strømstad, Poul. 'Artisan Travel and Technology Transfer to Denmark, 1750–1900', in: Kristine Bruland (ed.), *Technology Transfer and Scandinavian Industrialisation* (New York, NY 1991) 135–156.

Sundin, Bosse. *Den kupade handen: historien om människan och tekniken* (Stockholm 2006).

Sundin, Bosse. *Ingenjörsvetenskapens tidevarv: Ingenjörsvetenskapsakademin, Pappersmassekontoret, Metallografiska institutet och den teknologiska forskningen i början av 1900-talet* (Umeå 1981).

Sundt, Karl E. 'Af en stipendieindberetning fra tekniker Karl E. Sundt, vedrørende verktøimaskiner. 1897–98', *Teknisk ugeblad* 31 (1899) 347–350.

Surnevik, Per. 'Ingeniør Ludvig Thorsen i DSD ble mannen bak en epoke i lokalfarten på norskekysten', *Stavanger Aftenblad,* 31 July 1971.

Suul, Torgeir. *Arkitektur i 1000 år: en arkitekturguide for Trondheim* (Trondheim 1999).

Svartz Jos. and A. Norstrand, 'Westinghouse-kompaniets elektriska verkstäder i Pittsburgh', *Teknisk tidskrift. Afdelningen för mekanik och elektroteknik* 9 (1904) 151–156.

'Sven-Olof Ahlstedt', *Åbo underrättelser,* 21 March 1951.

'Svensk-amerikanska ingeniörers och arkitekters möte i Gefle', *Teknisk Tidskrift. Allmänna Afdelningen* (1901) 209.

Svensson, Thommy 'Japansk företagsledning och svenska bruk: en felande länk?', *Arkiv för studier i arbetarrörelsens historia* 33 (1986) 3–31.

T. R. H., 'Im memoriam. Ingenjören George A. Palmgren', *Hufvudstadsbladet,* 7 September 1927.

Tedebrand, Lars-Göran. *Västernorrland och Nordamerika 1875–1913: utvandring och återinvandring* (Stockholm 1972).

Tedebrand, Lars-Göran. 'Politik, väljare och valda efter 1888', in: Lars-Göran Tedebrand (ed.), *Sundsvalls Historia,* D.2 (Sundsvall 1997) 254–282.

Tedebrand, Lars-Göran. 'Remigration from America to Sweden', in: Lars-Göran Tedebrand, Roger Jacobsson, and Lars-Erik Edlund, *Historia och demografi: valda texter* (Umeå 1999) 1–27.

Teisch, Jessica B, 'Home is not so far away: Californian engineers in South Africa, 1868–1915.' *Australian Economic History Review* 45:2 (2005) 139–160.

Thörnberg, E. H. *Sverige i Amerika, Amerika i Sverige* (Stockholm 1938)

Thulesius, Olav. *The man who made the Monitor: A biography of John Ericsson, naval engineer* (Jefferson, NC 2007).

Thunholm, Lars-Erik. *Ivar Kreuger* (Stockholm 1995).

Torstendahl, Rolf. *Dispersion of engineers in a transitional society: Swedish technicians 1860–1940* (Uppsala 1975).

Trondheim ingeniørhøgskole 1912 - 1987: festskrift till jubileumsfeiringen 31. oktober 1987 (Trondheim 1987).

Tuomi, Timo. 'The call of the Mediterranean: Hilding Ekelund's great journeys of the 1920s', in: Timo Tuomi, Kristina Paatero, and Eija Rauske (eds.). *Hilding Ekelund: (1893–1984): arkkitehi=arkitekt=architect* (Helsinki 1997) 77–79.

Tvinnereim, Helga Stave. *Arkitektur i Ålesund 1904–1907: oppatbygginga av byen etter brannen 23. januar 1904* (Ålesund 1981).

Tvinnereim, Helga Stave. 'Sverre Pedersen—Utdypning (Norsk Biografisk Leksikon)—Store Norske Leksikon.' *Store Norske Leksikon.* https://nbl.snl.no/Sverre_Pedersen, 8 October 2018.

Tulkki, Pasi. *Valtion virka vai teollinen työ?: insinöörikoulutus sosiaalisena ilmiönä 1802–1939* (Turku 1996).

V. B. 'Krupps nye Jern- og Staalverk ved Rhinen: Friedrich Alfred Hütte', *Teknisk ugeblad* 14 (1908) 83–84.

V. M. 'Christiani & Nielsens 25 aars jubilæum 8/9 1904—8/9 1929', *Ingeniøren* 7 (1929) 77–78.

Venneslan, Christian. 'Electrification and industrialisation: An assessment of the industrial breakthrough in Norway', *Scandinavian Economic History Review* 57:2 (2009) 124–155.

Vestdal, Jón E., Stefán Bjarnason, and Steingrím Jónsson. *Verkfræðingatal: Æviágrip íslenzkra verkfræðinga og annarra félagsmanna verkfræðingafélags Íslands* (Reykjavík 1956).

Vi bygger vidare: byggnadsaktiebolaget Hallström & Nisses verksamhet under ett kvartssekel: en kort berättelse i ord och bild (Sundsvall 1955).

'Vid Karl Gustréns Grav Den 17.2.34.' *Åbo Underrättelser*, 28 February 1934.

Viklund, Roine. *Den första statsbaneelektrifieringen. Ett pionärprojekt i subarktisk miljö* (Gävle 2015).

Vinding, Povl. 'Carl Gimbel - Dansk Biografisk Leksikon', http://www.denstoredanske.dk/Dansk_Biografisk_Leksikon/Naturvidenskab_og_teknik/Ingeni%C3%B8r/Carl_Gimbel, 15 November 2017.

Virtanen, Keijo. *Settlement or return: Finnish emigrants (1860–1930) in the international overseas return migration movement* (Turku 1979).

W. H., 'Meddelanden från verldsutställningen i Filaldelfia. 33. William Sellers & Co's verkstäder och utställning af verktygsmaskiner', *Teknisk tidskrift.* (1876) 12–15, 36–37, 87–88, 159–160.

Waernberg, Jan. *Enoch Thulin - forskare, flygare, företagare* (Lund 2004).

Walaszek, Adam. 'Perserving or transforming role? Migrants and Polish territories in the era of mass migrations', in: Dirk Hoerder and Jörg Nagler (eds.), *People in transit: German migrations in comparative perspective, 1820–1930* (Cambridge 1995) 101–124.

Wale, Astrid. *Nyhet, nytte, framskritt: introduksjonen av lokale elektrisitetssystem 1877–1900: Trondheim i et nasjonalt og internasjonalt perspektiv* (Trondheim 2004).

Warren, Kenneth. *Big steel: The first century of the United States Steel Corporation, 1901–2001* (Pittsburgh, PA 2001).

Wegge, Simone A. 'Occupational self-selection of European emigrants: Evidence from nineteenth-century Hesse-Cassel', *European Review of Economic History* 6:3 (2002) 365–394.

Wengenroth, Ulrich. 'Germany: Competition abroad—cooperation at home, 1870–1990', in: Alfred D. Chandler, Franco Amatori, and Takashi Hikino (eds.), *Big business and the wealth of nations* (Cambridge 1997) 139–155.

Westman, C. Selim. 'Om verkstadsmetoder hos Ford Motor C:o', *Tekniska föreningens i Finland förhandlingar* 1 (1920) 37–42.

Widmark, Lawrence E. 'Engineers', in: Adolph B. Benson and Naboth Hedin (eds), *Swedes in America 1638–1938* (New Haven, CT 1938) 407–415.

Wiig, Jan. 'Gudbrand Gregersen Saági—utdypning (Norsk biografisk leksikon)—Store norske leksikon', *Store norske leksikon*, http://snl.no/.nbl_biografi/Gudbrand_Gregersen_Sa%c3%a1gi/utdypning, 15 November 2017.

Wise, George. *Willis R. Whitney, General Electric and the origins of U.S. industrial research* (New York, NY 1985).

Wohlert, Claus. 'The Introduction of the Bessermer process in Sweden', in: Kristine Bruland (ed.), *Technology transfer and Scandinavian industrialisation* (New York, NY 1991) 295–306.

Wuolle, Bernhard. *Den tekniska högskoleundervisningen i Finland 1849–1949* (Helsingfors 1949).

'Wuolle, Bernhard - Uppslagsverket Finland', http://www.uppslagsverket.fi/bin/view/Uppslagsverket/WuolleBernhard?template=highlightsearch&search=Wuolle%20%28tid.%20Wennerblom%29,%20Kustaa%20Bernhard%20%28f.%2020/4%201876%20Tfrs,%20d.%203/11%201962%20Hfrs%29,%20ingenj%F6r, 14 November 2017.

Wyman, Mark. *Round-trip to America: the immigrants return to Europe, 1880–1930* (Ithaca, NY 1993).

X. 'Patriotisme.' *Teknisk ugeblad*, 9 (1885) 43.

Zachariae, Hugo. 'Siams Jærnbanevæsen', *Ingeniøren*, 54 (1920) 28.

Index

Subjects

Architettura minore 209, 232, 307
Art Nouveau/Jugendstil 112, 123, 133, 135, 139, 149–150, 153–154, 200, 204, 210, 232, 234, 262, 267, 274, 302, 306, 307, 310, 314, 315, 317–319.

Baring crisis 117, 227, 236
Brain circulation 29
Brain drain 3, 13, 21, 29, 52
Brain gain 3, 21, 29

Development nationalism 238, 311

Female architects and engineers 59–60, 91–93, 199, 210, 224, 280, 281
Fordism 38, 41–42, 53, 82, 112, 122, 141, 143–144, 153, 157, 164–166, 171, 184–185, 297, 303, 305, 321.
Functionalism 189, 290, 319

Garden cities 26, 210, 234, 262–263, 285, 298, 306, 318

HISCLASS 69, 247

Industrial breakthrough 2

Labour-market mobility 8–9, 15, 36, 58, 102, 150, 243, 295, 300–303, 305, 307, 309–310

Mass production 57, 108, 125, 143–145, 147, 158, 166–169, 184–185, 258, 269.
Migration systems/mobility systems 16–18, 109, 124–126, 301–312.

National romanticism 115–116, 188–189
Neo-renaissance architecture 115, 133, 136, 153, 154, 290, 302, 317, 318
Nordic classicism 116, 209, 318
North-American manufacturing belt 158, 162–163, 184, 304
Norwegian Technical Society (Chicago) 181

Payroll migration (going abroad for domestic companies) 90–91, 100, 108–109, 126, 150–151, 154–155, 161–162, 189–190, 191–192, 194, 197, 201, 202, 203–204, 205–207, 211, 212, 214–215, 219, 221–223, 227, 228, 230, 231, 233–236, 301, 303, 306, 307
Placements abroad 8–9, 15, 22, 24, 36, 52, 106–107, 119, 121–122, 130, 162, 166, 184, 237, 250, 252–253, 259–260, 300–305, 307, 310, 313.

Rational workshop organisation 38, 41–42, 53, 56–58, 82, 112, 122, 141–144, 153, 157, 164–169, 171, 184–185, 189, 202, 268, 276–277, 279, 281, 292, 297, 303, 305, 321–322

Study trip/study travel 8–10, 15, 22, 36, 49–50, 55–56, 68–69, 71, 82, 84–86, 92, 93, 97, 99–102, 115–116, 119–123, 124–126, 128–129, 130, 142, 152, 153, 157, 160, 177, 183, 188, 190, 195–196, 200, 202, 203, 208–209, 211, 223–224, 226, 232, 234, 236, 237, 239–240, 247–248, 249, 250, 252, 253, 256, 259, 263, 267, 274, 276, 278, 280, 291, 293, 295, 296, 300–303, 306, 307–310, 312–313, 315, 316, 324
Swedish Engineers' Society of Chicago 178, 180–181.

Taylorism 38, 41–42, 53, 56, 58, 82, 112, 122, 141, 143–144, 153, 157, 164–167, 171, 184–185, 189, 202, 268, 276–277, 279, 281, 292, 297, 303, 305, 321–322

Welfarism, welfare capitalism 58, 141, 143–144, 145, 154, 157, 166, 171–172, 185, 253, 272, 303, 305.

Definitions

Architect 6

Denmark 11
Denmark's dependencies 12

Engineer 5–6

Finland 11
Finnish, Finland-Swedish 11

Migration 9

Nordic area 10
Norway 11–12

Placements abroad 8–9

Return migration 10

Scandinavia 11
Study trip/study travel 8–9
Sweden 11–12

Technology 7
Technology transfer 7
Traditional emigration 9
Transnational mobility 7–8

Companies and schools

A. J. Borsig 145, 314
Aarhus Oil Factory 219, 280
AGA 139, 197, 206, 228, 229, 268
Allgemeine Elektricitäts-Gesellschaft (AEG) 141, 143, 205, 233, 269, 314
Allis-Chalmers Mfg. Co 112, 167, 173, 184, 315
Anaconda Mining Co. 176, 229

Badische Anilin- und Soda- Fabrik (BASF) 147, 303
Bell Telephone Co. 202, 230, 233, 277
Bofors AB 219, 270
Boliden Mining Co. 170, 172, 176, 210, 271
Bolinders mekaniska verkstad 267, 321
Bruce, Peebles & Co. 112–113, 139, 142, 212, 234
Burmeister & Wain 69, 90, 222, 277

Carnegie Steel Co. 144, 158, 167, 169, 172, 184, 200, 305, 321
Crichton- Vulcan 215, 293

Danalith 197, 206, 223
Danish East Asiatic Co. 217–218, 219
Det Norske Veritas 90, 189, 197, 212
Dunderland Iron Ore Co. 170, 286, 321

Eidgenössische Technische Hochschule 14, 25, 38, 45, 46, 51, 89, 105, 113, 127, 128, 129, 131, 137–139, 152, 153, 257, 263, 267, 271, 276, 286, 302–303, 314
Electrolux 201, 206

Finnish National Railways 90, 215, 235, 322

General Electric Co. 1, 54, 112, 158, 162, 164, 170, 172–175, 179, 184, 264, 270, 272–273, 279, 305, 322
General Motors 201, 233
Götaverken 69, 172, 272
Guggenheim Exploration Co. 114, 123, 229, 236, 310

Illinois Steel Co. 175, 180

Jernbeton 286, 319
Julian Kennedy, Sahlin & Co. 201, 233

Kreuger & Toll 177, 191, 191n15, 201

L. M. Ericsson 197, 202, 204, 205, 206, 214, 219, 230, 233, 236, 268
Ludwig Loewe 143–144, 278
Luossavaara-Kiirunavaara AB (LKAB) 170, 271
Luth & Roséns Elektriska AB 198, 201, 267

Massachusettes Institute of Technology 53, 135, 165–166, 174n91, 184, 269, 304, 321

N. C. Monberg 151, 154, 189–190, 194, 277
Nielsen & Winther 90, 204, 205, 219, 277
Norsk Hydro 133, 145n108, 174n89, 197, 201, 206, 233, 283, 320, 322

Northern Pacific Railway Company 51, 182

Otis Elevator Co. 178n113, 182

Palmer's Shipbuilding Co. 156, 213

Rosenbergs mekaniske verksted 168, 284

Saabye & Lerche 206, 277
Siam Cement Co. 217–218
SKF 161–162, 197, 204, 206, 214, 228, 271–272, 303
Skånska Cementgjuteriet 135, 191, 214
Smith, Mygind & Hüttemeier 142n89, 277
Stavanger Steamship Co. 172, 284
Swedish Match Co. 204, 229

Tampere Linen and Iron Manufacturer 140, 287, 290
Technische-Hochschule Berlin-Charlottenburg 1, 48, 115, 122, 132–135, 139, 149, 153, 193, 211, 257, 274, 281, 283, 285, 289, 303, 314, 320
Technische Hochschule Darmstadt 113, 139, 153, 251, 303, 314
Technische Hochschule Dresden 49, 115, 135–136, 139, 149, 153, 257, 262, 276, 290, 303, 320
Technische Hochschule Hannover 139, 153, 283, 285, 303
Technische Hochschule Karlsruhe 139, 153, 277
Thomas B. Thrige 168, 279

Vulkan shipyard 142, 142n91

Westinghouse Electric Co. 112, 158, 162, 164, 173–175, 179, 184–185, 212, 257, 264, 305, 315, 322

Geographical terms

Aachen 139, 153, 201, 278, 284
Aalborg 60, 197m48, 276, 280–281
Äänekoski 289
Aarhus 1, 60, 64, 96, 96n59, 242, 280
Aberdeen 213
Adirondack 169–170, 297
Åland Islands 10, 11n25, 293
Ålesund 149, 210
Algeria 115, 223–224, 235, 310
Amsterdam 202–203, 307
Anaconda, MT 176, 229
Anholt 280
Ankara 206
Antwerp 50, 201–202, 233, 277, 307
Aosta 207
Arapuni 221
Archangel 215, 235, 287
Assens 278
Asuncion 228
Athens 37, 208
Auckland 221
Avesta 140, 270, 320

Baghdad 149
Baku 50, 214, 234, 306
Balen 201
Baltimore, MD 162
Bangkok 217–218, 220, 235
Barcelona 206, 207
Batavia (Djakarta) 220–221
Bavaria 116, 123, 140, 277, 303, 314
Beijing 219, 235
Belfast 113
Belleville, NJ 272
Bellingham, WA 183
Bengtsfors 168n48
Bergen 60, 64, 74, 78, 96, 101, 112, 148, 154, 163, 166, 183n155, 200, 213, 242, 281, 284–285, 299, 319, 321
Bergvik 168n48
Berlin 1, 49, 50, 113, 122, 123, 135, 138, 141, 143, 144, 145–146, 147, 149, 151, 153, 208, 263, 269, 274, 276, 278, 283, 285, 288, 303, 309, 322
Bielitz (Bielso-Biala) 151
Bilbao 207
Boden 270
Bodø 286
Boliden 210
Bologna 3, 26, 209
Bordeaux 199
Borneo 221
Bornholm 278, 299, 318
Borås 64, 80, 192
Boston, MA 112, 156, 158, 162, 163, 167, 178n114, 184, 305, 315

Breda 203
Bremen 50, 277
Bremerhaven 142, 154, 314
Breslau (Wroclaw) 179
Brisbane 221
Bristol 274
Brussels 197, 200, 201, 233, 307
Bucharest 205
Budapest 49, 204, 232, 233
Buenos Aires 114, 227, 290, 307
Buffalo, NY 162
Burcht 201
Burma 103
Bury 211

Caen 199
Camden, NJ 272
Canary Islands 228
Cape Town 224
Castile 17
Celebes 221
Cherbourg 197, 198, 199, 267
Chicago, IL 1, 8–9, 25, 42, 56, 57, 104, 108, 112, 134, 162–164, 171, 175, 176, 177–181, 184, 186, 263, 267, 270, 284, 288, 298, 304, 305, 314
Chuquicamata 114, 229, 236, 310
Cleveland, OH 162
Comodoro Rivadavia 227
Congo 114, 123, 224, 310
Constantinople (Istanbul) 205–206, 208, 233
Copenhagen 1, 6, 29, 31, 50, 61, 63–64, 66, 71, 77, 96, 97, 111, 115–116, 135, 140, 142, 142n89, 143n91, 145n107, 146, 151, 163, 174n89, 177n106, 188–189, 193, 195, 197n48, 202, 203, 217–218, 242, 244, 274–277, 290, 294, 299, 311, 318, 319, 320, 321, 322–323
Costa Rica 103, 225, 226, 291
Cuba 225, 226, 230

Dalnij 216–217
Detroit, MI 42, 162, 165n29, 177, 186, 202
Dettifoss 194
Deventer 203
Diarbekir 206
Djursholm 271
Dominican Republic 103, 231
Drammen 32, 61, 192, 283, 299, 320, 323
Dresden 49, 115, 135–136, 139, 149, 153, 257, 262, 276, 290, 303, 320
Duisburg 144
Dundee 210
Dunkirk 199
Düsseldorf 197, 277

Edinburgh 51, 113, 136, 142, 212, 213, 234
Egypt 115, 122, 223, 235, 310
Elsinore 135, 277
Eregli 206
Esbjerg 279
Eskilstuna 265, 269
Essen 144
Ethiopia 114, 224

Falkenberg 273
Fall River, MA 168
Falun 74, 145, 153, 270
Faroe Islands 10, 12
Fevzipaca 206
Fife 213, 234
Fiji 103, 221
Filipstad 74, 270
Firenze 209
Fiume (Rijeka) 205
Folldal 285
Fort Wayne, IN 181
Frankfurt-am-Main 1, 50, 116, 141, 147–148, 154, 303, 314
Frederikshavn 281, 321
Freiberg 113, 139, 153

Galapagos Islands 225
Gateshead 213
Gävle 162, 174n89, 270, 319
Geneva 200
Genoa 206, 209
Ghent 201
Glasgow 89, 113, 212–213, 234, 264, 277, 278, 285, 306, 315
Glencoe 225
Globe, AZ 176
Glomfjord 136, 286, 320
Gothenburg 51, 59, 60, 64, 74, 79–80, 88, 89, 96, 97, 101, 112, 129, 138, 140, 148, 154, 156, 162, 163, 172, 179, 190–191, 222, 235, 242, 269, 271–272, 281, 289, 291, 296, 319, 320, 321, 322
Göttingen 198n52

Grenoble 199
Grimsby 212, 286
Grängesberg 170, 170n61, 270, 321
Guam 103
Guatemala 226
Gudenå 280

Hague (the) 203, 220
Haiti 103, 231
Halmstad 148n126, 273
Hamar 285
Hamburg 1, 49, 50, 91, 141, 142, 149, 151, 154, 303, 314
Hämeenlinna 135–136, 290
Hamilton, ON 278
Hanko 292
Hankow 219
Harlu 289
Härnösand 64, 80, 178n114
Harrison, NY 175
Harstad 212, 286
Hartford, CT 167
Havana 225, 231
Hayange 199–200
Helsingborg 273–274, 320
Helsinki 1, 28, 49–50, 59, 61, 63, 66, 67, 71, 78n223, 89, 93, 130, 131–132, 133–134, 138, 145n107, 146, 148n126, 149, 150, 150n137, 153–154, 174, 177, 185, 188, 189, 191, 209, 256, 259, 262, 270, 291–292, 301, 308, 318, 319, 320, 321, 323
Hengelo 203
Hesse-Cassel 110, 159
Hissmofors 135, 271
Hobro 280, 320
Höchst-am-Main 141, 147, 154, 303
Hofors 270
Höganäs 273
Holbæk 277
Holeby 278
Holstebro 279
Holyoke, MA 169, 260
Honduras 103, 225
Honfleur 199
Hong Kong 219, 235
Horsens 66, 77, 90, 96, 242, 279, 321
Horten 74, 192n18, 321
Hull 156
Huskvarna 273
Hyvinkää 291, 320

Innsbruck 208
Inverurie 213
Ipswich 211
Isle of Wight 212
Ivigtut 195, 321
Izjevsk 216

Jaen 206
Jakobstad 289
Jamshedpur 223
Jarrow-upon-Tyne 156, 213
Jatibonico 230–231
Jena 133
Johannesburg 225, 236
Jönköping 273
Joutseno 289
Julianehåb 195
Jyväskylä 174n89, 289

Kalmar 156
Karlskoga 270
Karlskrona 143n91, 198, 273, 321
Karlstad 271
Kenya 224
Kiel 314
Kimberley 114, 225, 236
Kirkenes 286–287
Kiruna 270–271, 321, 323
Kobe 222
Køge 277, 321
Kokkola 289
Kolding 279, 322
Kolsva 270
Korea 103
Korsør 278
Koskis 169, 292
Kotka 148n126
Kramfors 168n48
Kristiania (Oslo) 32, 61, 64, 67, 71, 74, 93, 96, 104, 132–133, 140n83, 142, 142n89, 142n91, 145n107, 145n108, 146, 148n122, 154, 163, 165, 174–175, 175n92, 177, 188, 191, 192, 197, 199, 202, 203, 242, 281–283, 299, 314–315, 318, 319, 320, 321, 322, 323
Kristiansand 53, 113, 176, 185, 284, 320–321
Kristinehamn 173

Kuala Lumpur 220
Kuopio 289

La Franca 228
Lahti 134, 290
Lake Ladoga 114, 192, 289, 299
Lake Peipus 17
Lancaster 278
Landskrona 190, 273, 278
Lappeenranta 216, 289, 320
Le Creusot 199
Le Havre 199
Leiden 203
Leipzig 50, 140, 276
Leoben 113, 128, 130, 264
Liechtenstein 128
Liege 30, 39–40, 89, 113, 200–201, 233, 285, 307, 322
Lillehammer 285
Lillestrøm 199
Lima 229
Limhamn 274
Linköping 271
Lisbon 206
Liverpool 50, 156, 210, 211, 212
Lodz 204
Lojo 292
Løkken Verk 113–114, 285
London 1, 49, 50, 91, 93, 109, 196, 210, 211, 222, 225, 230, 234, 276, 277, 298, 306, 307, 308, 322
Loviisa 134, 174n89
Lowell, MA 168, 260
Ludvika 269
Ludwigshafen 147, 154, 303
Luleå 174n89, 270
Lund 135, 274
Luxembourg 199–200, 201
Lübeck 151

Madagascar 103, 114, 123, 170n61, 224, 236, 310
Madrid 17, 206
Makassar 221
Malaga 207
Malmberget 270, 321
Malmö 60, 76, 80, 96, 101, 111, 112, 135, 140n83, 242, 274–275, 319, 320, 321
Manchester 212
Mannheim 148, 151, 279
Mariager 280
Mariehamn 293
Marseille 197, 199
Melbourne 222
Meråker 285
Mexico City 230
Michigan 163n22, 181, 264, 297
Middelfart 165n30
Middlesbrough 113, 213
Mikkeli 290
Milan 206, 209
Milwaukee, WI 112, 162, 167, 173, 181, 184
Minneapolis, MN 163, 182
Minnesota 52, 108, 163, 163n22, 170, 182, 184, 264, 297
Mittweida 55, 140, 152, 192n18, 193
Molde 281
Monaco 199
Mongolia 103
Montevideo 228
Morocco 16, 92, 115, 122, 207, 223–224, 310
Moscow 17, 215, 235, 268
Mozambique 103, 224
Munich 28, 29, 113, 115, 139, 149, 153, 208, 262, 267, 303, 314
Mülhausen (Mulhause) 289
Mysore 223

Næstved 278, 279, 319, 320
Namibia (Southwest Africa) 103, 224
Namsos 286
Nancy 199
Naples 206, 209
Narni 207
Narvik 114, 270, 286
Nenthead 211
New Bedford, MA 168
New Caledonia 103
New Jersey 1, 88
New York, NY 1, 42, 92, 112, 116, 136, 158, 161, 162, 163, 163n22, 169n51, 177, 178n113, 182–183, 225
Newcastle-upon-Tyne 113, 156, 212–213, 264, 306, 315
Newport News, VA 171
Niagara Falls, NY/ON 42, 158, 160, 162, 272–273, 304
Niger 103, 224
Nigeria 103
Nizhny Novgorod 215

Norrköping 60, 61, 64, 80, 269, 320
Notodden 283, 320
Novaja Semlja 287
Nykøbing Falster 93, 278, 322
Ny-Ålesund 286

Odda 175n92, 284
Odense 60, 64, 65–66, 77, 96, 96n58, 140, 168, 242, 275–276, 279, 321
Odessa 17
Öland 156, 273
Örebro 168n48, 188–189, 270
Orijärvi 114, 192
Orkdal 285
Örnsköldsvik 271
Osaka 222
Östersund 179, 271, 320
Oulu 147, 154, 288, 319
Outukumpu 174n89

Panama 226, 236
Pargas 292
Paris 3, 9, 17, 18, 26, 28, 38–39, 49, 51, 55, 89, 91, 115, 137, 196–199, 211, 225, 232, 233, 234, 274, 277, 307, 315, 319, 322
Patagonia 227, 228
Perth 222
Petsamo 11, 71n204, 287
Philadelphia, PA 32, 95, 112, 145, 158, 162, 167, 184, 268, 274, 304, 305, 314–315
Pietermaritzburg 225
Pireus 208
Pisa 209
Pitkäranta 114, 192
Pittsburgh, PA 42, 53, 95, 112, 122, 158, 162, 168, 169, 174, 184–185, 202, 269, 276, 278, 305, 310, 315, 321
Plovdiv 205
Pompeii 208–209, 223
Pori 1, 142, 292
Porjus 173–174, 270, 319
Porsgrunn 213, 284
Porvoo 291
Præstø 278
Prague 3, 27, 51, 89, 131, 204, 232

Quebec 160, 166n34, 168n48

Rancagua 229
Renfrew 213
Reutlingen 140, 303
Reykjavik 193–194, 318, 320
Riga 203
Rio de Janeiro 228
Rjukan 133, 174n89, 175, 192, 283, 319, 320
Rome 7, 28, 195, 207, 208, 209
Rønne 278, 318
Rönnskär, Skelleftehamn 176
Røros 285
Roskilde 277, 320
Rostock 151, 154
Rugby 212

Saint Petersburg 1, 2, 17, 50, 61, 84, 86, 92, 102, 125, 145n106, 213–214, 215, 216, 234–235, 268, 292, 301, 302, 306, 322, 324
Sakhalin 216, 235, 322
Sakskøbing 278
Salo 148n126
San Francisco, CA 121, 219, 271
San Remo 206, 233, 270, cover illustration
Sandviken 95, 169, 176, 185, 321
Sao Paolo 228
Särkisalo 292
Sarpsborg 132, 212
Sassnitz 127, 154, 190
Sauda 284
Savonlinna 290
Schenectady, NY 112, 158, 162, 170–171, 173–174, 179–180, 273, 279
Senegal 103, 224
Seraing 201
Shanghai 219, 235, 308
Shawinigan, QC 160, 270, 304, 319
Sheffield 50, 211, 234, 306
Sicily 209
Siena 3, 209
Silesia 113, 123, 141, 264, 303
Silkeborg 148, 279, 320
Singapore 217, 220
Skellefteå 174n89, 176, 270, 321
Skien 74, 146, 284
Skövde 271
Slagelse 278
Smolensk 17
Smyrna 208
Söderhamn 271
Södertälje 268–269
Sokia 208

Solomon Islands 103
Sortavala 289
Soulom 197
South Charleston, WV 175
South Georgia 103, 228, 280
Southampton 31
St. Croix 12, 231, 278
St. John 12, 231
St. Louis, MO 9, 162, 304
St. Paul, MN 182
St. Thomas 12, 231
Stavanger 60, 96, 113, 168, 172, 207, 277, 284, 299, 321, 322
Steinkjer 286, 319
Stenlille 277
Stettin 50, 142, 154, 208, 314
Stockholm 1, 12, 16, 26, 28, 50, 64, 67, 71, 93, 96n59, 135n39, 139, 140, 140n83, 145, 145n107, 149, 154, 175, 177, 178, 178n113, 178n114, 188, 190, 198, 206, 230, 242, 262, 267–268, 278, 282, 287, 289, 292, 294, 311, 318, 319, 320, 321, 322, 323
Stord 284
Strasbourg 199
Struer 279
Strömstad 271
Sudan 224
Suez Canal 218
Sulitjelma 114, 286
Sumatra 221
Sunderland 113, 212–213, 273
Sundsvall 135, 177n106, 271, 319
Svalbard 192, 286, 299
Svelgfoss 136
Svendborg 279
Sydney 221, 222
Syracuse, NY 177
Szeged 204–205

Tallinn (Reval) 203, 287
Tampere 1, 11n24, 60, 61, 78n223, 135–136, 140, 168n47, 169n51, 191, 287–288, 290, 291, 299, 318, 319, 320, 321, 323, 324
Tervakoski 290, 320
Tientsin 219
Timgad 223, 235
Timor 221
Tokyo 222
Torino 207, 322
Tornio 288
Toronto, ON 158
Trail, BC 183
Trelleborg 127, 154–155, 189–190, 273
Trento 208
Trinidad and Tobago 231
Trollhättan 272–273, 319
Tromsø 286
Trondheim 1, 60, 61, 64, 67, 87, 96, 101, 133, 142n91, 148n122, 153, 163, 188, 189, 192, 193, 198, 212, 224, 228, 242, 250, 259, 285–286, 287, 294, 311, 319, 321, 322
Tula 216
Tunisia 115, 122, 223–224, 310
Turda 205
Turkey 1, 68, 198, 205–206, 223, 324
Turku 11n25, 60, 96, 150, 150n137, 154, 177, 190, 191, 292–293
Tynset 285

Umeå 174n89, 178n114, 270
Uppsala 174n89, 178n114, 280
Utrecht 203
Uusikapunkki 191, 292

Vaasa 174n89, 177, 191, 289, 319, 322
Varberg 273
Vargon 168n48
Varkaus 289
Västerås 143, 153, 189, 265, 269, 271, 323
Vejle 275, 279, 321
Venice 209, 267, 319
Vera Cruz 225
Verona 209
Verviers 201
Vienna 17, 49, 51, 89, 115, 122, 128, 131, 150, 203, 204, 314
Viipuri 60, 96, 192, 230, 289–290, 299, 319
Visby 271
Vladivostok 216–217, 235, 322

Warsaw 204, 268
Weihenstephan 140
Wellington 221, 235
Westman Islands 194
Wintherthur 145, 153
Wisconsin 163n22, 166

Yaroslavl 215, 234
Yellow River 219

Zurich 14, 25, 38, 45–46, 49, 51, 81, 89, 105, 113, 127, 128, 129, 131–132, 137–139, 152, 153, 257, 263, 267, 271, 276, 286, 302–303, 314

Persons

Aalto, Alvar 190
Adler-Nissen, Christian 278
Aga, Ole Waldemar 219
Ahlberg, Hakon 116
Ahlrot, Georg 274
Ahlstedt, Sven-Olof 142, 291
Airo, A. E. 148n126
Alexanderson, Ernst F. W. 57, 157, 173, 179, 185, 306
Allesen-Vernø, Cay 279
Alsaker, Alfred 181
Ambjørn, Gustaf 269
Andersen, Folmer 231, 236
Andersen, Hans Niels 217–218
Andersin, Johannes 169, 290, 320
Andersin, Olle 223
Andersson, Bror 170, 270
Angelo, Aage Rørbye 138, 276, 318
Appelberg, Axel 170, 175
Arneberg, Olaf 203
Aschan, Ossian 134, 291
Asplund, Gunnar 116, 188, 267
Åstrand, Edward 148
Åström, Bruno 181
Åström, Carl S. 145n107
Åström, Karl S. 291–292
Aus, Gunvald 183

Bache-Wiig, Jens 52
Bachke, Fritz Morten Anker 189
Bagge, P. F. 205
Barfoed, Gunner Falck 278
Behrens, Peter 143
Berg, Ernst 179
Berg, Eskil 179
Berger, Bernt 183
Bergh, Axel 140n83
Bergius, Eino 199
Bergman, Rudolf 174n89
Berlage, Hendrik Petrus 202, 307
Berle, Kort 183
Beskow, Jakob 273–274
Bie, Jacob Anker 280
Birk, Louis 182
Birkeland, Richard 198
Birn, Valdemar 279
Björkman, Axel 176–177
Björnbom, Thor 174n89
Blache, Hans 277
Blom, Aage Gram 202
Blume, Gustaf 168n46
Boberg, Ferdinand 177–178, 319
Bodin, Victor 178n113
Boeck-Hansen, Christian 280
Bolinder, Erik August 267–268
Borch,Martin 189
Borg, Elsi 92–93, 224
Borgstedt, Martin 116, 122, 160, 186, 263, 298, 310
Brag, Albin 139, 267, 319
Bragstad, Ole S. 52
Bratager, Stork Johan 182
Brathole, A. A. O. 142n89
Bro, Viggo 142n89
Brun, Leif 202
Bruun, Sigurd 177n206
Brænne, Gunnar 148n122
Buur, Victor 191

Candelin, Georg 290
Carlsen, Carl-Erik 165n30
Carlsen, Peter 151
Carlson, Birger 148
Carlson, Fredrik 135
Carlsson, Hugo 269–270
Carlsund, Anton 214
Christiani, Rudolf 149, 197, 263, 277
Clausen, Christian 227
Coolidge, William D. 171
Curie, Marie 199

Dahlander, Magnus 188–189
Dahlander, Robert 267
Dahlbäck, Gunnar 173–174
Dalen, Gustaf 139, 153, 268
Danielson, Ernst 173n86
de Laval, Gustaf 268
Dekke, Christian Stoltz 213, 285
Dorenfelt, Margot 199
Dreyer, Jørgen 148, 279

Eck, Carl 272
Eckbo, Olaf 212
Eckermann, August von 198
Edström, J. Sigfrid 57, 138–139, 141, 142, 143, 146, 164, 168, 173, 174, 184–185, 264, 269, 271, 272, 302, 308, 321.
Ekelund, Hilding 93, 188, 207–210, 223, 232, 235, 262, 307, 319
Emanuelsson, Gösta 224
Engberg, Gunnar 272n59
Engberg, Jens F. 166–168, 171–172, 185, 279, 305, 321.
Engel, Walter 165n30
Engels, Hubert 136
Englesson, Elov 173
Enwald, Bertha 92
Ericson, John (Chicago's city engineer) 180
Ericson, John (USS Monitors' constructor) 57, 180
Eriksson, Karl-Erik 171, 269
Esselius, Eric 169, 270, 321
Estlander, Jakob 214
Ewe, Nils August 274, 319
Eyde, Sam 74, 132–133, 251, 281, 296, 323

Fæster, Knud 195
Fagerberg, Georg 170, 271
Faith-Ell, Oscar 192
Falch, Søren Johan 283
Faleide, Isak Rasmussen 181
Falkenberg, Andreas 285
Falkman, Oskar 171, 271, 323
Fiehn, Berrit 93
Finne, Henrik 284
Fischer, Albrecht 195
Fisk, Gustaf 223
Fleron, Emil 151
Fogelberg, Ivar 148n126
Folin, Thorild 168n48
Forchhammer, Herluf 1, 4, 151, 280, 324
Ford, Henry 141, 164–165, 251
Forsberg, Erik August 268, 322
Forsberg, Uno 161, 271
Forsman, Fritz 174n89
Fougner, Nicolay 53
Fredriksson, Frans 135, 274
Fredriksson, Robert 174n89
Frick, Otto 194
Friis, Johannes 278
Fritsch, Alvar 148n126

Gadelius, Knut 222, 235
Ganz, Abraham 204, 233
Gaudi, Antoni 207, 232, 306, 310
Gerlow, Poul 145–146, 276
Gill, Sonja 93
Gimbel, Carl 219
Gleditsch, Ellen 199
Goursat, Èdouard 198
Granholm, Johannes Christensen 135
Gregersen, Gudbrand 204–205, 233
Grill, Anthony 148n126
Grut, Torben 267
Grønningsæther, Anton Martin 53, 176, 284, 304, 321
Guldahl, Axel 133, 286
Gustrén, Karl 292
Gylphe, John M. 292

Haavardsholm, Nils 284
Hagemann, Gunnar 231
Hahl, Bernhard 289
Hahr, Erik 143
Håkansson, Harald 173n86
Hållden, Arvid 168n48
Hallgren, Johan 274
Hallström, Per 177n206
Hammar, Hugo 156, 163, 172, 184, 213, 272, 305
Hänschell, Theodora 280
Hansen, Harald 199
Hansen, Niels Christian 278
Hauan, Haakon 1, 4, 133, 324
Haussmann, Georges Eugène 196
Haustrup, N. J. 55
Hedèn, Ernst A. 272
Hedman, Ernst Gustaf 215
Heggstad, Olav 136, 285
Heikinheimo, Mikko 146, 291
Heje, Kolbjørn 138, 285
Hellman, Oscar 141n84
Helweg-Larsen, Harald 276
Hennebique, Francois 197, 232
Henriksen, Einar Skjølberg 174n89
Herfeldt, Trygve Olsen 201
Heroult, Paul 207
Hertzsprung, Ejnar 203

Hesser, Helmer 214
Hietanen, Toivo 148n126
Hilbert, David 198n52
Hindrum, Hans 228
Hjelmman, Leonard 134, 291
Hjelt, Paul 198
Hjort, Thomas 175n92
Hlìðdal, Guðmundur J. 193, 320
Holm, Georg 169
Holm, Ove 174n89
Holmberg, Emil 134, 291
Holmsen, Holm 212, 286
Holst, Malthe Conrad 280
Holst-Grubbe, Gudrun 92–93
Holtan, Eugen 148, 199
Holwech, Randi 91–92
Hornborg, Signe 59
Horta, Victor 200
Houm, Ole 200–201
Hubendick, Edward 267
Hybinette, Victor 176, 284

Ignatius, Pentti 168n47
Illstrup, Carl 182
Isleifson, Jon Hallson 193

Jahn, Carl 216–217, 235
Jebsen, Gustav 138
Jensen, Harald 140, 291
Jensen, Søren Valeur 161
Jenssen, Kolbjørn Thorvaldsen 221
Jentoft, Harald 286–287
Jentoft, Johan 286–287
Jentoft, Ole 287
Johansson, John 169n51
Jonassen, Even 142n89
Jørgensen, Alfred 189
Jørgensen, Johanne Wille 281
Julin, Albert von 169, 185, 292
Jung, Bertel 150, 291, 292

Kaarbø, Reidar Darre 212, 286
Kahn, Julius 177, 186, 201
Kahrs, Wilhelm 147–148, 285
Kampmann, Per 206
Karén, Uuno Aleksanteri 288
Karhi, Iivari 147, 288
Kärne, Göran 174n89
Kempe, Tor 149
Kerstens, Thorvald 278
Kielland, Christofer Kahrs 133, 283, 320
Kielland, Jacob de Rytter 145n107
Kiljander, Elna 93
Kinck, Johan 136
Kirk, Niels Pedersen 194
Kittler, Erasmus 139
Kjems, Frode 276
Klein, Felix 198n52
Knudsen, Finn Christian 213
Knudsen, Gunnar 97
Knudtzen, Harald 286–287
Knudtzen, Ole 287
Kolbjørnsen, Arne 231
Konstantin-Hansen, Karl Kristian 279
Koren, Wilhelm 224
Körner, Julius 1, 4, 270, 324.
Krabbe, Adolf Fredrik 183
Krabbe, Thorvald 194
Krantz, Einar 194
Kreuger, Ivar 177, 225, 236
Kristjansson, Thòrarinn 194
Kuhlefelt-Ekelund, Eva 93, 188, 207, 208–210, 232, 262, 307, 319
Kure, Per 192, 192n18
Kyrklund, Harald 134, 291

la Cour, Jens Lassen 138–139
Lagerroos, Thor 177, 289
Larsen, Orla 161
Larsen, Poul Sehested 276, 280
Laurell, Bror M. C. 168n48, 222
Laval, Gustaf de 268
Lavonius, Magnus 140, 291
Lavonius, Robert 292
Lehmann, Alfred 140, 276
Lehmkuhl, Joakim 165, 185, 281, 321
Lepsøe, Robert Kock 183
Levanto, Kaarlo Ilmari 174n89
Lewerentz Siegfried 116, 188
Lie, Ragnvald 136
Lilius, Karl 145n108
Liljekvist, Rudolf 213, 234
Lindeman, Thorvald 133
Lindqvist, David L. 182
Lindqvist, Selim A. 149, 291, 319
Ljungzell, Nils 142n91
Lobnitz, Henry C. 213
Löfgren, Johan 142n91

Lønberg-Holm, Åge 143n91
Lorentzen, Herbert Oscar 151, 154
Lund, Bernt 286
Lundbohm, Hjalmar 271
Lundqvist, Emil 168, 173, 174, 225, 236, 269
Luth, John 198, 267, 319
Lyman, Reginald 279

Madsen, Edgar 277
Magnusson, Ivar 95, 270, 321
Mäklin, Hugo 174, 291
Markelin, Jenny 59
Mattson, Guss 138
Melin, Carl Oskar 274, 319
Meyer, Viggo 279
Michalsen, Eystein 177, 281, 319
Michelsen, Waldemar 230–231
Moberg, Wilhelm 9
Monberg, Niels Christiansen 151
Müller-Breslau, Heinrich 132

Nielsen, Aage 277
Nilsson, Emanuel 217
Nobel, Alfred 206–207, 213, 223, 234, 270, cover illustration
Nobel, Carl 214
Nobel, Immanuel 214
Nobel, Ludvig 50, 214–215, 234, 292, 306
Nobel, Robert 50, 214–215, 234, 292, 306
Nobell, Årad Svensson 135
Norberg-Schulz, Thomas 283
Nordberg, Bruno 181
Nordenfeldt, Per 196, 211
Nordenfeldt, Thorsten 196, 211, 234, 306, 307
Nørgaard, Oscar 284
Norstedt, Claes Oskar 290
Nyberg, Otto Fridolf 288
Nyquist, Viktor Theodor 228
Nyrop, Martin 115–116, 188

Obelitz, Axel 221
Öfverholm, Ivan 174n86
Olson, Erik 140
Olsson, Axel 215
Orell, Napoleon 140n83
Örn, Gustaf 192
Osness, Johan 133, 286, 319
Östberg, Ragnar 267, 319
Östlund, Erik 230
Oustad, Krisfoffer Olsen 182
Övergaard, Yngve 221

Paavola, Werner 219
Palén, Paul 176, 271, 323
Palmcrantz, Helge 196
Pauss, Sigurd 142n91
Pedersen, Sverre 285
Personne, Emil 206
Persson, Nils 114, 192
Petersson, Lambert 135–136, 290, 318
Petrelius, Albert 133
Picard, Èmile 198
Piltz, Gottlieb 268
Piper, Oscar 279
Pohjanpalo, Soini 289
Poincarè, Henri 198
Preus, A. W. 207

Ranko, K. E 204
Rasmussen, Henrik 278
Rasmussen, Ludvig 278
Reimer, Justus L. 285
Rice Jr., Edwin W 173
Richardson, Henry Hobson 25, 178, 186, 304, 317
Rich-Christensen, Christian 280
Richelieu, Andreas du Plessis de 218
Rockefeller, John D. 171
Rølvaag, Ole Edvart 9
Rørvik, Johannes 174n89
Rost, Helge 230
Rotheim, Eric Jacobsen 191
Rua, Engebret 223
Ryselin, Verner 290–291

Saarinen, Eliel 181, 245, 306
Sachs, Josef 177
Sadolin, Knud 231
Sahlin, Axel 201
Sahlin, Olof 223
Samberg, Johan Vilhelm 289
Sandberg, Vera 59, 92
Sandwall, John 273
Sarlin, Emil 292
Schmidt, Wilhelm 192